CHEMICAL ENGINEERING METHODS AND TECHNOLOGY

ALGEBRAIC CHEMISTRY

APPLICATIONS AND ORIGINS

CHEMICAL ENGINEERING METHODS AND TECHNOLOGY

Additional books in this series can be found on Nova's website under the Series tab.

Additional E-books in this series can be found on Nova's website under the E-book tab.

MATHEMATICS RESEARCH DEVELOPMENTS

Additional books in this series can be found on Nova's website under the Series tab.

Additional E-books in this series can be found on Nova's website under the E-book tab.

CHEMICAL ENGINEERING METHODS AND TECHNOLOGY

ALGEBRAIC CHEMISTRY APPLICATIONS AND ORIGINS

CYNTHIA KOLB WHITNEY

New York

Copyright © 2013 by Nova Science Publishers, Inc.

All rights reserved. No part of this book may be reproduced, stored in a retrieval system or transmitted in any form or by any means: electronic, electrostatic, magnetic, tape, mechanical photocopying, recording or otherwise without the written permission of the Publisher.

For permission to use material from this book please contact us:
Telephone 631-231-7269; Fax 631-231-8175
Web Site: http://www.novapublishers.com

NOTICE TO THE READER

The Publisher has taken reasonable care in the preparation of this book, but makes no expressed or implied warranty of any kind and assumes no responsibility for any errors or omissions. No liability is assumed for incidental or consequential damages in connection with or arising out of information contained in this book. The Publisher shall not be liable for any special, consequential, or exemplary damages resulting, in whole or in part, from the readers' use of, or reliance upon, this material. Any parts of this book based on government reports are so indicated and copyright is claimed for those parts to the extent applicable to compilations of such works.

Independent verification should be sought for any data, advice or recommendations contained in this book. In addition, no responsibility is assumed by the publisher for any injury and/or damage to persons or property arising from any methods, products, instructions, ideas or otherwise contained in this publication.

This publication is designed to provide accurate and authoritative information with regard to the subject matter covered herein. It is sold with the clear understanding that the Publisher is not engaged in rendering legal or any other professional services. If legal or any other expert assistance is required, the services of a competent person should be sought. FROM A DECLARATION OF PARTICIPANTS JOINTLY ADOPTED BY A COMMITTEE OF THE AMERICAN BAR ASSOCIATION AND A COMMITTEE OF PUBLISHERS.

Additional color graphics may be available in the e-book version of this book.

Library of Congress Cataloging-in-Publication Data

Whitney, Cynthia Kolb, 1941-
Algebraic chemistry : applications and origins / Cynthia Kolb Whitney (The Journal Galilean Electrodynamics, Arlington, MA, USA).
  pages cm
Includes bibliographical references and index.
ISBN: 978-1-62257-861-0 (hardcover)
1. Molecular dynamics--Mathematics. 2. Ionization constants--Measurement. 3. Quantum chemistry. I. Title.
 QD96.M65W44 2013
 541.01'512--dc23
                                             2012031714

Published by Nova Science Publishers, Inc. † New York

Contents

Preface	vii
Introduction	1
Part I. Chemistry as Numerical Regularities	
Prolog to Part I	13
Chapter 1 Ionization Potentials of Atoms	19
Chapter 2 Ionization Potentials of Ions	29
Chapter 3 Ions and States of Matter	159
Chapter 4 Singular Elements	195
Chapter 5 Typical Molecules	207
Chapter 6 Important Reactions	237
Chapter 7 Catalysis of Chemical Reactions	263
Chapter 8 Electro-Chemistry in Power Generation	279
Part II. Chemistry as Quantum Mechanics	
Prolog to Part II	297
Chapter 9 Hydrogen as the Prototypical Atom	299
Chapter 10 General Charge Pairs	307
Chapter 11 Electron Rings and Structures Thereof	317
Chapter 12 Explosions and Explanations	327
Part III. Quantum Mechanics as Electrodynamics	
Prolog to Part III	345
Chapter 13 Photons and Maxwell's Equations	347
Chapter 14 On the Invariance of Maxwell's Equations	359
Conclusion	367
Index	373

PREFACE

Algebraic chemistry is based on numerical patterns observed in readily available data about ionization potentials of atoms, and on a physical model that interpolates and extrapolates from that data to situations for which data is not readily available, such as ionization potentials for atoms that are already ionized, or corresponding energy increments involved in adding, rather than subtracting, electrons. This Book presents an approach to chemistry that permits numerical evaluation of many chemical scenarios without use of much computation power. Everything here can be worked out with a hand calculator. The approach thus makes numerical analysis of scenarios in chemistry feasible for students, or up-coming researchers, or retirees, who work with minimal financial support.

INTRODUCTION

1. THE BACKGROUND

Chemistry certainly has its deepest roots in antiquity, tracing to a number of very prescient individuals, especially individuals in ancient Greece. For example, Democritus was prescient, with his 'indivisible' bits of matter, which foretell of atoms. And Aristotle was prescient with his four 'elements', which foretell the main macroscopic states of matter. And Archimedes was prescient, applying reason to various practical problems involving change, which foretell of people in modern times studying chemical reactions.

An even stronger connection for Chemistry is to medieval Alchemy, whose practitioners were specifically interested in particular transformations, *i.e.*, from lead to gold. But medieval Alchemy wasn't any great step toward science. It was all about resurrection of the ancient beliefs, preservation of secrecy about the ancient knowledge, application of various mystical incantations, magical transmutation of seemingly immutable substances, and, all too often, a bit of profitable fraud.

To be fair, the medieval Europeans were quite right in seeking out the wisdom of wise men from the ancient world. But, probably on account of having just emerged from a Dark Age, and having rediscovered a previous Golden Age, they were overly deferential to the ancients. They were quite wrong in that. No body of knowledge is ever complete, or completely right. So there is always need for on-going progress. And there can be real progress only when knowledge can be reviewed and updated with additions and corrections. That time came in the 17th century.

My personal hero of the 17th century era is Sir Isaac Newton. We know him mostly as the first true 'Mathematical Physicist'. But on occasion, he also pursued Alchemy (and no doubt thereby learned some skill, or at least some skepticism, valuable for his later calling as the master of his nation's mint). That contact with Alchemy ties Newton to what would ultimately become Modern Chemistry.

Newton's work is all about reliance on experiments and observations, rather than ancient authority. It is also about elucidating natural phenomena, even without a pre-determined technological objective (pure research). It exemplifies the 'virtuous cycle' of scientific development, which goes: 1) collect quantitative data; 2) recognize numerical patterns in the data; 3) develop a mathematical theory about the data; 4) predict new results to test with new experiments or observations; and, eventually, 5) publish results widely.

My next personal hero is James Clerk Maxwell. His development of electromagnetic theory makes him the giant of the 19th century. His work figures prominently in this book because his set of four coupled field equations constitute the point of reference for the development of a mathematical theory about Algebraic Chemistry. And how else could it be? Chemistry is, after all, about events that are fundamentally electrical in nature.

And of course my other personal hero of the 19th century is Dmitri Mendeleyev. By his time, constituents of matter had been characterized as different 'elements'. But the information about the characteristics of different elements was a total mess. Mendeleyev cleaned up the mess. He was the all-time champion in the art of 'Pattern Recognition'. The Periodic Table he worked out was a thing of value for anticipating the existence of missing elements, even before we knew anything much about atoms. And today, in the development of Algebraic Chemistry, a new exercise in basic Pattern Recognition is the first step. And the pattern made manifest here is again a thing of value, this time for chemical engineering purposes, even before we try to develop a deep understanding of it.

In modern times, the atom has been accepted as a real thing, and recognized as divisible into particles: its electrons and its nucleus. And the nucleus in turn has particles: protons and neutrons. Today we have a mathematical theory about the electrons, the atoms, the molecules, and matter in just about any state or form. It is Quantum Mechanics (QM).

Are we done? No, we are not done. We are in the midst of trying to make QM tell us all that we want to know about Chemistry, without necessarily burning out our super-computers. Effectively addressing that task appears to require a short digression into Philosophy.

Within the modern discipline of Philosophy of Science, there exists a clear distinction between the concept of 'Model' and the concept of 'Theory'. Models are utilitarian tools for anticipating phenomena and calculating their particulars, whereas Theories are deep explanations for phenomena, aiming to be the actual physical truth of what goes on. Models are not necessarily based in any established Theory; they can be based on empiricism: data patterns observed, rules of thumb developed, relationships described with simple math. So, while eminently practical for everyday use, Models are humble and unpretentious tools. Theories are more formal; they are rigorously based on established physical principles, and/or mathematical axioms, and/or or postulates. They may involve deep mathematics, and they may require elaborate computer calculations. But because Theories purport to describe what is actually, physically, going on, they are highly valued, even if they are sometimes inconveniently cumbersome to apply.

Quantum Mechanics is quite like that. It is clearly stated, but it can be hard to use for problems at the scales that Chemistry often needs. Chemistry can involve atom counts that are large for QM calculations, but small for simple statistical characterizations, and therefore altogether awkward to handle.

2. This Book

This Book is meant for a wide audience, from youth to old age. It represents an act of contrition on my part, concerning my youthful attitude about Chemistry. I thought it looked like a huge collection of facts to be memorized. I was wrong. I see now that Chemistry does not at all require memorization, because everything is connected to everything else. There is a pattern to the data. Once you see the numerical regularities, you can calculate any specific datum that you may need to know. This Book illustrates a method to do just that. The math required is not at all deep. It is mostly Arithmetic. The deepest thing required is a square root operation. Only that square root operation justifies calling the method 'Algebraic', rather than just 'Arithmetic'.

The Book follows the path that Chemistry has often taken in the past: first empirical facts and a numerical Model useful for applications, then an effort to build a well-founded Theory. It has three Parts, corresponding to Applications, followed by Origins at two levels: Model and Theory. Part I is an exercise in pattern recognition, followed by exploitation in all sorts of chemical scenarios. Part II explains the pattern in terms of a Model for atoms that is Newtonian in body and Maxwellian in spirit. Part III uncovers some reasons why, historically, the QM Theory of atoms wasn't developed in this way in the first place.

2.1. About Part I

The first Chapter of Part I cites a lot of data about ionization potentials of individual elements, and exposes relationships among data points for all elements and all orders of ionization. This is a Pattern. It implies a model sufficient to express the ionization potential of any order, whether measured yet or not, for any atom, whether discovered or synthesized yet or not. The next Chapter gives simple formulae for interpolating between data points, and extrapolating beyond data points, to model potentials for further ionization of atoms that are already ionized. This kind of extension allows a user to make a quantitative assessment of any atom in any state of ionization, whether positive or negative. The next Chapter suggests a correlation between energies of ionization states and the physical states of matter: solid, liquid, gas, and plasma. The next Chapter uses this information to analyze some elements that are particularly interesting on account of the physical state they present, or the reactions they exhibit (or don't exhibit). The next Chapter uses ionization potentials to make numerical assessments of many common molecules. The next Chapter analyzes a number of example reactions, all surrounding the theme of hydrocarbon combustion, and including subjects such as fuel additives and catalytic converters. Those subjects lead to the larger subject of catalysis in general, treated some more in the next Chapter. The final Chapter of Part I of this Book analyzes one proposed means of generating electrical power *without* hydrocarbon combustion.

2.2. About Part II

Part II of the book delves into Chemistry as Quantum Mechanics (QM), inasmuch as QM is the source of some of the parameters in the numerical model developed and used in Part I. QM has already had several incarnations in the twentieth century, and may now be ready to accommodate another one, if a new one can do a little bit more about explaining the Periodic Table (PT) of the elements. Present-day QM *describes* the PT, but it doesn't *explain* it. The model developed here starts anew. It suggests a view of atoms that is hierarchical, with the electrons forming a subsystem that has an overall orbit around the nucleus, plus internal relationships among the electrons only. The first Chapter revisits Hydrogen, as that atom has always been the starting point for the QM of all atoms. The second Chapter deepens the analysis of Hydrogen to explain some otherwise unexplained spectral features, and introduces some other two-charge systems: the electron-positron system called 'positronium', and a ring system composed of two electrons. The next Chapter shows that electron rings with larger electron counts can build all the electron populations of all the elements. The algorithm for doing this involves pairs of rings creating regions of magnetic confinement for geometrically smaller rings. The algorithm is basically fractal. It explains why the single-electron states of QM fill in the order that they do through the Periodic Table. The last Chapter of Part II revisits higher-order ionization potentials, to explain them more, and suggest their potential practical utility in analyzing chemical explosions.

2.3. About Part III

Part III dips into Quantum Mechanics as Electrodynamics. There are two areas that seem to merit attention. One concerns the connection between Maxwell fields and photons. It has been widely supposed that photons are something very different from Maxwell fields. But actually, it is entirely possible to model photons in terms of Maxwell fields. It has also been widely assumed that Maxwell's equations require that coordinate transformations be Lorentz transformations, and so mandate Einstein's Special Relativity Theory. But actually Maxwell's equations tell us nothing whatsoever about what kind of coordinate transformation Nature requires. Both of these Part III Chapters are somewhat mathematical, and so can be viewed as steps in the development of a more conveniently applicable Theory for use in Chemistry.

2.4. About So Much More

There are many interesting and important topics that are in the history and current practice of Chemistry, but are not among the topics treated in the present Book. The very best way I know to find out about some of them is through the series of Chemistry lectures by Prof. Don Sadoway in the "open courseware" video collection produced by the Massachusetts Institute of Technology, and available online at ocw.mit.edu/courses. Look under "Materials Science and Engineering", course 3-091sc, "Introduction to Solid State Chemistry".

3. THE CHARACTERS

The story inherent in Chemistry is nothing short of a good Drama. The plot is all about the advantageous giving and taking of electrons. Electrons make the world go 'round. Electrons are like money. The treasure they purchase is an overall lower energy state, and hence enhanced stability. Stability is the goal sought by all the characters in the drama.

The Characters are the elements. The elements form endlessly varied alliances, and wars, all in pursuit of lower energy states. Sometimes they even go nuclear, and kill themselves or even each other. (But of course, they then transmute to altogether different elements, so this story is mostly Comedy, and not much Tragedy.)

In preparation for the story, here is the 'Dramatis Personae':

$_1$H : Hydrogen, a keystone element, and the prototype for all physical and chemical models;

$_2$He : Helium, the sunny, the super-fluid, the He-Ne laser component, the most noble;

$_3$Li : Lithium, the lightest of the alkaline metals, the battery component, the medicinal;

$_4$Be : Beryllium, the rather dangerous to work on;

$_5$B : Boron, the cleanser;

$_6$C : Carbon, the keystone element, the basis of Earthling life;

$_7$N : Nitrogen, the funny, the peace maker, the protein builder;

$_8$O : Oxygen, the incendiary, the hot head, the spark of animal life;

$_9$F : Fluorine, the light, the halogen, the hole-filler;

$_{10}$Ne : Neon, the He-Ne laser component, the aloof noble;

$_{11}$Na : Sodium, the dangerous alkaline metal that make the safe table salt;

$_{12}$Mg : Magnesium, the light-weight metal, when powdered, it is good for flash-bangs;

$_{13}$Al : Aluminum, the light, the abundant, the good conductor of electrons and heat;

$_{14}$Si : Silicon, a keystone element, the basis of present-day technological life;

$_{15}$P : Phosphorus, the element that has several forms: OK white, OK red, toxic yellow;

$_{16}$S : Sulfur, the stinker, but also the constituent of sulfa drugs;

$_{17}$Cl : Chlorine, the most abundant halogen, the constituent of bleach, cleanser, table salt;

$_{18}$Ar : Argon, another aloof noble;

$_{19}$K : Potassium, the alkaline metal, the salt sometimes better for you than Sodium's;

$_{20}$Ca : Calcium, the maker of milk and stone, tooth and bone;

$_{21}$Sc : Scandium, the elusive, long known only by its salts.

$_{22}$Ti : Titanium, the difficult as metal, but wonderful white as oxide;

$_{23}$V : Vanadium, the most hard, but abundant, and used in many alloys;

$_{24}$Cr : Chromium, the peculiar, the very hard, very shiny metal on you car accents;

$_{25}$Mn : Manganese, the additive, used with others, to make iron into steel;

$_{26}$Fe : Iron, the interloper metal defining a historical age that we still live in;

$_{27}$Co : Cobalt, a keystone metal, a metal for defining our historical age;

$_{28}$Ni : Nickel, the cheap, used in coinage since the invention of coins;

$_{29}$Cu : Copper, the expensive, the good electrical conductor, the good heat conductor;

$_{30}$Zn : Zinc, the plentiful, galvanizer of iron, the negative plate in a dry cell, and more;

$_{31}$Ga : Gallium, the easy to melt, the hard to vaporize, the odd: expands upon solidifying;

$_{32}$Ge : Germanium, the material of some transistors;

$_{33}$As : Arsenic, the constituent of many poisonous compounds;

$_{34}$Se : Selenium, the element that builds up in the artificially watered desert 'garden';

$_{35}$Br : Bromine, the halogen, one of two liquid elements, both dangerous;

$_{36}$Kr : Krypton, the super-man substance;

$_{37}$Rb : Rubidium, the alkaline metal sometimes used in photoelectric cells;

$_{38}$Sr : Strontium, the sometimes-radioactive substitute for Calcium in bone;

$_{39}$Y : Yttrium, the rare earth that comes up way before any others in the PT ;

$_{40}$Zr : Zirconium, the maker of bling, the maker of windows for thermal neutrons;

$_{41}$Nb : Niobium, the peculiar, the 'daughter' of Tantalum;

$_{42}$Mo : Molybdenum, the peculiar, the additive that helps make steel;

$_{43}$Tc : Technetium, the metal long expected and late found, the companion of Platinum;

$_{44}$Ru : Ruthenium, the peculiar, the sometimes catalyst, the sometimes alloy component;

$_{45}$Rh : Rhodium, the peculiar, a keystone element, a good alloy or plate material;

$_{46}$Pd : Palladium, the peculiar, the catalyst, the negative electrode (cathode);

$_{47}$Ag : Silver, the peculiar, the good catalyst;

$_{48}$Cd : Cadmium, a constituent of many paint pigments;

$_{49}$In : Indium, the shiny relative of Aluminum;

$_{50}$Sn : Tin, the roofing material, the copper-pot lining material, the tin can, the alloy;

$_{51}$Sb : Antimony, the rare, but known since ancient times;

$_{52}$Te : Tellurium, the versatile non-metal used in glass, electrical equipment, dyestuffs;

$_{53}$I : Iodine, the halogen with medical uses;

$_{54}$Xe : Xenon, another aloof noble.

$_{55}$Cs : Cesium, the surprising, the alkali metal that is softer than beeswax;

$_{56}$Ba : Barium, the maker of many useful compounds;

$_{57}$La : Lanthanum, the peculiar, the rare earth that might be in your glasses;

$_{58}$Ce : Cerium, the peculiar, the rare-earth metal so malleable and ductile that a knife cuts it;

$_{59}$Pr : Praseodymium, the leek-green rare-earth metal;

$_{60}$Nd : Neodymium, the not-so-rare, the yellow rare earth metal that makes red salts.

$_{61}$Pm : Promethium, the rare earth metal that was long predicted and late found;

$_{62}$Sm : Samarium, the rare and little used rare earth.

$_{63}$Eu : Europium, one of the really rare of the rare earth elements.

$_{64}$Gd : Gadolinium, the peculiar, the rare, rare earth;

$_{65}$Tb : Terbium, the rare, rare earth;

$_{66}$Dy : Dysprosium, the rare-earth metal not used for much;

$_{67}$Ho : Holmium, the rare, rare earth;

$_{68}$Er : Erbium, the other rare earth metal not used for much.

$_{69}$Tm : Thulium, the close companion of Erbium

$_{70}$Yb : Ytterbium, the rare earth generally found mixed with other rare earths

$_{71}$Lu : Lutetium, the rare earth;

$_{72}$Hf : Hafnium, the neutron rich, the companion of Zirconium;

$_{73}$Ta : Tantalum, the tough metal in lab equipment, medical devices, *etc.*

$_{74}$W : Tungsten, the metal with the highest melting point, the light-bulb filament;

$_{75}$Re : Rhenium, the other companion of Platinum, the possible catalyst;

$_{76}$Os : Osmium, the metal used with Iridium to make a really tough alloy.

$_{77}$Ir : Iridium, the tell-tale clue about the asteroid, the putative killer of dinosaurs;

$_{78}$Pt : Platinum, the peculiar, the catalyst, the ultimate metal, the best credit card;

$_{79}$Au : Gold, the peculiar, the incorruptible, the reliably much desired;

$_{80}$Hg : Mercury, the runny, the other of two liquid elements, both dangerous;

$_{81}$Tl : Thallium, the element with spy-drama poisonous salts, discovered by spectroscopy;

$_{82}$Pb : Lead, the bad heavy metal, the good radiation shield;

$_{83}$Bi : Bismuth, the metal with the lowest heat conductivity;

$_{84}$Po : Polonium, the very rare radioactive element discovered very early;

$_{85}$At : Astatine, the heaviest halogen;

$_{86}$Rn : Radon, the aloof noble, but radioactive, gas that may be in your basement;

$_{87}$Fr : Francium, the heaviest alkali metal;

$_{88}$Ra : Radium, the radioactive element that might be on your antique watch face;

$_{89}$Ac : Actinium, the peculiar, the companion of Uranium and Radium;

$_{90}$Th : Thorium, the peculiar, the metal convertible to Uranium;

$_{91}$Pa : Protactinium, the peculiar, the element that decays to Actinium;

$_{92}$U : Uranium, the peculiar, the politically and literally dangerous;

$_{93}$Np : Neptunium, the peculiar, the first trans-Uranic element;

$_{94}$Pu : Plutonium, the second, and even-more dangerous, trans-Uranic element;

$_{95}$Am : Americium, the product of bombarding Uranium with Helium ions;

$_{96}$Cm : Curium, the peculiar, the product of bombarding Polonium with Helium ions;

$_{97}$Bk : Berkelium, the product of bombarding Americium with Helium ions;

$_{98}$Cf : Californium, the product of bombarding Curium with Helium ions.

$_{99}$Es : Einsteinium, the heavy (254);

$_{100}$Fm : Fermium, the elusive;

$_{101}$Md : Mendelevium, the elusive;

$_{102}$No : Nobelium, the elusive top keystone element;

$_{103}$Lr : Lawrencium, the product of bombarding Californium with Helium ions;

$_{104}$Rf : Rutherfordium, the elusive;

$_{105}$Db : Dubnium, the heavy (262);

$_{106}$Sg : Seborgium, the heavy (263);

$_{107}$Bh : Bohrium, the heavy (262);

$_{108}$Hs : Hassium, the heavy (265);

$_{109}$Mt : Meitnerium, the heavy (266);

$_{110}$Ds : Darmstadtium, the heavy (271);

$_{111}$Rg : alias 'Unununium', the heavy (272);

Extras:

$_{112}$?? : alias Ununbium, the presently unknown;

$_{113}$?? : alias Ununtrium, the presently unknown;

$_{114}$?? : alias Unnilquadium, the presently unknown;

$_{115}$?? : alias Ununpentium, the presently unknown;

$_{116}$?? : alias Ununhexium, the presently unknown;

$_{117}$?? : alias Ununseptium, the presently unknown;

$_{118}$?? : alias Ununoctium, the presently unknown.

That is quite a long list of characters. So it is helpful to sort them in some way. The society of Chemical Elements does divide into various social classes. The classes are defined with reference to an economy based on the trading of electrons for energy. Electrons are like cash, and energy is like a commodity. Both are totally fungible. Atoms trade in electrons and energy because they aspire to some recognized status that has something to do with sense of wealth: either as a 'noble gas' (possessing a balanced amount of energy and number of electrons, and so not seeking to engage in further trading), or else as a member of a 'priestly class' (not needing to possess any worldly electrons at all).

Most neutral atoms are in neither of these desired classes. So most neutral atoms need to acquire or give up electrons to achieve their aspirations. Some are in the warrior class (alkali metals, who throw an electron whenever possible). Some are among the gentry (halogens, who seek to receive an electron whenever possible). Some are in the working class (adaptable, giving or receiving electrons as requested). Some are technologically important (keystone elements). Some exhibit electrons out of place (peculiar). Some are hard to classify at all (troublesome). You will get to know them all in this book.

Electron trading creates ions, and hence ionic bonds. Other names are often attributed to particular bonding situations in Chemistry, but when you come right down to it, they all involve ions. The ions form molecules. The energy involved in molecule formation might be negative, meaning the molecule forms automatically, or it might be positive, meaning the molecule forms only with assistance. You will be able to tell which is which. Algebraic Chemistry is about quantifying the energy of molecule formation in the most convenient way possible. Enjoy it.

Part I. Chemistry as Numerical Regularities

PROLOG TO PART I

The ancient and medieval history of humanity does feature a lot of people practicing a lot of arcane numerology. The Pythagoreans of ancient Greece exemplify the idea. Vladimir Ginzberg has written several wonderful books [1] that contain many fascinating stories from this period of history. In the western hemisphere, the ancient Mayans developed a cosmic calendar, terminating we originally thought in 2012 (coincidentally the target publication date for the present book), but now known to go immeasurably further (so you can relax and read). Similar numerical investigations occurred in the vast area that is today China. In medieval Europe, Jewish communities devoted much intellectual effort to the idea of Kabala. And of course the medieval Alchemists played with numbers and formulae. Paul Strathern has told the most fascinating stories about them in [2]. He continued to the near-modern times of Lavoisier and Priestly. Joe Jackson treated those near-modern times again, a little later and little longer, in [3].

In [4], Eric Scerri remarks on a somewhat discomfiting modern attitude: the imputation of 'numerology' can be delivered and/or experienced as an insult. So here I feel the need to explain why the first and biggest Part of this Book is called "Chemistry as Numerical Regularities". The information developed here is definitely not a demonstration of ancient numerology or magical incantation; it is instead a demonstration of modern pattern recognition and exploitation. And although Part I focuses on a Model emphasizing numerical regularities, the Book goes beyond Model and toward Theory in Parts II and III.

I am a physicist, now borrowing the treasure trove of information that Chemistry possesses. In Physics, the early course of development went: 1) data collection, exemplified by Copernicus; 2) pattern recognition, exemplified by Kepler; 3) mathematical formulation, exemplified by Newton; 4) more data collection. The development of Physics continues in the same way today, with all three types of activity now operating in parallel. The most vulnerable part of the development cycle is the new data collection, because it has become so expensive. Just witness my own country's abandonment of its super-conducting super-collider.

In Chemistry, the early course of development has gone similarly: 1) data collection, exemplified by many, many early metallurgists, beer, wine, and bread makers, dye masters, ceramicists, *etc.*; 2) pattern recognition, exemplified by the work of giants like Lavoisier and Priestly, Mendeleyev and Dalton; 3) mathematical formulation, currently most exemplified by efforts in Quantum Chemistry (QC); and 4) more data collection.

Modern Chemistry is most proud to be now operating at the Newton level of mathematical formulation. But there are still important opportunities at the Kepler level of numerical pattern recognition. Unfortunately though, such efforts in pattern recognition do sometimes get branded as 'numerology'. That charge reminds us of the historical tie between modern Chemistry and medieval Alchemy, and therefore discourages modern efforts at numerical pattern recognition in Chemistry. Those efforts are thus the most vulnerable part of the overall development cycle for Chemistry. That is why Part I of this book is entirely devoted to the recognition of, documentation of, and exploitation of, numerical patterns in chemical data.

The data featured consists of ionization potentials. So the book is focusing on the electron populations of atoms. The main idea is that simple addition and subtraction of electrons can account for situations that otherwise invite various purpose-built terminology, such as 'electron sharing', 'covalent bonding', 'perfect covalent bonding', 'general-fraction covalent bonding', 'polar bonding', 'semi-polar bonding', *etc.*, *etc.*, *etc.* The objective is to reveal that Chemistry is really less complicated than it currently appears to be.

Part I, Chapter 1, investigates all the ionization potentials of all atoms reported in the literature. It demonstrates an underlying pattern that relates the reported ionization potentials of any order to those of first order. The significance of this relationship is that we never have to look up, or, worse yet, measure, information about any higher-order ionization potentials. We can infer all such information from the first-order ionization potentials.

As for those, they follow a simple pattern that depends on the ratio of nuclear charge to nuclear mass for each element, the nominal quantum numbers of the single-electron states being filled at that element, a parameter that characterizes the lengths of the period in the Periodic Table, plus a handful of integer ratios.

So we don't really have to look up first-order ionization potentials either; we can easily generate them all, just given the first one, the ionization potential of Hydrogen. As for that one, it is easy to predict if you know Planck's constant. (But if you read further on, through Part II and to Part III of the book, you will find that even Planck's constant is not necessary; everything traces ultimately to Maxwell's equations).

Part I, Chapter 2, constructs a large number of ionization potentials of positive and negative ions of many elements – all the ones that this author has yet had reason to study. This is a big Chapter, and it will grow bigger over time as more and more interesting problems are addressed. It is a developing handy reference work. The algorithm applied in this work amounts to interpolation between, and extrapolation from, the ionization potentials of the neutral atoms, provided in exhaustive detail in Chapter 1. The point is: there exists a grid of bits of information so comprehensive and complete, with so many algebraic relationships between those bits of information, that practically any other bit of information that you may wish to possess can be computed. Since all molecules are made of atoms in various states of ionization, this computation scheme can assist assessment of compounds of arbitrary complexity, and then the reactions among the molecules.

This capacity promises to help make Chemistry more quantitative, in a more physically meaningful way, than it currently seems to be. But to be trusted, the model has to be shown to comport with known facts. So the next several Chapters are devoted to showing that this model conforms to known facts.

Part I, Chapter 3, discusses a plausible relationship between ions and states of matter. It proposes a correlation between the ionic configurations of material substances and their macroscopic states – solid, liquid, gas, or plasma. Those top-level characterizations of states of matter again reflect the deep roots of Chemistry in antiquity. Aristotle proposed four basic 'elements': earth, air, fire, and water. That was a Model, and what is on offer today is really the same Model, just re-interpreted as 'states of matter', as distinct from 'elements', and re-ordered into solid, liquid, gas, and plasma.

Part I, Chapter 4, looks at a handful of especially interesting elements, the physical, and chemical, states that they exhibit, and a the sorts of compounds that they do, or do not, make. The noble gases are so interesting because, unlike any other category of elements, they resist making reactions of any kind. A few other elements are peculiar in that they present themselves as liquids under the conditions that human beings are familiar with.

Part I, Chapter 5, looks at various hydrocarbon molecules. The hydrocarbons come in linear chains, in cyclic loops, and in branching forms. The geometry of the molecule can be reflected in its ionic configuration. Some molecules allow more than one ionic configuration, and sometimes a different ionic configuration makes not only for a different geometry but also a drastically different behavior.

But just confirming known facts is not enough; it is important to shed light on situations where the facts may be presently uninvestigated, or investigated but still unclear, or even hotly disputed. Those situations seem to come up with many new technology developments. So the next several Chapters are devoted to such situations.

Part I, Chapter 6, examines a number of interesting chemical reactions in which the hydrocarbons are involved. Issues include hydrocarbon combustion, fuel additives, and catalytic converters. Incomplete combustion can create the need for a fuel additive and/or a catalytic converter. Algebraic Chemistry can explain why incomplete combustion happens, what a fuel additive needs to do, how a catalytic converter can remedy a residual problem, and especially how to do all these jobs a little better.

Part I, Chapter 7, addresses the existence of some reactions that may be desired, but just do not transpire spontaneously. They need catalysis. But it hasn't always been clear what catalysis scheme would work best in a given problem, because it hasn't always been clear *how* any catalysis scheme works. But a reaction with catalysis is just a sequence of several reactions, and the techniques of Algebraic Chemistry illustrated in this Book can address all of the reactions. This approach can make the phenomenon of catalysis a little less mysterious than it currently seems to be. It can also expose any weaknesses in previous interpretations of particular catalysis scenarios, and offer better interpretations.

Part I, Chapter 8, illustrates how one may use the techniques illustrated earlier to discover some new and surprising information about technologies that have been attempted in advance of having a scientific base sufficient for a real understanding of the problems involved. The technology illustrated is so-called 'Cold Fusion'. Except for a hoped-for nuclear fusion event at the end, it is electrochemistry all the way. Ionic configuration matters, molecular geometry matters, catalysis matters – all issues to which Algebraic Chemistry can speak.

ABOUT PATTERNS

The ultimate pattern in Chemistry is the Periodic Table (PT). Readers of this Book will need to refer to the PT. So it is provided at the end of this Prolog. Although many new elements have been discovered since the time of Mendeleyev, the format he settled on is still standard. There are columns, corresponding to so-called Chemical Groups, which are composed of elements with similar chemical properties.

The PT is now bigger than it was in Mendeleyev's day, and consequently a bit awkward to display. Originally, the PT had 18 columns. In more recent times, the Lanthanide and Actinide elements have been discovered and incorporated, so now there are 32 columns, with 14 of them having only two elements each. That fact makes the display awkward.

Mendeleyev was a man who spent a lot of time playing the game known in the USA as Solitaire. You can see that personal fact reflected in the form of PT he established: columns like 'suites' of elements, descending in numerical order down his Table. Today there also exist many alternative display formats for the PT. Dr. Mark Leach has collected many of them in [5], and I believe each one similarly reflects the soul of its creator.

REFERENCES

[1] Ginsberg, V. Unified Spiral Field and Matter, Helicola Press, Pittsburgh, PA, 1999; The Unification of Strong, Gravitational and Electric Forces, Helicola Press, Pittsburgh, PA, 2003; Prime Elements of Ordinary Matter, Dark Matter and Dark Energy, Helicola Press, Pittsburgh, PA, 2006; Prime Elements of Ordinary Matter, Dark Matter and Dark Energy – Beyond Standard Model and String Theory, Helicola Press, Pittsburgh, PA, 2007.
[2] Strathern, P. Mendeleyev's Dream, Hamish Hamilton, London, 2000.
[3] Jackson, J. A World on Fire, Viking, New York, 2005.
[4] Scerry, E. Explaining the periodic table, and the role of chemical triands, see Section entitled 'Possible objections to such use of triads', Foundations of Chemistry DOI 10.1007/s10698-010-9082-9, published online 28 January 2010.
[5] Leach, M., http://www.meta-synthesis.com/webbook/35_pt/pt_database.php.

APPENDIX: THE PERIODIC TABLE, MENDELEYEV STYLE

$_1$H						$_2$He
$_3$Li	$_4$Be			5		$_{10}$Ne
$_{11}$Na	$_{12}$Mg			col-		$_{18}$Ar
$_{19}$K	$_{20}$Ca		10 col-	umns		$_{36}$Kr
$_{37}$Rb	$_{38}$Sr		umns of	of 6		$_{54}$Xe
$_{55}$Cs	$_{56}$Ba	14 columns of	4 rows	rows		$_{86}$Rn
$_{87}$Fr	$_{88}$Ra	2 rows each	each	each		$_{118}$??

The fourteen columns of two rows each are:

$_{57}$La $_{58}$Ce $_{59}$Pr $_{60}$Nd $_{61}$Pm $_{62}$Sm $_{63}$Eu
$_{89}$Ac $_{90}$Th $_{91}$Pa $_{92}$U $_{93}$Np $_{94}$Pu $_{95}$Am

and

$_{64}$Gd $_{65}$Tb $_{66}$Dy $_{67}$Ho $_{68}$Er $_{69}$Tm $_{70}$Yb
$_{96}$Cm $_{97}$Bk $_{98}$Cf $_{99}$Es $_{100}$Fm $_{101}$Md $_{102}$No

The ten columns of four rows each are:

$_{21}$Sc $_{22}$Ti $_{23}$V $_{24}$Cr $_{25}$Mn
$_{39}$Y $_{40}$Zr $_{41}$Nb $_{42}$Mo $_{43}$Tc
$_{71}$Lu $_{72}$Hf $_{73}$Ta $_{74}$W $_{75}$Re
$_{103}$Lr $_{104}$Rf $_{105}$Db $_{106}$Sg $_{107}$Bh

and

$_{26}$Fe $_{27}$Co $_{28}$Ni $_{29}$Cu $_{30}$Zn
$_{44}$Ru $_{45}$Rh $_{46}$Pd $_{47}$Ag $_{48}$Cd
$_{76}$Os $_{77}$Ir $_{78}$Pt $_{79}$Au $_{80}$Hg
$_{108}$Hs $_{109}$Mt $_{110}$?? $_{111}$?? $_{112}$??

The five columns of six rows each are:

$_{5}$B $_{6}$C $_{7}$N $_{8}$O $_{9}$F
$_{13}$Al $_{14}$Si $_{15}$P $_{16}$S $_{17}$Cl
$_{31}$Ga $_{32}$Ge $_{33}$As $_{34}$Se $_{35}$Br
$_{49}$In $_{50}$Sn $_{51}$Sb $_{52}$Te $_{53}$I
$_{81}$Tl $_{82}$Pb $_{83}$Bi $_{84}$Po $_{85}$At
$_{113}$?? $_{114}$?? $_{115}$?? $_{116}$?? $_{117}$??

Chapter 1

IONIZATION POTENTIALS OF ATOMS

ABSTRACT

This Chapter shows that all the higher-order ionization potentials are linear combinations of the constituents of first-order ionization potentials. The coefficients involve only the ionization order itself and the square of the ionization order. This is a potentially handy fact to know, since higher-order ionization potentials can be difficult to measure reliably in the laboratory or to calculate with the traditional techniques of quantum mechanics.

It is a demonstrable fact that the first-order ionization potentials fall into a simple pattern. The present Chapter documents what this pattern is. It involves the usual quantum numbers, the principal, or orbital, n, the angular momentum l, and the spin s, of the single-electron states filled, and another parameter N that gives the length of the periods in the Periodic Table as $2N^2$.

INTRODUCTION

The basic data set used throughout this whole Book is the set of first-order ionization potentials of all the elements. Any other information about any individual element that may be required to address matters such as chemical bonding, *etc.*, is derived from the set of first-order ionization potentials.

For example, the Part I Prolog mentioned the subject of 'Chemical Groups'. Numbers are traditionally assigned to different chemical Groups, and the Group Number is involved in the expression currently used to compute 'electronegativity', a parameter meant to characterize the ability of an atom to attract electrons for a covalent bond. So the numerical value characterizing a Chemical Group matters, and the proper Group assignment for a troublesome element is a subject to be argued over.

But Algebraic Chemistry posits that all bonds are ultimately ionic in nature, and bases all numerical characterizations on ionization potentials. So it does not rely on a value of 'electronegativity' to make calculations about a covalent bond. So we simply skip that whole argument.

Ionization potentials of atoms are a type of data that is accessible experimentally in a number of ways. One can shoot beams: photon, electron, atomic, or molecular; one can document optical spectra: emission, or absorption; one can do electrolysis, or perhaps use other techniques as well. The various possibilities having been exploited for over a century, there is now available quite a large body of data about ionization potentials of atoms.

Any situation where a lot of data exists naturally invites an exercise in 'data mining'. But how shall we mine the data in this case? We need to have a model in mind. A model is developed in detail in Part II of this Book. What you need to know now is just the overview of the basic concept.

In the early 20th century, the Hydrogen atom was the prototype atom for the development of the Quantum Mechanics (QM) of atoms. And so it is here also. The model detailed later says the basic ingredients of an atom are a positively charged nucleus and a negatively charged population of electrons. In the Hydrogen atom, there is just one electron, which at the beginning of the 20th century was viewed as a point particle; *i.e.*, localized.

We begin similarly here, supposing the electrons of an atom to form a 'sub-system', interacting among themselves as individuals, but remaining at least somewhat localized, and so interacting collectively with the nucleus like an orbiting body, albeit with Z body parts, where Z is the nuclear charge number, equal to the electron count for a neutral atom. (The question of *how* an electron population could ever be at all localized, given that, in our macro world, same-sign charges repel each other, is deferred to Part II of this Book. See Chapter 2 of Part II.)

The proposed localization of the electron population suggested that a pattern, if one existed, would emerge when the raw data on ionization potentials were scaled by M/Z, where M is the nuclear mass number. The rationale was that M/Z scaling would convert raw ionization-potential data from being element-specific into being population-generic. (The question of *why* the ratio M/Z should function in this way is also deferred to Part II if this Book. See Chapter 1 of Part II.) After scaling the raw data by M/Z, a pattern in the data about ionization potentials did indeed emerge. The present Chapter documents this pattern. The empirical fact that this pattern exists is quite separate from any theoretical reason for why it exists. The pattern makes a Model, not a Theory. Every reader can learn the Model, and learn how to use it, even without a Theory. And for those readers who do want to know about the Theory, that still-developing information is given in Part II and in Part III of this Book.

1. OBSERVED BEHAVIOR OF IONIZATION POTENTIALS OF ALL ORDERS

Figure 1 depicts the behavior of the M/Z-scaled ionization potentials, called 'IP's', for all elements (nuclear charge $Z=1$ to $Z=118$ shown). It includes not only first-order IP's, but also IP's of higher ionization order, IO, up to 7, the limit for readily available data. That being the limit, Figure 1 has less than 400 out of approximately 5000 desired data points. This author would certainly welcome any additional data that readers might possess or know about.

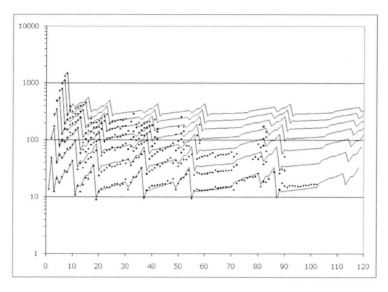

Figure 1. Ionization potentials, scaled appropriately and modeled algebraically.

The most striking fact revealed by Figure 1 is that, apart from its first point, the curve for *any* IO is similar to the curve for $IO = 1$, but shifted right by $IO - 1$ elements, and muted in amplitude, more and more muted as IO increases. This shift behavior means that each successive element reveals one additional bit of information about *all* subsequent elements. This fact speaks to the universality of chemical information: basic information about any one element can be inferred reliably from basic information about other elements.

The Hydrogen-based model used here invites the division of each ionization potential $IP_{1,Z}$ into two parts, one being $IP_{1,1}$ for the generic Hydrogen-like collective interaction with the nucleus, and the other being the increment $\Delta IP_{1,Z} = IP_{1,Z} - IP_{1,1}$ for the element specific electron-electron interactions. Figure 2 uses these entities to relate the IP's of higher ionization order, with $IO > 1$, to the first-order IP's, with $IO = 1$.

$$IP_{IO,IO} = 2 \times IP_{1,1} \times IO^2$$

$$IP_{IO,IO+1} = IP_{1,1} \times 2 \times IO^2 + \frac{1}{2} \times IP_{1,1} \times IO + \frac{1}{2} \Delta IP_{1,2} \times IO$$

$$IP_{IO,IO+2} = \frac{1}{2} IP_{1,1}(IO^2 + IO) + \frac{1}{2} \Delta IP_{1,3} \times (IO^2 + IO)$$

$$IP_{IO,IO+3} = \frac{1}{2} IP_{1,1}(IO^2 + IO) + \frac{1}{2} \Delta IP_{1,3}(IO^2 - IO) + \Delta IP_{1,4} IO, \text{ etc.}$$

Figure 2. Behavior of higher-order IP's.

The '*etc.*' means that the pattern established at $IP_{IO,IO+3}$ continues from there on. Indeed, even $IP_{IO,IO+2}$ is a special case of the general pattern revealed in $IP_{IO,IO+3}$. The first term in $IP_{IO,IO+3}$ is universal, involving only $IP_{1,1}$. The second term is period specific, involving $\Delta IP_{1,3}$, or $\Delta IP_{1,11}$, $\Delta IP_{1,19}$, $\Delta IP_{1,37}$, $\Delta IP_{1,55}$, $\Delta IP_{1,87}$ as needed. The third term is element specific, generally involving $\Delta IP_{1,Z}$ in place of $\Delta IP_{1,4}$.

The weakest area for the IP model fit to data is in the IP's for $IO = 2$ and $IO = 3$ elements in the sixth and seventh periods. The existence of this weak area may mean that the model needs further development. But the existence of the area of weaker fit may just mean that the experimental data are difficult to obtain there, or confusing to interpret. For example, observe that the reported second-order and third-order IP's there are very close to 2 and 3 times the corresponding first-order IP's. It is possible that the physical process that produced those data points was coincident production of multiple atoms each singly ionized, instead of production of a single atom multiply ionized, as witnessed everywhere else by all the IO^2 factors that fit the data available. The behavior of higher-order IP's is intriguing. It demands more explanation, and it invites some practical application. All of this is deferred to Part II, Chapter 4. The behavior of higher-order IP's is introduced here to make just one important point: absolutely *everything* is related to *everything* else. There is one grand pattern to it all.

And everything depends, first of all, on the behavior of first-order ionization potentials. This behavior is detailed next.

2. DETAILS ON BEHAVIOR OF FIRST-ORDER IONIZATION POTENTIALS

Figure 3 below characterizes the overall pattern of first-order IP's in terms of simple integers and ratios thereof. Multiple routes through the periods are available; bold font indicates the route actually taken in constructing Figure 1. The most startling fact is that the rise on *every* period is the same: total rise $= 7/2$.

For sub-periods, the incremental rises can be expressed in the form

incremental rise = total rise × fraction ,

For first sub-periods, the fractions are empirical numbers; for subsequent sub-periods, they fit the empirical formula:

$$\text{fraction} = \left[(2l+1)/N^2\right]\left[(N-l)/l\right],$$

The value of l in the pattern is the nominal angular momentum quantum number of the electron state being filled. The word 'nominal' is needed because nineteen 'peculiar' elements depart from nominal in their actual electron populations.

	$_1$H,	$IP_{1,1}$	→ 7/2 →	$IP_{1,2}$,	$_2$He	
7/8	↓	↓	← 1/4 ←	↵	↓	7/8
	$_3$Li,	$IP_{1,3}$	→ 7/2 →	$IP_{1,10}$,	$_{10}$Ne	
7/8	↓	↓	← 1/4 ←	↵	↓	7/8
	$_{11}$Na,	$IP_{1,11}$	→ 7/2 →	$IP_{1,18}$,	$_{18}$Ar	
7/8	↓	↓	← 1/4 ←	↵	↓	7/8
	$_{19}$K,	$IP_{1,19}$	→ 7/2 →	$IP_{1,36}$,	$_{36}$Kr	
1	↓	↓	← 2/7 ←	↵	↓	1
	$_{37}$Rb,	$IP_{1,37}$	→ 7/2 →	$IP_{1,54}$,	$_{54}$Xe	
1	↓	↓	← 2/7 ←	↵	↓	1
	$_{55}$Cs,	$IP_{1,55}$	→ 7/2 →	$IP_{1,86}$,	$_{86}$Rn	
1	↓	↓	← 2/7 ←	↵	↓	1
	$_{87}$Fr,	$IP_{1,87}$	→ 7/2 →	$IP_{1,118}$,	$_{118}$??	

Figure 3. First-order IP's: map of main highways through the periods.

Figure 4 quantifies the detailed pattern of first-order IP's. On each period rise, there are sub-period rises keyed to the traditional 'angular momentum' quantum number l, and to a non-traditional parameter N that goes $1,2,2,3,3,4,4$ for periods 1 through 7. The parameter N gives the number of elements in a period as $2N^2$. From Figure 4, it can also be characterized as counting the number of different l-values occurring in a period.

N	l	fraction	l	fraction	l	fraction	l	fraction
1	0	1						
2	0	1/2	1	3/4				
2	0	1/3	1	3/4				
3	0	1/4	2	5/18	1	2/3		
3	0	1/4	2	5/18	1	2/3		
4	0	1/4	3	7/48	2	5/16	1	9/16
4	0	1/4	3	7/48	2	5/16	1	9/16

Figure 4. First-order IP's: map of local roads through the periods.

The detailed characterization of the behavior of first-order IP's reiterates the same point that the study of higher-order IP's made before; namely, that there exists a certain level of universality about chemical information. Basic information about any one element can be inferred reliably from basic information about other elements.

CONCLUSION

This Chapter has revealed a lot of algebraic relationships among data points representing ionization potentials. You have seen that data about ionization potentials of order 1 to 7 makes the case that everything is related to everything else. The first-order ionization potential of each successive element reveals information about the higher-order ionization potentials of *all* subsequent elements. This fact speaks to the universality of chemical information: basic information about any one element can be inferred reliably from basic information about other elements.

For now, the information about the IP's of all orders has to be regarded as purely empirical. It is revealed by, but not explained by, the scaling of the ionization-potential raw data by the factor M/Z. Nor is it explained by any widely available theory. The need for explanation, or at least interpretation, is revisited in Part II, Chapter 4.

But first, the utility of the information is explored. The rest of Part I is about common industrial or domestic chemical reactions, which are gentle, or at least controllable. Exchanges of electrons occur one at a time, so higher-order ionization potentials aren't directly involved. But a principle drawn from their behavior is involved. Observe that the population-generic ΔIP's from all elements provided valid population-generic information about all elements of higher Z. That means the core information is dependent only on electron count, and not dependent on anything about the nucleus. This same principle is used to work out the energy requirements of electron exchanges that go one-at-a time. The actual numerical information about higher-order ionization potentials is potentially useful for more vigorous chemical reactions, where exchanges of electrons happen all at once. These are super-fast events, often occurring not as reactions among several molecules, but rather as detonations of individual molecules, so the participant atoms are very close at hand. One of these is investigated in Part II, Chapter 4. The present Chapter concludes with an Appendix that collects all the numerical data about first-order ionization potentials of all elements, including some that have not been found or created yet. Each $IP_{1,Z}$ is presented, and with it is the increment $\Delta IP_{1,Z}$. The $IP_{1,1}$ and the $\Delta IP_{1,Z}$ are the data used again and again throughout the Book. All data are expressed in electon volts, eV.

ACKNOWLEDGMENTS

The basic numerical data displayed in Figure 1 and detailed in the following Appendix has been presented in graphical and/or tabular form in many talks and papers, in many conferences and journals, and I thank them all for listening to it and/or publishing it. Special thanks go to *Foundations of Physics* for giving the first journal exposure back in 2007. My title then was *Relativistic Dynamics in Basic Chemistry* [vol. 37, nos. 4/5, pp. 788-812]. The formulation at that time did not make the separation of $IP_{1,Z}$ into $IP_{1,1}$ and $\Delta IP_{1,Z}$, so the formulae given then were different from those given now. Figure 1, however, looked exactly the same. Another special thanks goes to *International Journal of Molecular Sciences*, where Figure 1 was first reconstructed using the new formulation, in a paper for a Special Issue entitled *Atoms in Molecules and in Nanostructures* (2012, vol. 13), organized by Prof. M.V. Putz. The title of that paper is *On the Several Molecules and Nanostructures of Water* (pp. 1066-1094).

APPENDIX: BASIC DATA ON FIRST-ORDER IONIZATION POTENTIALS OF ATOMS

Periods 1, 2 and 3

Element	Charge Z	Mass M	Ionization Potential	IP = Ionization Potential × M/Z	Model IP	Model ΔIP
H	1	1.008	13.610	13.718	14.250	0
He	2	4.003	24.606	49.244	49.875	35.625
Li	3	6.941	5.394	12.480	12.469	−1.781
Be	4	9.012	9.326	21.011	23.327	9.077
B	5	10.811	8.309	17.966	17.055	2.805
C	6	12.011	11.266	22.551	21.570	7.320
N	7	14.007	14.544	29.101	27.281	13.031
O	8	15.999	13.631	27.260	27.281	13.031
F	9	18.998	17.438	36.810	34.504	20.254
Ne	10	20.180	21.587	43.562	43.641	29.391
Na	11	22.990	5.145	10.753	10.910	−3.340
Mg	12	24.305	7.656	15.506	16.565	2.315
Al	13	26.982	5.996	12.444	14.923	0.673
Si	14	28.086	8.154	16.357	18.874	4.624
P	15	30.974	10.498	21.677	23.871	9.621
S	16	32.066	10.373	20.790	23.871	9.621
Cl	17	35.453	12.977	27.063	30.192	15.942
Ar	18	39.948	15.778	35.017	38.186	23.936

Period 4

Element	Charge Z	Mass M	Ionization Potential	IP = Ionization Potential × M / Z	Model IP	Model ΔIP
K	19	39.098	4.346	8.944	9.546	−4.704
Ca	20	40.078	6.120	12.265	13.057	−1.193
Sc	21	44.956	6.546	14.013	13.057	−1.193
Ti	22	47.867	6.826	14.851	13.638	−0.612
V	23	50.942	6.743	14.934	14.244	−0.006
Cr	24	51.996	6.774	14.676	14.877	0.627
Mn	25	54.938	7.438	16.345	15.539	1.289
Fe	26	55.845	7.873	16.911	15.539	1.289
Co	27	58.933	7.863	17.163	16.229	1.980
Ni	28	58.693	7.645	16.026	16.951	2.701
Cu	29	63.546	7.728	16.934	17.705	3.455
Zn	30	65.390	9.398	20.485	18.492	4.242
Ga	31	69.723	6.006	13.509	14.494	0.244
Ge	32	72.610	7.905	17.936	17.860	3.610
As	33	74.922	9.824	22.303	22.007	7.757
Se	34	78.960	9.761	22.669	22.007	7.757
Br	35	79.904	11.826	26.998	27.116	12.866
Kr	36	83.800	14.015	32.623	33.412	19.162

Period 5

Element	Charge Z	Mass M	Ionization Potential	IP = Ionization Potential × M / Z	Model IP	Model ΔIP
Rb	37	85.468	4.180	9.657	9.546	−4.704
Sr	38	87.620	5.695	13.132	13.057	−1.193
Y	39	88.906	6.390	14.567	13.057	−1.193
Zr	40	91.224	6.846	15.614	13.638	−0.612
Nb	41	92.906	6.888	15.608	14.244	−0.006
Mo	42	95.940	7.106	16.232	14.877	0.627
Tc	43	98.000	7.282	16.597	15.539	1.289
Ru	44	101.070	7.376	16.942	15.539	1.289
Rh	45	102.906	7.469	17.080	16.230	1.980
Pd	46	106.420	8.351	19.319	16.951	2.701
Ag	47	107.868	7.583	17.403	17.705	3.455
Cd	48	112.411	9.004	21.087	18.492	4.242
In	49	114.818	5.788	13.563	14.494	0.244
Sn	50	118.710	7.355	17.462	17.860	3.610
Sb	51	121.760	8.651	20.655	22.007	7.757
Te	52	127.600	9.015	22.120	22.007	7.757
I	53	126.904	10.456	25.037	27.116	12.866
Xe	54	131.290	12.137	29.508	33.412	19.162

Periods 6 and 7 follow

Element	Charge Z	Mass M	Ionization Potential	IP = Ionization Potential × M / Z	Model IP	Model ΔIP
Cs	55	132.905	3.900	9.425	9.546	−4.704
Ba	56	137.327	5.218	12.796	13.057	−1.192
La	57	138.906	5.581	13.600	12.393	−1.857
Ce	58	140.116	5.477	13.232	12.583	−1.667
Pr	59	140.908	5.425	12.957	12.776	−1.474
Nd	60	144.240	5.498	13.217	12.972	−1.278
Pm	61	145.000	5.550	13.192	13.171	−1.079
Sm	62	150.360	5.633	13.660	13.374	−0.876
Eu	63	151.964	5.674	13.687	13.579	−0.671
Gd	64	157.250	6.141	15.089	13.579	−0.671
Tb	65	158.925	5.851	14.305	13.787	−0.463
Dy	66	162.500	5.934	14.609	13.999	−0.251
Ho	67	164.930	6.027	14.836	14.213	−0.037
Er	68	167.260	6.110	15.029	14.431	0.181
Tm	69	168.934	6.183	15.137	14.653	0.403
Yb	70	170.040	6.255	15.463	14.878	0.627
Lu	71	174.967	5.436	13.395	17.860	3.610
Hf	72	178.490	7.054	17.487	18.755	4.505
Ta	73	180.948	7.894	19.568	19.696	5.446
W	74	183.840	7.988	19.844	20.684	6.434
Re	75	186.207	7.884	19.574	21.721	7.471
Os	76	190.230	8.714	21.811	21.721	7.471
Ir	77	192.217	9.129	22.788	22.811	8.560
Pt	78	195.076	9.025	22.571	23.955	9.705
Au	79	196.967	9.232	23.019	25.156	10.906
Hg	80	200.530	10.446	26.184	26.418	12.168
Tl	81	204.383	6.110	15.417	16.515	2.265
Pb	82	207.200	7.427	18.768	19.696	5.446
Bi	83	208.980	7.293	18.361	23.490	9.240
Po	84	209.000	8.423	20.958	23.490	9.240
At	85	210.000			28.015	13.765
Rn	86	222.000	10.757	27.769	33.412	19.164

Element	Charge Z	Mass M	Ionization Potential	IP = Ionization Potential × M / Z	Model IP	Model ΔIP
Fr	87	223.000			9.546	−4.704
Ra	88	226.000	5.280	13.560	13.057	−1.193
Ac	89	227.000	6.950	17.727	12.393	−1.857
Th	90	232.038	6.089	15.699	12.583	−1.667
Pa	91	231.036	5.892	14.959	12.776	−1.474
U	92	238.029	6.203	16.050	12.972	−1.277
Np	93	237.000	6.276	15.994	13.171	−1.079
Pu	94	244.000	6.068	15.752	13.374	−0.876
Am	95	243.000	5.996	15.337	13.579	−0.671
Cm	96	247.000	6.027	15.507	13.579	−0.671
Bk	97	247.000	6.234	15.875	13.787	−0.463
Cf	98	251.000	6.307	16.154	13.999	−0.251
Es	99	252.000	6.421	16.345	14.213	−0.037
Fm	100	257.000	6.504	16.716	14.431	0.181
Md	101	258.000	6.587	16.827	14.653	0.403
No	102	259.000	6.660	16.911	14.877	0.627
Lf	103				17.859	3.610
Rf	104				18.755	4.505
Db	105				19.696	5.446
Sg	106				20.684	6.434
Bh	107				21.721	7.471
Hs	108				21.721	7.471
Mt	109				22.811	8.561
Uun	110				23.955	9.705
Uuu	111				25.156	10.906
Uub	112				26.418	12.168
???	113				16.515	2.265
???	114				17.019	2.769
???	115				23.490	9.240
???	116				23.490	9.240
???	117				28.015	13.765
???	118				33.412	19.162

Chapter 2

IONIZATION POTENTIALS OF IONS

ABSTRACT

This Chapter takes the next step in characterizing the ionized species that play roles in Chemistry. With ionization potentials of neutral atoms in hand, we need to infer from those the ionization potentials involved in changing the ionization states of all manner of ions. This task includes not only the stripping of more electrons off an already singly ionized atom, but also the addition of arbitrarily many electrons to a neutral atom. This Chapter works out a rational way to do this, based upon, and extrapolating from, the M/Z scaling that brought the first-order ionization potentials into focus as IP's, and based upon the separation into $IP_{1,1}$, for the gross interaction between the nucleus and the whole electron population, and $\Delta IP_{1,Z}$, for interactions just among the electrons. The extrapolation is based on the interpretation that a factor of Z for a neutral atom becomes a factor of $\sqrt{Z_p Z_e}$ for an ion with Z_p proton count and Z_e electron count.

INTRODUCTION

The preceding Chapter revealed some empirical facts about the ionization potentials of neutral atoms. Some of the facts are not yet fully explained, but all of them are nevertheless useful. The first of these facts is that, when multiplied by nuclear charge Z, divided by nuclear mass M, the available data reveal a regular pattern that embraces all ionization potentials of all orders for which data is available. The scope of the applicability of the pattern means that chemical information is really universal. A fact about any one element maps into a fact about any other element.

The second important fact is the Z/M scaling itself. It serves to expose even more regularity in the Periodic Table than was previously appreciated. The regularity is not just bevavioral - it is numerical. For example, the total rise of first-order ionization potential over every period is the same: a factor of $7/2$. And within a period, there are sub-periods keyed to the nominal quantum numbers of single electron states being successively filled, along with another parameter N, which provides the total period lengths, $2N^2$.

The third important fact is that when data are plotted on a log scale versus Z, the ionization potentials present themselves as a sequence of straight lines. Such straight lines mean power laws. We can reasonably anticipate that the ionization potentials of already-existing ions will exhibit power laws too.

We can also reasonably anticipate that similar, but slightly more general, patterns describe ions as well. We know already about the pattern exhibited by higher-order ionization potentials. That appeared to reflect rather vigorous 'all-at-once' ionization scenarios. There has to exist something like it for gentler, 'one-at-a-time' ionization scenarios. The purpose of the present Chapter is to show what this pattern is.

The Chapter concludes with a quite long Appendix that provides copious numerical results for handy reference throughout the remainder of this book.

1. MODEL DEVELOPMENT

Recall that we separated $IP_{1,1}$ and $\Delta IP_{1,Z}$. For first-order ionizations, the IP's scale with Z/M. This means that the constituent parts, $IP_{1,1}$ and $\Delta IP_{1,Z}$, both scale with Z/M too. But for an already ionized atom, the modifications to Z/M will be different for the two parts, $IP_{1,1}$ and $\Delta IP_{1,Z}$.

The M appropriate for an $IP_{1,1}$ remains the nuclear mass M. And the M appropriate for a ΔIP is also the nuclear mass, even though ΔIP's are about electron-electron interactions, and *not* electron-nucleus interactions. Why? The reason is as follows. $\Delta IP_{1,Z}$ is the minimum energy required to take an electron just outside of the electron subsystem. The $\Delta IP_{1,Z}$ depends on both the natural radius ΔR of the electron sub-system, and on the radius R_0 at which the electron subsystem center orbits in the $1/R$ potential well created by the nucleus. The energy required to take the electron out of the electron subsystem scales as $\Delta R / R_0$, and R_0 is linear with M.

In general, the Z appropriate for either an $IP_{1,1}$ or for $\Delta IP_{1,Z}$ is not the nuclear proton count Z_p. Instead, the electron count Z_e is involved in some way. The electron count Z_e and the proton count Z_p are the same Z only for a neutral atom.

In the case of $IP_{1,1}$, the appropriate Z is the geometric mean of Z_p and Z_e, or square root of their product, $\sqrt{Z_p Z_e}$. Why is the product involved? It is because the electron – nucleus interaction is a Coulomb interaction, and so involves the product. Why is the square root applied? Clearly, the expression then reduces to the correct expression when $Z_e = Z_p \equiv Z$. More detailed mathematical support for this assertion comes later in this Book, in Part II, Chapter 1.

In general, the Z appropriate for $\Delta IP_{1,Z}$ does not at all involve the proton count Z_p, but rather just the electron count Z_e. This is because $\Delta IP_{1,Z}$ is about electron-on-electron interactions. The proton count Z_p is not relevant. Only Z_e can matter.

Recall that $\Delta IP_{1,Z}$ is usually positive. This means that the electron's best escape route usually departs from the greatest possible distance from the nucleus, $R_0 + \Delta R$. But $\Delta IP_{1,Z}$ is sometimes negative. That happens, for example, for alkali metals. The negative $\Delta IP_{1,Z}$ suggests that ΔR is actually larger than R_0, and the electron's best escape route departs from the other side of the nucleus, at distance $\Delta R - R_0$ away from the nucleus.

2. Symbolic Formulae

The formulae are essentially the same for every element, so let us use the symbol '$_Z E$' for an arbitrary element, so we can write the formulae in a symbolic way.

First consider the transition $_Z E \rightarrow {}_Z E^+$. It definitely takes an energy investment of $IP_{1,Z} \times Z / M_Z$, where the factors of Z and $1/M_Z$ restore the population-generic information $IP_{1,Z}$ to element-specific information. This energy investment corresponds to a potential wall to be gotten over. The wall has two parts, $IP_{1,1} \times Z / M_Z$ and $\Delta IP_{1,Z} \times Z / M_Z$. The transition $_Z E \rightarrow {}_Z E^+$ may also consume some heat, or generate some heat, as the remaining $Z-1$ electrons form new relationships, not necessarily instantaneously. This process constitutes adjustment to the rock pile, or the ditch, on the other side of the potential wall. It is represented by a term $-\Delta IP_{1,Z-1} \times (Z-1)/M_Z$, where the factors of $(Z-1)$ and $1/M_Z$ restore the population-generic information $-\Delta IP_{1,Z-1}$ to element-specific information tailored for $_Z E$. Thus altogether,

$_Z E \rightarrow {}_Z E^+$ takes:

$$IP_{1,1} \times Z/M_Z + \Delta IP_{1,Z} \times Z/M_Z - \Delta IP_{1,Z-1} \times (Z-1)/M_Z.$$

Now consider removal of a second electron, $_Z\mathrm{E}^+ \to {_Z\mathrm{E}^{2+}}$.

Being already stripped of one of its electrons, the $_Z\mathrm{E}^+$ system has less internal Coulomb attraction than neutral $_Z\mathrm{E}$ has. So the factor of Z multiplying $IP_{1,1}$ for $_Z\mathrm{E} \to {_Z\mathrm{E}^+}$ has to change to something smaller. Since Coulomb attraction generally reflects the product of the number of positive charges (here Z) and the number of negative charges (here $Z-1$), the reduced factor is $\sqrt{Z \times (Z-1)}$. Given this factor,

$_Z\mathrm{E}^+ \to {_Z\mathrm{E}^{2+}}$ takes:

$$IP_{1,1} \times \sqrt{Z \times (Z-1)}/M_Z$$
$$+\Delta IP_{1,Z-1} \times (Z-1)/M_Z - \Delta IP_{1,Z-2} \times (Z-2)/M_Z .$$

Observe that putting the steps $_Z\mathrm{E} \to {_Z\mathrm{E}^+}$ and $_Z\mathrm{E}^+ \to {_Z\mathrm{E}^{2+}}$ together, the terms involving $\Delta IP_{1,Z-1}$ cancel, leaving that altogether,

$_Z\mathrm{E} \to {_Z\mathrm{E}^{2+}}$ takes:

$$IP_{1,1} \times Z/M_Z + IP_{1,1}\sqrt{Z \times (Z-1)}/M_Z$$
$$+\Delta IP_{1,Z} \times Z/M_Z - \Delta IP_{1,Z-2} \times (Z-2)/M_Z$$

This reduction to just two terms involving ΔIP's is typical of all sequential ionizations, of however many steps.

Observe too that in the cumulative, we have *two* terms in $IP_{1,1}$, $IP_{1,1} \times Z/M_Z$ and $IP_{1,1}\sqrt{Z \times (Z-1)}/M_Z$. If we went further in stripping electrons, we would have IO terms in $IP_{1,1}$. This feature is somewhat reminiscent of the higher-order IP model, which had terms linear in ionization order, IO.

From the above, it should be clear how to proceed with stripping however many more electrons you may be interested in removing.

For another example, think about adding an electron to $_Z\mathrm{E}$. The problem is similar to removing an electron from $_{Z+1}\mathrm{E}$, but in reverse. So,

$_Z E \to {}_Z E^-$ takes:

$$-IP_{1,1} \times \sqrt{Z \times (Z+1)}/M_Z - \Delta IP_{1,Z+1}(Z+1)/M_Z + \Delta IP_{1,Z} Z/M_Z.$$

Going one step further, the problem of adding another electron to $_Z E^-$ is similar to removing an electron from $_{Z+2} E$, but in reverse. So,

$_Z E^- \to {}_Z E^{2-}$ takes:

$$-IP_{1,1} \times \sqrt{Z \times (Z+2)}/M_Z \\ -\Delta IP_{1,Z+2} \times (Z+2)/M_Z + \Delta IP_{1,Z+1} \times (Z+1)/M_Z.$$

Observe that putting the steps $_Z E \to {}_Z E^-$ and $_Z E^- \to {}_Z E^{2-}$ together, the terms involving $\Delta IP_{1,Z+1}$ cancel, leaving that altogether,

$_Z E \to {}_Z E^{2-}$ takes:

$$-IP_{1,1} \times \sqrt{Z \times (Z+1)}/M_Z - IP_{1,1} \times \sqrt{Z \times (Z+2)}/M_Z \\ +\Delta IP_{1,Z} \times Z/M_Z - \Delta IP_{1,Z+2} \times (Z+2)/M_Z.$$

From the above, it should be clear how to proceed with adding however many more electrons you may be interested in adding.

For future reference, here is a summary of formulae for example element $_Z E$:

$_Z E \to {}_Z E^+$:
$$IP_{1,1} \times Z/M_Z + \Delta IP_{1,Z} \times Z/M_Z - \Delta IP_{1,Z-1} \times (Z-1)/M_Z$$

$_Z E^+ \to {}_Z E^{2+}$:
$$IP_{1,1} \times Z/M_Z + IP_{1,1} \sqrt{Z \times (Z-1)}/M_Z \\ +\Delta IP_{1,Z} \times Z/M_Z - \Delta IP_{1,Z-2} \times (Z-2)/M_Z$$

etc.

$_ZE \to {_Z}E^-$:

$-IP_{1,1} \times \sqrt{Z \times (Z+1)}/M_Z - \Delta IP_{1,Z+1}(Z+1)/M_Z + \Delta IP_{1,Z}Z/M_Z$

$_ZE^- \to {_Z}E^{2-}$:

$-IP_{1,1} \times \sqrt{Z \times (Z+2)}/M_Z$

$-\Delta IP_{1,Z+2} \times (Z+2)/M_Z + \Delta IP_{1,Z+1} \times (Z+1)/M_Z$

etc.

CONCLUSION

You now have the information necessary to evaluate the energy requirement for any ionization state of any element. But of course there is a lot of calculation involved. Appendix 1 is included to save you some work in the future. It documents many elements, and it treats some of them to a high degree of ionization. I have chosen the elements and the extent of evaluation based on what data you will need later in the Book, what elements I happen to mention later in the Book, and what elements I just felt personally curious about. There are enough elements and ionization states treated here to thoroughly cement your appreciation for the pattern involved. There are always three steps: 1) Write the formulae, using the summary of formulae for the example element $_ZE$; 2) Insert the data, using the Appendix to Chapter 1; 3) Evaluate the formulae, using a hand calculator.

AN INVITATION TO READERS

Many of you have on-going work involving elements other than the ones featured below. You can develop more information of a similar kind for those elements, and it can be integrated into later editions of this Book, with a credit line for you.

APPENDIX: FORMULAE AND EVALUATIONS FOR SEQUENTIAL IONIZATIONS OF SELECTED ELEMENTS

1. Hydrogen

Write Formulae:

(Note: $\Delta IP_{1,1} \equiv 0$, and $\Delta IP_{1,0}$ does not exist.)

$_1H \to {_1}H^+$: $IP_{1,1} \times 1/M_1 + \Delta IP_{1,1} \times 1/M_1 - \Delta IP_{1,0} \times 0/M_1$

$_1H \rightarrow {_1}H^-$:

$-IP_{1,1} \times \sqrt{1 \times 2}/M_1 - \Delta IP_{1,2} \times 2/M_1 + \Delta IP_{1,1} \times 1/M_1$

$_1H^- \rightarrow {_1}H^{2-}$:

$-IP_{1,1} \times \sqrt{1 \times 3}/M_1 - \Delta IP_{1,3} \times 3/M_1 + \Delta IP_{1,2} \times 2/M_1$

$_1H^{2-} \rightarrow {_1}H^{3-}$:

$-IP_{1,1} \times \sqrt{1 \times 4}/M_1 - \Delta IP_{1,4} \times 4/M_1 + \Delta IP_{1,3} \times 3/M_1$

Insert Data:

$_1H \rightarrow {_1}H^+$: $14.250 \times 1/1.008 + 0 - 0$

$_1H \rightarrow {_1}H^-$:

$-14.250 \times 1.4142/1.008 - 35.625 \times 2/1.008 + 0 \times 1/1.008$

$_1H^- \rightarrow {_1}H^{2-}$:

$-14.250 \times 1.7321/1.008 - (-1.781) \times 3/1.008 + 35.625 \times 2/1.008$

$_1H^{2-} \rightarrow {_1}H^{3-}$:

$-14.250 \times 2/1.008 - 9.077 \times 4/1.008 + (-1.781) \times 3/1.008$

Evaluate Formulae:

$_1H \rightarrow {_1}H^+$: $14.1369 + 0 - 0 = 14.1369$

$_1H \rightarrow {_1}H^-$: $-19.9924 - 70.6845 + 0 = -90.6769$

$_1H^- \rightarrow {_1}H^{2-}$: $-24.4865 + 5.3006 + 70.6845 = 51.4986$

$_1H^{2-} \rightarrow {_1}H^{3-}$: $-28.2738 - 36.0198 - 5.3006 = -69.5942$

2. Helium

Write Formulae:

(Note: $\Delta IP_{1,1} \equiv 0$.)

$_2\text{He} \to {}_2\text{He}^+$: $IP_{1,1} \times 2 / M_2 + \Delta IP_{1,2} \times 2 / M_2 - \Delta IP_{1,1} \times 1 / M_2$

$_2\text{He}^+ \to {}_2\text{He}^{2+}$:
$IP_{1,1} \times \sqrt{2 \times 1}/M_2 - \Delta IP_{1,1} \times 1/M_2 - \Delta IP_{1,0} \times 0 / M_2$

$_2\text{He} \to {}_2\text{He}^-$:
$-IP_{1,1} \times \sqrt{2 \times 3}/M_2 - \Delta IP_{1,3} \times 3 / M_2 + \Delta IP_{1,2} \times 2 / M_2$

$_2\text{He}^- \to {}_2\text{He}^{2-}$:
$-IP_{1,1} \times \sqrt{2 \times 4}/M_2 - \Delta IP_{1,4} \times 4 / M_2 + \Delta IP_{1,3} \times 3 / M_2$

Insert Data:

$_2\text{He} \to {}_2\text{He}^+$:
$\quad -14.250 \times 2.4495/4.003 - (-1.781) \times 3 / 4.003 + 35.625 \times 2 / 4.003$

$_2\text{He}^+ \to {}_2\text{He}^{2+}$:
$14.250 \times 1.4142/4.003 + 0 \times 1 / 4.003 - 0 \times 0 / 4.003$

$_2\text{He} \to {}_2\text{He}^-$:
$\quad -14.250 \times 2.4495/4.003 - (-1.781) \times 3 / 4.003 + 35.625 \times 2 / 4.003$

$_2\text{He}^- \to {}_2\text{He}^{2-}$:
$\quad -14.250 \times 2.8284/4.003 - 9.077 \times 4 / 4.003 + (-1.781) \times 3 / 4.003$

Evaluate Formulae:

$_2\text{He} \to {_2\text{He}^+}$: $7.1197 + 17.7992 - 0 = 24.9189$

$_2\text{He}^+ \to {_2\text{He}^{2+}}$: $5.0343 + 0 + 0 = 5.0343$

$_2\text{He} \to {_2\text{He}^-}$: $-8.7198 + 1.3347 + 17.7992 = 10.4141$

$_2\text{He}^- \to {_2\text{He}^{2-}}$: $-10.0686 - 9.0702 + -1.3347 = -20.4735$

3. Lithium

Write Formulae:

$_3\text{Li} \to {_3\text{Li}^+}$: $IP_{1,1} \times 3 / M_3 + \Delta IP_{1,3} \times 3 / M_3 - \Delta IP_{1,2} \times 2 / M_3$

$_3\text{Li}^+ \to {_3\text{Li}^{2+}}$:
$IP_{1,1} \times \sqrt{3 \times 2}/M_3 + \Delta IP_{1,2} \times 2 / M_3 - \Delta IP_{1,1} \times 1 / M_3$

$_3\text{Li}^{2+} \to {_3\text{Li}^{3+}}$:
$IP_{1,1} \times \sqrt{3 \times 1}/M_3 + \Delta IP_{1,1} \times 1 / M_3 - \Delta IP_{1,0} \times 0 / M_3$

$_3\text{Li} \to {_3\text{Li}^-}$:
$-IP_{1,1} \times \sqrt{3 \times 4}/M_3 - \Delta IP_{1,4} \times 4 / M_3 + \Delta IP_{1,3} \times 3 / M_3$

$_3\text{Li}^- \to {_3\text{Li}^{2-}}$:
$-IP_{1,1} \times \sqrt{3 \times 5}/M_3 - \Delta IP_{1,5} \times 5 / M_3 + \Delta IP_{1,4} \times 4 / M_3$

$_3\text{Li}^{--} \to {_3\text{Li}^{3-}}$:
$-IP_{1,1} \times \sqrt{3 \times 6}/M_3 - \Delta IP_{1,6} \times 6 / M_3 + \Delta IP_{1,5} \times 5 / M_3$

Insert Data:

$_3$Li → $_3$Li$^+$:
$14.250 \times 3 / 6.941 + (-1.781) \times 3 / 6.941 - 35.625 \times 2 / 6.941$

$_3$Li$^+$ → $_3$Li^{2+}: $14.250 \times 2.449/6.941 + 35.625 \times 2 / 6.941$

$_3$Li^{2+} → $_3$Li^{3+}:
$14.250 \times 1.7321/6.941 + 0 \times 1 / 6.941 - 0 \times 0 / 6.941$

$_3$Li → $_3$Li$^-$:
$-14.250 \times 3.464/6.941 - 9.077 \times 4 / 6.941 + (-1.781) \times 3 / 6.941$

$_3$Li$^-$ → $_3$Li^{2-}:
$-14.250 \times 3.8729/6.941 - 2.805 \times 5 / 6.941 + 9.077 \times 4 / 6.941$

$_3$Li^{--} → $_3$Li^{3-}:
$-14.250 \times 4.2426/6.941 - 7.320 \times 6 / 6.941 + 2.805 \times 5 / 6.941$

Evaluate Formulae:

$_3$Li → $_3$Li$^+$: $6.1591 + -0.7698 - 10.2651 = -4.8758$

$_3$Li$^+$ → $_3$Li^{2+}: $5.0278 + 10.2651 = 15.2929$

$_3$Li^{2+} → $_3$Li^{3+}: $3.5560 + 0 - 0 = 3.5560$

$_3$Li → $_3$Li$^-$: $-7.1117 - 5.2309 - 0.7698 = -13.1124$

$_3$Li$^-$ → $_3$Li^{2-}: $-7.9511 - 2.0206 + 5.2309 = -4.7408$

$_3$Li^{2-} → $_3$Li^{3-}: $-8.7101 - 6.3276 + 2.0206 = -13.0171$

4. Beryllium

Write Formulae:

$_4\text{Be} \rightarrow {_4\text{Be}^+}$: $IP_{1,1} \times 4/M_4 + \Delta IP_{1,4} \times 4/M_4 - \Delta IP_{1,3} \times 3/M_4$

$_4\text{Be}^+ \rightarrow {_4\text{Be}^{2+}}$:
$IP_{1,1} \times \sqrt{4 \times 3}/M_4 + \Delta IP_{1,3} \times 3/M_4 - \Delta IP_{1,2} \times 2/M_4$

$_4\text{Be}^{2+} \rightarrow {_4\text{Be}^{3+}}$:
$IP_{1,1} \times \sqrt{4 \times 2}/M_4 + \Delta IP_{1,2} \times 2/M_4 - \Delta IP_{1,1} \times 1/M_4$

$_4\text{Be}^{3+} \rightarrow {_4\text{Be}^{4+}}$:
$IP_{1,1} \times \sqrt{4 \times 1}/M_4 + \Delta IP_{1,1} \times 1/M_4 - \Delta IP_{1,0} \times 2/M_4$

$_4\text{Be} \rightarrow {_4\text{Be}^-}$:
$-IP_{1,1} \times \sqrt{4 \times 5}/M_4 - \Delta IP_{1,5} \times 5/M_4 + \Delta IP_{1,4} \times 4/M_4$

$_4\text{Be}^- \rightarrow {_4\text{Be}^{2-}}$:
$-IP_{1,1} \times \sqrt{4 \times 6}/M_4 - \Delta IP_{1,6} \times 6/M_4 + \Delta IP_{1,5} \times 5/M_4$

$_4\text{Be}^{2-} \rightarrow {_4\text{Be}^{3-}}$:
$-IP_{1,1} \times \sqrt{4 \times 7}/M_4 - \Delta IP_{1,7} \times 7/M_4 + \Delta IP_{1,6} \times 6/M_4$

$_4\text{Be}^{3-} \rightarrow {_4\text{Be}^{4-}}$:
$-IP_{1,1} \times \sqrt{4 \times 8}/M_4 - \Delta IP_{1,8} \times 8/M_4 + \Delta IP_{1,7} \times 7/M_4$

Insert Data:

$_4Be \to {}_4Be^+$:
$14.250 \times 4 / 9.012 + 9.077 \times 4 / 9.012 - (-1.781) \times 3 / 9.012$

$_4Be^+ \to {}_4Be^{2+}$:
$14.250 \times 3.4641 / 9.012 + (-1.781) \times 3 / 9.012 - 35.625 \times 2 / 9.012$

$_4Be^{2+} \to {}_4Be^{3+}$:
$14.250 \times 2.8284 / 9.012 + 35.625 \times 2 / 9.012 - 0 \times 1 / 9.012$

$_4Be^{3+} \to {}_4Be^{4+}$:
$14.250 \times 2 / 9.012 + 0 \times 1 / 9.012 - 0 \times 2 / 9.012$

$_4Be \to {}_4Be^-$:
$-14.250 \times 4.472 / 9.012 - 2.805 \times 5 / 9.012 + 9.077 \times 4 / 9.012$

$_4Be^- \to {}_4Be^{2-}$:
$-14.250 \times 4.899 / 9.077 - 7.320 \times 6 / 9.077 + 2.805 \times 5 / 9.077$

$_4Be^{2-} \to {}_4Be^{3-}$:
$-14.250 \times 5.2915 / 9.012 - 13.031 \times 7 / 9.012 + 7.320 \times 6 / 9.012$

$_4Be^{3-} \to {}_4Be^{4-}$:
$-14.250 \times 5.6569 / 9.012 - 13.031 \times 8 / 9.012 + 13.031 \times 7 / 9.012$

Evaluate Formulae:

$_4Be \to {}_4Be^+$: $6.3249 + 4.02885 + 0.5929 = 10.9466$

$_4Be^+ \to {}_4Be^{2+}$: $5.4775 - 0.5929 - 7.9061 = -6.7203$

$_4Be^{2+} \to {_4}Be^{3+}$: $4.4723 + 7.9061 - 0 = 12.3784$

$_4Be^{3+} \to {_4}Be^{4+}$: $3.1625 + 0 - 0 = 3.1625$

$_4Be \to {_4}Be^-$: $-7.0712 - 1.556 + 4.0289 = -4.5983$

$_4Be^- \to {_4}Be^{2-}$: $-7.6909 - 4.8386 + 1.5451 = -10.9844$

$_4Be^{2-} \to {_4}Be^{3-}$: $-8.3671 - 10.1217 + 4.8735 = -13.6153$

$_4Be^{3-} \to {_4}Be^{4-}$: $-8.9448 - 11.5677 + 10.1217 = -10.3908$

5. Boron

Write Formulae:

$_5B \to {_5}B^+$: $IP_{1,1} \times 5 / M_5 + \Delta IP_{1,5} \times 5 / M_5 - \Delta IP_{1,4} \times 4 / M_5$

$_5B^+ \to {_5}B^{2+}$:
$IP_{1,1} \times \sqrt{5 \times 4} / M_5 + \Delta IP_{1,4} \times 4 / M_5 - \Delta IP_{1,3} \times 3 / M_5$

$_5B^{2+} \to {_5}B^{3+}$:
$IP_{1,1} \times \sqrt{5 \times 3} / M_5 + \Delta IP_{1,3} \times 3 / M_5 - \Delta IP_{1,2} \times 2 / M_5$

$_5B^{3+} \to {_5}B^{4+}$:
$IP_{1,1} \times \sqrt{5 \times 2} / M_5 + \Delta IP_{1,2} \times 2 / M_5 - \Delta IP_{1,1} \times 1 / M_5$

$_5B^{4+} \to {_5}B^{5+}$:
$IP_{1,1} \times \sqrt{5 \times 1} / M_5 + \Delta IP_{1,1} \times 1 / M_5 - \Delta IP_{1,0} \times 1 / M_5$

$_5B \to {}_5B^-$:
$-IP_{1,1} \times \sqrt{5 \times 6}/M_5 - \Delta IP_{1,6} \times 6 / M_5 + \Delta IP_{1,5} \times 5 / M_5$

$_5B^- \to {}_5B^{2-}$:
$-IP_{1,1} \times \sqrt{5 \times 7}/M_5 - \Delta IP_{1,7} \times 7 / M_5 + \Delta IP_{1,6} \times 6 / M_5$

$_5B^{2-} \to {}_5B^{3-}$:
$-IP_{1,1} \times \sqrt{5 \times 8}/M_5 - \Delta IP_{1,8} \times 8 / M_5 + \Delta IP_{1,7} \times 7 / M_5$

$_5B^{3-} \to {}_5B^{4-}$:
$-IP_{1,1} \times \sqrt{5 \times 9}/M_5 - \Delta IP_{1,9} \times 9 / M_5 + \Delta IP_{1,8} \times 8 / M_5$

$_5B^{4-} \to {}_5B^{5-}$:
$-IP_{1,1} \times \sqrt{5 \times 10}/M_5 - \Delta IP_{1,10} \times 10 / M_5 + \Delta IP_{1,9} \times 9 / M_5$

Insert Data:

$_5B \to {}_5B^+$:
$14.250 \times 5 / 10.811 + 2.805 \times 5 / 10.811 - 9.077 \times 4 / 10.811$

$_5B^+ \to {}_5B^{2+}$:
$14.250 \times 4.4721/10.811 + 9.077 \times 4 / 10.811 - (-1.781) \times 3 / 10.811$

$_5B^{2+} \to {}_5B^{3+}$:
$14.250 \times 3.8730/10.811 + (-1.781) \times 3 / 10.811 - 35.625 \times 2 / 10.811$

$_5B^{3+} \to {}_5B^{4+}$:
$14.250 \times 3.1623/10.811 + 35.625 \times 2 / 10.811 - 0 \times 1 / 10.811$

$_5B^{4+} \to {_5}B^{5+}$:

$14.250 \times 2.2361/10.811 + 0 \times 1/10.811 - 0 \times 1/10.811$

$_5B \to {_5}B^-$:

$-14.250 \times 5.4772/10.811 - 7.320 \times 6/10.811 + 2.805 \times 5/10.811$

$_5B^- \to {_5}B^{2-}$:

$-14.250 \times 5.9161/10.811 - 13.031 \times 7/10.811 + 7.320 \times 6/10.811$

$_5B^{2-} \to {_5}B^{3-}$:

$-14.250 \times 6.3245/10.811 - 13.031 \times 8/10.811 + 13.031 \times 7/10.811$

$_5B^{3-} \to {_5}B^{4-}$:

$-14.250 \times 6.708/10.811 - 20.254 \times 9/10.811 + 13.031 \times 8/10.811$

$_5B^{4-} \to {_5}B^{5-}$:

$-14.250 \times 7.0711/10.811 - \Delta IP_{1,10} \times 10/10.811 + \Delta IP_{1,9} \times 9/10.811$

Evaluate Formulae:

$_5B \to {_5}B^+$: $6.5905 + 1.2973 - 3.3584 = 4.5294$

$_5B^+ \to {_5}B^{2+}$: $5.8947 + 3.3584 + 0.4942 = 9.7473$

$_5B^{2+} \to {_5}B^{3+}$: $5.1050 - 0.4942 - 6.5905 = -1.9797$

$_5B^{3+} \to {_5}B^{4+}$: $4.1682 + 6.5905 - 0 = 10.7587$

$_5B^{4+} \to {_5}B^{5+}$: $2.9474 + 0 - 0 = 2.9474$

$_5B \to {_5}B^-$: $-7.2195 - 4.0625 + 1.2973 = -9.9847$

$_5B^- \to {}_5B^{2-}$: $-7.7980 - 8.4374 + 4.0625 = -12.1729$

$_5B^{2-} \to {}_5B^{3-}$: $-8.3363 - 9.6428 + 8.4374 = -9.5417$

$_5B^{3-} \to {}_5B^{4-}$: $-8.8418 - 16.8612 + 9.6428 = -16.0602$

$_5B^{4-} \to {}_5B^{5-}$: $-9.3204 - 27.1862 + 16.8612 = -19.6454$

6. Carbon

Write Formulae:

$_6C \to {}_6C^+$: $IP_{1,1} \times 6 / M_6 + \Delta IP_{1,6} \times 6 / M_6 - \Delta IP_{1,5} \times 5 / M_6$

$_6C^+ \to {}_6C^{2+}$:
$IP_{1,1} \times \sqrt{6 \times 5} / M_6 + \Delta IP_{1,5} \times 5 / M_6 - \Delta IP_{1,4} \times 4 / M_6$

$_6C^{2+} \to {}_6C^{3+}$:
$IP_{1,1} \times \sqrt{6 \times 4} / M_6 + \Delta IP_{1,4} \times 4 / M_6 - \Delta IP_{1,3} \times 3 / M_6$

$_6C^{3+} \to {}_6C^{4+}$:
$IP_{1,1} \times \sqrt{6 \times 3} / M_6 + \Delta IP_{1,3} \times 3 / M_6 - \Delta IP_{1,2} \times 2 / M_6$

$_6C \to {}_6C^-$:
$-IP_{1,1} \times \sqrt{6 \times 7} / M_6 - \Delta IP_{1,7} \times 7 / M_6 + \Delta IP_{1,6} \times 6 / M_6$

$_6C^- \to {}_6C^{2-}$:
$-IP_{1,1} \times \sqrt{6 \times 8} / M_6 - \Delta IP_{1,8} \times 8 / M_6 + \Delta IP_{1,7} \times 7 / M_6$

$_6C^{2-} \to {}_6C^{3-}$:

$-IP_{1,1} \times \sqrt{6 \times 9}/M_6 - \Delta IP_{1,9} \times 9/M_6 + \Delta IP_{1,8} \times 8/M_6$

$_6C^{3-} \to {}_6C^{4-}$:

$-IP_{1,1} \times \sqrt{6 \times 10}/M_6 - \Delta IP_{1,10} \times 10/M_6 + \Delta IP_{1,9} \times 9/M_6$

Insert Data:

$_6C \to {}_6C^+$:
$14.250 \times 6/12.011 + 7.320 \times 6/12.011 - 2.805 \times 5/12.011$

$_6C^+ \to {}_6C^{2+}$:
$14.250 \times 5.4772/12.011 + 2.805 \times 5/12.011 - 9.077 \times 4/12.011$

$_6C^{2+} \to {}_6C^{3+}$:
$14.250 \times 4.8990/12.011 + 9.077 \times 4/12.011 - (-1.781) \times 3/12.011$

$_6C^{3+} \to {}_6C^{4+}$:
$14.250 \times 4.2426/12.011 + (-1.781) \times 3/12.011 - 35.625 \times 2/12.011$

$_6C \to {}_6C^-$:
$-14.250 \times 6.4807/12.011 - 13.031 \times 7/12.011 + 7.320 \times 6/12.011$

$_6C^- \to {}_6C^{2-}$:
$-14.250 \times 6.9282/12.011 - 13.031 \times 8/12.011 + 13.031 \times 7/12.011$

$_6C^{2-} \to {}_6C^{3-}$:
$-14.250 \times 7.3485/12.011 - 20.254 \times 9/12.011 + 13.031 \times 8/12.011$

$_6C^{3-} \to {}_6C^{4-}$:
$-14.250 \times 7.7460/12.011 - 29.391 \times 10/12.011 + 20.254 \times 9/12.011$

Evaluate Formulae:

$_6C \to {}_6C^+$: $7.1185 + 3.6566 - 1.1677 = 9.6074$

$_6C^+ \to {}_6C^{2+}$: $6.4982 + 1.1678 - 3.0229 = 4.6431$

$_6C^{2+} \to {}_6C^{3+}$: $5.8122 + 3.0229 + 0.4448 = 9.2799$

$_6C^{3+} \to {}_6C^{4+}$: $5.0335 - 0.4448 - 5.9321 = -1.3434$

$_6C \to {}_6C^-$: $-7.6888 - 7.5945 + 3.6566 = -11.6267$

$_6C^- \to {}_6C^{2-}$: $-8.2197 - 8.6794 + 7.5945 = -9.3046$

$_6C^{2-} \to {}_6C^{3-}$: $-8.7184 - 15.1766 + 8.6794 = -15.2156$

$_6C^{3-} \to {}_6C^{4-}$: $-9.1900 - 24.4701 + 15.1766 = -18.4835$

7. Nitrogen

Write Formulae:

$_7N \to {}_7N^+$: $IP_{1,1} \times 7 / M_7 + \Delta IP_{1,7} \times 7 / M_7 - \Delta IP_{1,6} \times 6 / M_7$

$_7N^+ \to {}_7N^{2+}$:
$IP_{1,1} \times \sqrt{7 \times 6} / M_7 + \Delta IP_{1,6} \times 6 / M_7 - \Delta IP_{1,5} \times 5 / M_7$

$_7N^{2+} \to {}_7N^{3+}$:
$IP_{1,1} \times \sqrt{7 \times 5} / M_7 + \Delta IP_{1,5} \times 5 / M_7 - \Delta IP_{1,4} \times 4 / M_7$

$_7N^{3+} \to {_7}N^{4+}$:

$IP_{1,1} \times \sqrt{7 \times 4}/M_7 + \Delta IP_{1,4} \times 4/M_7 - \Delta IP_{1,3} \times 3/M_7$

$_7N \to {_7}N^-$:

$-IP_{1,1} \times \sqrt{7 \times 8}/M_7 - \Delta IP_{1,8} \times 8/M_7 + \Delta IP_{1,7} \times 7/M_7$

$_7N^- \to {_7}N^{2-}$:

$-IP_{1,1} \times \sqrt{7 \times 9}/M_7 - \Delta IP_{1,9} \times 9/M_7 + \Delta IP_{1,8} \times 8/M_7$

$_7N^{2-} \to {_7}N^{3-}$:

$-IP_{1,1} \times \sqrt{7 \times 10}/M_7 - \Delta IP_{1,10} \times 10/M_7 + \Delta IP_{1,9} \times 9/M_7$

$_7N^{3-} \to {_7}N^{4-}$:

$-IP_{1,1} \times \sqrt{7 \times 11}/M_7 - \Delta IP_{1,11} \times 11/M_7 + \Delta IP_{1,10} \times 10/M_7$

Insert Data:

$_7N \to {_7}N^+$:

$14.250 \times 7/14.007 + 13.031 \times 7/14.007 - 7.320 \times 6/14.007$

$_7N^+ \to {_7}N^{2+}$:

$14.250 \times 6.481/14.007 + 7.320 \times 6/14.007 - 2.805 \times 5/14.007$

$_7N^{2+} \to {_7}N^{3+}$:

$14.250 \times 5.916/14.007 + 2.805 \times 5/14.007 - 9.077 \times 4/14.007$

$_7N^{3+} \to {_7}N^{4+}$:

$14.250 \times 5.292/14.007 + 9.077 \times 4/14.007 - (-1.781) \times 3/14.007$

$_7N \to {_7N^-}$:

$-14.250 \times 7.483/14.007 - 13.031 \times 8/14.007 + 13.031 \times 7/14.007$

$_7N^- \to {_7N^{2-}}$:

$-14.250 \times 7.937/14.007 - 20.254 \times 9/14.007 + 13.031 \times 8/14.007$

$_7N^{2-} \to {_7N^{3-}}$:

$-14.250 \times 8.367/14.007 - 29.391 \times 10/14.007 + 20.254 \times 9/14.007$

$_7N^{3-} \to {_7N^{4-}}$:

$-14.250 \times 7.7460/12.011 - 29.391 \times 10/12.011 + 20.254 \times 9/12.011$

Evaluate Formulae:

$_7N \to {_7N^+}$: $7.1214 + 6.5122 - 3.1356 = 10.498$

$_7N^+ \to {_7N^{2+}}$: $6.5934 + 3.1356 - 1.0013 = 8.7277$

$_7N^{2+} \to {_7N^{3+}}$: $6.0186 + 1.0013 - 2.5921 = 4.4278$

$_7N^{3+} \to {_7N^{4+}}$: $5.3838 + 2.5921 + 0.3815 = 8.3574$

$_7N \to {_7N^-}$: $-7.6128 - 7.4426 + 6.5122 = -8.5432$

$_7N^- \to {_7N^{2-}}$: $-8.0747 - 13.0139 + 7.4426 = -13.646$

$_7N^{2-} \to {_7N^{3-}}$: $-8.5122 - 29.9831 + 13.0139 = -25.4814$

$_7N^{3-} \to {_7N^{4-}}$: $-8.9272 + 2.6230 + 20.9831 = 14.6789$

8. Oxygen

Write Formulae:

$_8O \to {_8O^+}$: $IP_{1,1} \times 8 / M_8 + \Delta IP_{1,8} \times 8 / M_8 - \Delta IP_{1,7} \times 7 / M_8$

$_8O^+ \to {_8O^{2+}}$:
$IP_{1,1} \times \sqrt{8 \times 7} / M_8 + \Delta IP_{1,7} \times 7 / M_8 - \Delta IP_{1,6} \times 6 / M_8$

$_8O^{2+} \to {_8O^{3+}}$:
$IP_{1,1} \times \sqrt{8 \times 6} / M_8 + \Delta IP_{1,6} \times 6 / M_8 - \Delta IP_{1,5} \times 5 / M_8$

$_8O^{3+} \to {_8O^{4+}}$:
$IP_{1,1} \times \sqrt{8 \times 5} / M_8 + \Delta IP_{1,5} \times 5 / M_8 - \Delta IP_{1,4} \times 4 / M_8$

$_8O \to {_8O^-}$:
$-IP_{1,1} \times \sqrt{8 \times 9} / M_8 - \Delta IP_{1,9} \times 9 / M_8 + \Delta IP_{1,8} \times 8 / M_8$

$_8O^- \to {_8O^{2-}}$:
$-IP_{1,1} \times \sqrt{8 \times 10} / M_8 - \Delta IP_{1,10} \times 10 / M_8 + \Delta IP_{1,9} \times 9 / M_8$

$_8O^{2-} \to {_8O^{3-}}$:
$-IP_{1,1} \times \sqrt{8 \times 11} / M_8 - \Delta IP_{1,11} \times 11 / M_8 + \Delta IP_{1,10} \times 10 / M_8$

$_8O^{3-} \to {_8O^{4-}}$:
$-IP_{1,1} \times \sqrt{8 \times 12} / M_8 - \Delta IP_{1,12} \times 12 / M_8 + \Delta IP_{1,11} \times 11 / M_8$

Insert Data:

$_8O \to {_8O^+}$:
$14.250 \times 8 / 15.999 + 13.031 \times 8 / 15.999 - 13.031 \times 7 / 15.999$

$_8O^+ \to {_8O^{2+}}$:

$14.250 \times 7.4833/15.999 + 13.031 \times 7/15.999 - 7.320 \times 6/15.999$

$_8O^{2+} \to {_8O^{3+}}$:

$14.250 \times 6.9282/15.999 + 7.320 \times 6/15.999 - 2.805 \times 5/15.999$

$_8O^{3+} \to {_8O^{4+}}$:

$14.250 \times 6.3246/15.999 + 2.805 \times 5/15.999 - 9.077 \times 4/15.999$

$_8O \to {_8O^-}$:

$-14.250 \times 8.4853/15.999 - 20.254 \times 9/15.999 + 13.031 \times 8/15.999$

$_8O^- \to {_8O^{2-}}$:

$-14.250 \times 8.9443/15.999 - 29.391 \times 10/15.999 + 20.254 \times 9/15.999$

$_8O^{2-} \to {_8O^{3-}}$:

$-14.250 \times 9.3808/15.999 - (-3.340) \times 11/15.999 + 29.391 \times 10/15.999$

$_8O^{3-} \to {_8O^{4-}}$:

$-14.250 \times 9.7980/15.999 - 2.315 \times 12/15.999 + (-3.340) \times 11/15.999$

Evaluate Formulae:

$_8O \to {_8O^+}$: $7.1254 + 6.5159 - 5.7014 = 7.9399$

$_8O^+ \to {_8O^{2+}}$: $6.6652 + 5.7014 - 2.7452 = 9.6214$

$_8O^{2+} \to {_8O^{3+}}$: $6.1708 + 2.7452 - 0.8766 = 8.0394$

$_8O^{3+} \to {_8O^{4+}}$: $5.6332 + 0.8766 - 2.2694 = 4.2404$

$_8O \to {_8O^-}$: $-7.5577 - 11.3936 + 6.5159 = -12.4354$

$_8O^- \rightarrow {_8O^{2-}}$: $-7.9665 - 18.3705 + 11.3936 = -14.9434$

$_8O^{2-} \rightarrow {_8O^{3-}}$: $-8.3553 + 2.2964 + 18.375 = 12.3116$

$_8O^{3-} \rightarrow {_8O^{4-}}$: $-8.7269 - 1.7364 - 2.2964 = -12.7597$

9. Fluorine

Write Formulae:

$_9F \rightarrow {_9F^+}$: $IP_{1,1} \times 9 / M_9 + \Delta IP_{1,9} \times 9 / M_9 - \Delta IP_{1,8} \times 8 / M_9$

$_9F^+ \rightarrow {_9F^{2+}}$:
$IP_{1,1} \times \sqrt{9 \times 8}/M_9 + \Delta IP_{1,8} \times 8 / M_9 - \Delta IP_{1,7} \times 7 / M_9$

$_9F^{2+} \rightarrow {_9F^{3+}}$:
$IP_{1,1} \times \sqrt{9 \times 7}/M_9 + \Delta IP_{1,7} \times 7 / M_9 - \Delta IP_{1,6} \times 6 / M_9$

$_9F \rightarrow {_9F^-}$:
$-IP_{1,1} \times \sqrt{9 \times 10}/M_9 - \Delta IP_{1,10} \times 10 / M_9 + \Delta IP_{1,9} \times 9 / M_9$

$_9F^- \rightarrow {_9F^{2-}}$:
$-IP_{1,1} \times \sqrt{9 \times 11}/M_9 - \Delta IP_{1,11} \times 11 / M_9 + \Delta IP_{1,10} \times 10 / M_9$

$_9F^{2-} \rightarrow {_9F^{3-}}$:
$-IP_{1,1} \times \sqrt{9 \times 12}/M_9 - \Delta IP_{1,12} \times 12 / M_9 + \Delta IP_{1,11} \times 11 / M_9$

Insert Data:

$_9F \rightarrow {_9F^+}$:
$14.250 \times 9 / 18.998 + 20.254 \times 9 / 18.998 - 13.031 \times 8 / 18.998$

$_9F^+ \to {_9}F^{2+}$:

$14.250 \times 8.4852/18.998 + 13.031 \times 8/18.998 - 13.031 \times 7/18.998$

$_9F^{2+} \to {_9}F^{3+}$:

$14.250 \times 7.9373/18.998 + 13.031 \times 7/18.998 - 7.320 \times 6/18.998$

$_9F \to {_9}F^-$:

$-14.250 \times 9.4868/18.998 - 29.391 \times 10/18.998 + 20.254 \times 9/18.998$

$_9F^- \to {_9}F^{2-}$:

$-14.250 \times 9.9499/18.998 - (-3.340) \times 11/18.998 + 29.391 \times 10/18.998$

$_9F^{2-} \to {_9}F^{3-}$:

$-14.250 \times 10.3923/18.998 - 2.315 \times 12/18.998 + (-3.340) \times 11/18.998$

Evaluate Formulae:

$_9F \to {_9}F^+$: $6.7507 + 9.5950 - 5.4873 = 10.8584$

$_9F^+ \to {_9}F^{2+}$: $6.3646 + 5.4873 - 4.8014 = 7.0505$

$_9F^{2+} \to {_9}F^{3+}$: $5.9536 + 4.8014 - 2.3118 = 8.4432$

$_9F \to {_9}F^-$: $-7.1158 - 15.4706 + 9.5950 = -12.9914$

$_9F^- \to {_9}F^{2-}$: $-7.4632 + 1.9339 + 15.4706 = 9.9413$

$_9F^{2-} \to {_9}F^{3-}$: $-7.7950 - 1.4623 - 1.9339 = -11.1912$

10. Neon

Write Formulae:

$_{10}Ne \rightarrow {}_{10}Ne^{+}$:

$IP_{1,1} \times 10 / M_{10} + \Delta IP_{1,10} \times 10 / M_{10} - \Delta IP_{1,9} \times 9 / M_{10}$

$_{10}Ne^{+} \rightarrow {}_{10}Ne^{2+}$:

$IP_{1,1} \times \sqrt{10 \times 9} / M_{10} + \Delta IP_{1,9} \times 9 / M_{10} - \Delta IP_{1,8} \times 8 / M_{10}$

$_{10}Ne^{2+} \rightarrow {}_{10}Ne^{3+}$:

$IP_{1,1} \times \sqrt{10 \times 8} / M_{10} + \Delta IP_{1,8} \times 8 / M_{10} - \Delta IP_{1,7} \times 7 / M_{10}$

$_{10}Ne^{3+} \rightarrow {}_{10}Ne^{4+}$:

$IP_{1,1} \times \sqrt{10 \times 7} / M_{10} + \Delta IP_{1,7} \times 7 / M_{10} - \Delta IP_{1,6} \times 6 / M_{10}$

$_{10}Ne \rightarrow {}_{10}Ne^{-}$:

$-IP_{1,1} \times \sqrt{10 \times 11} / M_{10} - \Delta IP_{1,11} \times 11 / M_{10} + \Delta IP_{1,10} \times 10 / M_{10}$

$_{10}Ne^{-} \rightarrow {}_{10}Ne^{2-}$:

$-IP_{1,1} \times \sqrt{10 \times 12} / M_{10} - \Delta IP_{1,12} \times 12 / M_{10} + \Delta IP_{1,11} \times 11 / M_{10}$

$_{10}Ne^{2-} \rightarrow {}_{10}Ne^{3-}$:

$-IP_{1,1} \times \sqrt{10 \times 13} / M_{10} - \Delta IP_{1,13} \times 13 / M_{10} + \Delta IP_{1,12} \times 12 / M_{10}$

$_{10}Ne^{3-} \rightarrow {}_{10}Ne^{4-}$:

$-IP_{1,1} \times \sqrt{10 \times 14} / M_{10} - \Delta IP_{1,14} \times 14 / M_{10} + \Delta IP_{1,13} \times 13 / M_{10}$

Insert Data:

$_{10}Ne \rightarrow {}_{10}Ne^+$:
$14.250 \times 10 / 20.180 + 29.391 \times 10 / 20.180 - 20.254 \times 9 / 20.180$

$_{10}Ne^+ \rightarrow {}_{10}Ne^{2+}$:
$14.250 \times 9.4868 / 20.180 + 20.254 \times 9 / 20.180 - 13.031 \times 8 / 20.180$

$_{10}Ne^{2+} \rightarrow {}_{10}Ne^{3+}$:
$14.250 \times 8.9443 / 20.180 + 13.031 \times 8 / 20.180 - 13.031 \times 7 / 20.180$

$_{10}Ne^{3+} \rightarrow {}_{10}Ne^{4+}$:
$14.250 \times 8.3666 / 20.180 + 13.031 \times 7 / 20.180 - 7.320 \times 6 / 20.180$

$_{10}Ne \rightarrow {}_{10}Ne^-$:
$-14.250 \times 10.4881 / 20.180 - (-3.340) \times 11 / 20.180 + 20.391 \times 10 / 20.180$

$_{10}Ne^- \rightarrow {}_{10}Ne^{2-}$:
$-14.250 \times 10.9545 / 20.180 - 2.315 \times 12 / 20.180 + (-3.340) \times 11 / 20.180$

$_{10}Ne^{2-} \rightarrow {}_{10}Ne^{3-}$:
$-14.250 \times 11.4018 / 20.180 - 0.673 \times 13 / 20.180 + 2.315 \times 12 / 20.180$

$_{10}Ne^{3-} \rightarrow {}_{10}Ne^{4-}$:
$-14.250 \times 11.8322 / 20.180 - 4.624 \times 14 / 20.180 + 0.6373 \times 13 / 20.180$

Evaluate Formulae:

$_{10}Ne \rightarrow {}_{10}Ne^+$: $7.0614 + 14.5644 - 9.0330 = 12.5928$

$_{10}Ne^+ \rightarrow {}_{10}Ne^{2+}$: $6.6991 + 9.0330 - 5.1659 = 10.5662$

$_{10}Ne^{2+} \rightarrow {}_{10}Ne^{3+}$: $6.3160 + 5.1659 - 4.5202 = 6.9617$

$_{10}Ne^{3+} \to {}_{10}Ne^{4+}$: $5.9080 + 4.5202 - 2.1764 = 8.2518$

$_{10}Ne \to {}_{10}Ne^{-}$: $-7.4061 + 1.8206 + 10.1046 = 4.5191$

$_{10}Ne^{-} \to {}_{10}Ne^{2-}$: $-7.7355 - 1.3766 - 1.8206 = -10.9327$

$_{10}Ne^{2-} \to {}_{10}Ne^{3-}$: $-8.0513 - 0.4335 + 1.3766 = -7.1082$

$_{10}Ne^{3-} \to {}_{10}Ne^{4-}$: $-8.3552 - 3.2079 + 0.4106 = -11.1525$

11. Sodium

Write Formulae:

$_{11}Na \to {}_{11}Na^{+}$:
$IP_{1,1} \times 11 / M_{11} + \Delta IP_{1,11} \times 11 / M_{11} - \Delta IP_{1,10} \times 10 / M_{11}$

$_{11}Na^{+} \to {}_{11}Na^{2+}$:
$IP_{1,1} \times \sqrt{11 \times 10} / M_{11} + \Delta IP_{1,10} \times 10 / M_{11} - \Delta IP_{1,9} \times 9 / M_{11}$

$_{11}Na^{2+} \to {}_{11}Na^{3+}$:
$IP_{1,1} \times \sqrt{11 \times 9} / M_{11} + \Delta IP_{1,9} \times 9 / M_{11} - \Delta IP_{1,8} \times 8 / M_{11}$

$_{11}Na^{3+} \to {}_{11}Na^{4+}$:
$IP_{1,1} \times \sqrt{11 \times 8} / M_{11} + \Delta IP_{1,8} \times 8 / M_{11} - \Delta IP_{1,7} \times 7 / M_{11}$

$_{11}Na \to {}_{11}Na^{-}$:
$-IP_{1,1} \times \sqrt{11 \times 12} / M_{11} - \Delta IP_{1,12} \times 12 / M_{11} + \Delta IP_{1,11} \times 11 / M_{11}$

$_{11}\text{Na}^- \to {}_{11}\text{Na}^{2-}$:

$-IP_{1,1} \times \sqrt{11 \times 13}/M_{11} - \Delta IP_{1,13} \times 13/M_{11} + \Delta IP_{1,12} \times 12/M_{11}$

$_{11}\text{Na}^{2-} \to {}_{11}\text{Na}^{3-}$:

$-IP_{1,1} \times \sqrt{11 \times 14}/M_{11} - \Delta IP_{1,14} \times 14/M_{11} + \Delta IP_{1,13} \times 13/M_{11}$

$_{11}\text{Na}^{3-} \to {}_{11}\text{Na}^{4-}$:

$-IP_{1,1} \times \sqrt{11 \times 15}/M_{11} - \Delta IP_{1,15} \times 15/M_{11} + \Delta IP_{1,14} \times 14/M_{11}$

Insert Data:

$_{11}\text{Na} \to {}_{11}\text{Na}^+$:
$14.250 \times 11/22.990 + (-3.340) \times 11/22.990 - 29.391 \times 10/22.990$

$_{11}\text{Na}^+ \to {}_{11}\text{Na}^{2+}$:
$14.250 \times 10.4881/22.990 + 29.391 \times 10/22.990 - 20.254 \times 9/22.990$

$_{11}\text{Na}^{2+} \to {}_{11}\text{Na}^{3+}$:
$14.250 \times 9.9499/22.990 + 20.254 \times 9/22.990 - 13.031 \times 8/22.990$

$_{11}\text{Na}^{3+} \to {}_{11}\text{Na}^{4+}$:
$14.250 \times 9.3808/22.990 + 13.031 \times 8/22.990 - 13.031 \times 7/22.990$

$_{11}\text{Na} \to {}_{11}\text{Na}^-$:
$-14.250 \times 11.4891/22.990 - 2.315 \times 12/22.990 + (-3.340) \times 11/22.990$

$_{11}\text{Na}^- \to {}_{11}\text{Na}^{2-}$:
$-14.250 \times 11.9583/22.990 - 0.673 \times 13/22.990 + 2.315 \times 12/22.990$

$_{11}\text{Na}^{2-} \to {}_{11}\text{Na}^{3-}$:
$-14.250 \times 12.4097/22.990 - 4.624 \times 14/22.990 + 14.923 \times 13/22.990$

$_{11}Na^{3-} \to {}_{11}Na^{4-}$:

$-14.250 \times 12.8452/22.990 - 9.621 \times 15/22.990 + 4.624 \times 14/22.990$

Evaluate Formulae:

$_{11}Na \to {}_{11}Na^{+}$: $6.8181 - 1.5981 - 12.7843 = -7.5643$

$_{11}Na^{+} \to {}_{11}Na^{2+}$: $6.5009 + 12.7843 - 7.9289 = 11.3563$

$_{11}Na^{2+} \to {}_{11}Na^{3+}$: $6.1673 + 7.9289 - 4.5345 = 9.5617$

$_{11}Na^{3+} \to {}_{11}Na^{4+}$: $5.8145 + 4.5345 - 3.9677 = 6.3813$

$_{11}Na \to {}_{11}Na^{-}$: $-7.1213 - 1.2084 - 1.5981 = -9.9278$

$_{11}Na^{-} \to {}_{11}Na^{2-}$: $-7.4122 - 0.3806 + 1.2084 = -6.5844$

$_{11}Na^{2-} \to {}_{11}Na^{3-}$: $-7.6920 - 2.8158 + 8.4384 = -2.0694$

$_{11}Na^{3-} \to {}_{11}Na^{4-}$: $-7.9619 - 6.2773 + 2.8158 = -11.4234$

12. Magnesium

Write Formulae:

$_{12}Mg \to {}_{12}Mg^{+}$:
$IP_{1,1} \times 12/M_{12} + \Delta IP_{1,12} \times 12/M_{12} - \Delta IP_{1,11} \times 11/M_{12}$

$_{12}Mg^{+} \to {}_{12}Mg^{2+}$:
$IP_{1,1} \times \sqrt{12 \times 11}/M_{12} + \Delta IP_{1,11} \times 11/M_{12} - \Delta IP_{1,10} \times 10/M_{12}$

$_{12}Mg \to {}_{12}Mg^-$:
$$-IP_{1,1} \times \sqrt{12 \times 13}/M_{12} - \Delta IP_{1,13} \times 13/M_{12} + \Delta IP_{1,12} \times 12/M_{12}$$

$_{12}Mg^- \to {}_{12}Mg^{2-}$:
$$-IP_{1,1} \times \sqrt{12 \times 14}/M_{12} - \Delta IP_{1,14} \times 14/M_{12} + \Delta IP_{1,13} \times 13/M_{12}$$

Insert Data:

$_{12}Mg \to {}_{12}Mg^+$:
$14.250 \times 12/24.305 + 2.315 \times 12/24.305 - (-3.340) \times 11/24.305$

$_{12}Mg^+ \to {}_{12}Mg^{2+}$:
$14.250 \times 11.4891/24.305 + (-3.340) \times 11/24.305 - 29.391 \times 10/24.305$

$_{12}Mg \to {}_{12}Mg^-$:
$-14.250 \times 12.4900/24.305 - 0.673 \times 13/24.305 + (-3.340) \times 12/24.305$

$_{12}Mg^- \to {}_{12}Mg^{2-}$:
$-14.250 \times 12.9615/24.305 - 4.624 \times 14/24.305 + 0.673 \times 13/24.305$

Evaluate Formulae:

$_{12}Mg \to {}_{12}Mg^+$: $7.0356 + 1.1430 + 1.5116 = 9.6902$

$_{12}Mg^+ \to {}_{12}Mg^{2+}$: $6.4493 - 1.5116 - 12.0926 = -7.1549$

$_{12}Mg \to {}_{12}Mg^-$: $-7.3229 - 0.3600 - 1.6490 = -9.3319$

$_{12}Mg^- \to {}_{12}Mg^{2-}$: $-7.5993 - 2.6635 + 0.3600 = -9.9028$

13. Aluminum

Write Formulae:

$_{13}Al \rightarrow {}_{13}Al^{+}$:

$IP_{1,1} \times 13 / M_{13} + \Delta IP_{1,13} \times 13 / M_{13} - \Delta IP_{1,12} \times 12 / M_{13}$

$_{13}Al^{+} \rightarrow {}_{13}Al^{2+}$:

$IP_{1,1} \times \sqrt{13 \times 12} / M_{13} + \Delta IP_{1,12} \times 12 / M_{12} - \Delta IP_{1,11} \times 11 / M_{13}$

$_{13}Al^{2+} \rightarrow {}_{13}Al^{3+}$:

$IP_{1,1} \times \sqrt{13 \times 11} / M_{13} + \Delta IP_{1,11} \times 11 / M_{13} - \Delta IP_{1,10} \times 10 / M_{13}$

$_{13}Al^{3+} \rightarrow {}_{13}Al^{4+}$:

$IP_{1,1} \times \sqrt{13 \times 10} / M_{13} + \Delta IP_{1,10} \times 10 / M_{13} - \Delta IP_{1,9} \times 9 / M_{13}$

$_{13}Al^{4+} \rightarrow {}_{13}Al^{5+}$:

$IP_{1,1} \times \sqrt{13 \times 9} / M_{13} + \Delta IP_{1,9} \times 9 / M_{13} - \Delta IP_{1,8} \times 8 / M_{13}$

$_{13}Al^{5+} \rightarrow {}_{13}Al^{6+}$:

$IP_{1,1} \times \sqrt{13 \times 8} / M_{13} + \Delta IP_{1,8} \times 8 / M_{13} - \Delta IP_{1,7} \times 7 / M_{13}$

$_{13}Al \rightarrow {}_{13}Al^{-}$:

$-IP_{1,1} \times \sqrt{13 \times 14} / M_{13} - \Delta IP_{1,14} \times 14 / M_{13} + \Delta IP_{1,13} \times 13 / M_{13}$

$_{13}Al^{-} \rightarrow {}_{13}Al^{2-}$:

$-IP_{1,1} \times \sqrt{13 \times 15} / M_{13} - \Delta IP_{1,15} \times 15 / M_{13} + \Delta IP_{1,14} \times 14 / M_{13}$

$_{13}Al^{2-} \to {}_{13}Al^{3-}$:
$-IP_{1,1} \times \sqrt{13 \times 16}/M_{13} - \Delta IP_{1,16} \times 16/M_{13} + \Delta IP_{1,15} \times 15/M_{13}$

$_{13}Al^{3-} \to {}_{13}Al^{4-}$:
$-IP_{1,1} \times \sqrt{13 \times 17}/M_{13} - \Delta IP_{1,17} \times 17/M_{13} + \Delta IP_{1,16} \times 16/M_{13}$

$_{13}Al^{4-} \to {}_{13}Al^{5-}$:
$-IP_{1,1} \times \sqrt{13 \times 18}/M_{13} - \Delta IP_{1,18} \times 18/M_{13} + \Delta IP_{1,17} \times 17/M_{13}$

$_{13}Al^{5-} \to {}_{13}Al^{6-}$:
$-IP_{1,1} \times \sqrt{13 \times 19}/M_{13} - \Delta IP_{1,19} \times 19/M_{13} + \Delta IP_{1,18} \times 18/M_{13}$

Insert Data:

$_{13}Al \to {}_{13}Al^{+}$:
$14.250 \times 13/26.982 + 0.673 \times 13/26.982 - 2.315 \times 12/26.982$

$_{13}Al^{+} \to {}_{13}Al^{2+}$:
$14.250 \times 12.490/26.982 + 2.315 \times 12/26.982 - (-3.340) \times 11/26.982$

$_{13}Al^{2+} \to {}_{13}Al^{3+}$:
$14.250 \times 11.958/26.982 + (-3.340) \times 11/26.982 - 29.391 \times 10/26.982$

$_{13}Al^{3+} \to {}_{13}Al^{4+}$:
$14.250 \times 11.402/26.982 + 29.391 \times 10/26.982 - 20.254 \times 9/26.982$

$_{13}Al^{4+} \to {}_{13}Al^{5+}$:
$14.250 \times 10.817/26.982 + 20.254 \times 9/26.982 - 13.031 \times 8/26.982$

$_{13}Al^{5+} \to {}_{13}Al^{6+}$:
$14.250 \times 10.198/26.982 + 13.031 \times 8/26.982 - 13.031 \times 7/26.982$

$_{13}Al \rightarrow {}_{13}Al^-$:
$-14.250 \times 13.491/26.982 - 4.624 \times 14/26.982 + 0.673 \times 13/26.982$

$_{13}Al^- \rightarrow {}_{13}Al^{2-}$:
$-14.250 \times 13.964/26.982 - 9.621 \times 15/26.982 + 4.624 \times 14/26.982$

$_{13}Al^{2-} \rightarrow {}_{13}Al^{3-}$:
$-14.250 \times 14.422/26.982 - 9.621 \times 16/26.982 + 9.621 \times 15/26.982$

$_{13}Al^{3-} \rightarrow {}_{13}Al^{4-}$:
$-14.250 \times 14.866/26.982 - 15.942 \times 17/26.982 + 9.621 \times 16/26.982$

$_{13}Al^{4-} \rightarrow {}_{13}Al^{5-}$:
$-14.250 \times 15.297/26.982 - 23.936 \times 18/26.982 + 15.942 \times 17/26.982$

$_{13}Al^{5-} \rightarrow {}_{13}Al^{6-}$:
$-14.250 \times 15.716/26.982 - (-4.704) \times 19/26.982 + 23.936 \times 18/26.982$

Evaluate Formulae:

$_{13}Al \rightarrow {}_{13}Al^+$: $6.8657 + 0.3243 - 1.0296 = 6.1604$

$_{13}Al^+ \rightarrow {}_{13}Al^{2+}$: $6.5963 + 1.0296 + 1.3616 = 8.9875$

$_{13}Al^{2+} \rightarrow {}_{13}Al^{3+}$: $6.3154 - 1.3616 - 10.8928 = -5.9390$

$_{13}Al^{3+} \rightarrow {}_{13}Al^{4+}$: $6.0217 + 10.8928 - 6.7558 = 10.1587$

$_{13}Al^{4+} \rightarrow {}_{13}Al^{5+}$: $5.7128 + 6.7558 - 3.8636 = 8.6050$

$_{13}Al^{5+} \rightarrow {}_{13}Al^{6+}$: $5.3859 + 3.8636 - 3.3807 = 5.8688$

$_{13}Al \to {}_{13}Al^-$: $-7.1250 - 2.3992 + 0.3243 = -9.1999$

$_{13}Al^- \to {}_{13}Al^{2-}$: $-7.3748 - 5.3486 + 2.3992 = -10.3242$

$_{13}Al^{2-} \to {}_{13}Al^{3-}$: $-7.6167 - 5.7051 + 5.3486 = -7.9732$

$_{13}Al^{3-} \to {}_{13}Al^{4-}$: $-7.8512 - 10.0443 + 5.7051 = -12.1904$

$_{13}Al^{4-} \to {}_{13}Al^{5-}$: $-8.0788 - 15.9680 + 10.0443 = -14.0025$

$_{13}Al^{5-} \to {}_{13}Al^{6-}$: $-8.3001 + 3.3124 + 15.9680 = 10.9803$

14. Silicon

Write Formulae:

$_{14}Si \to {}_{14}Si^+$:
$IP_{1,1} \times 14 / M_{14} + \Delta IP_{1,14} \times 14 / M_{14} - \Delta IP_{1,13} \times 13 / M_{14}$

$_{14}Si^+ \to {}_{14}Si^{2+}$:
$IP_{1,1} \times \sqrt{14 \times 13} / M_{14} + \Delta IP_{1,13} \times 13 / M_{14} - \Delta IP_{1,12} \times 12 / M_{14}$

$_{14}Si^{2+} \to {}_{14}Si^{3+}$:
$IP_{1,1} \times \sqrt{14 \times 12} / M_{14} + \Delta IP_{1,12} \times 12 / M_{14} - \Delta IP_{1,11} \times 11 / M_{14}$

$_{14}Si^{3+} \to {}_{14}Si^{4+}$:
$IP_{1,1} \times \sqrt{14 \times 11} / M_{14} + \Delta IP_{1,11} \times 11 / M_{14} - \Delta IP_{1,10} \times 10 / M_{14}$

$_{14}Si \to {}_{14}Si^-$:
$-IP_{1,1} \times \sqrt{14 \times 15} / M_{14} - \Delta IP_{1,15} \times 15 / M_{14} + \Delta IP_{1,14} \times 14 / M_{14}$

$_{14}Si^- \to {}_{14}Si^{2-}$:

$-IP_{1,1} \times \sqrt{14 \times 16}/M_{14} - \Delta IP_{1,16} \times 16/M_{14} + \Delta IP_{1,15} \times 15/M_{14}$

$_{14}Si^{2-} \to {}_{14}Si^{3-}$:

$-IP_{1,1} \times \sqrt{14 \times 17}/M_{14} - \Delta IP_{1,17} \times 17/M_{14} + \Delta IP_{1,16} \times 16/M_{14}$

$_{14}Si^{3-} \to {}_{14}Si^{4-}$:

$-IP_{1,1} \times \sqrt{14 \times 18}/M_{14} - \Delta IP_{1,18} \times 18/M_{14} + \Delta IP_{1,17} \times 17/M_{14}$

Insert Data:

$_{14}Si \to {}_{14}Si^+$:
$14.250 \times 14/28.086 + 4.624 \times 14/28.086 - 0.673 \times 13/28.086$

$_{14}Si^+ \to {}_{14}Si^{2+}$:
$14.250 \times 13.4907/28.086 + 0.673 \times 13/28.086 - 2.315 \times 12/28.086$

$_{14}Si^{2+} \to {}_{14}Si^{3+}$:
$14.250 \times 12.9615/28.086 + 2.315 \times 12/28.086 - (-3.340) \times 11/28.086$

$_{14}Si^{3+} \to {}_{14}Si^{4+}$:
$14.250 \times 12.4097/28.086 + (-3.340) \times 11/28.086 - 29.391 \times 10/28.086$

$_{14}Si \to {}_{14}Si^-$:
$-14.250 \times 14.4914/28.086 - 9.621 \times 15/28.086 + 4.624 \times 14/28.086$

$_{14}Si^- \to {}_{14}Si^{2-}$:
$-14.250 \times 14.9666/28.086 - 9.621 \times 16/28.086 + 9.621 \times 15/28.086$

$_{14}Si^{2-} \to {}_{14}Si^{3-}$:
$-14.250 \times 15.4242/28.086 - 15.942 \times 17/28.086 + 9.621 \times 16/28.086$

$_{14}Si^{3-} \rightarrow {}_{14}Si^{4-}$:
$-14.250 \times 15.8745/28.086 - 23.936 \times 18/28.086 + 15.942 \times 17/28.086$

Evaluate Formulae:

$_{14}Si \rightarrow {}_{14}Si^{+}$: $7.1032 + 2.3049 - 0.3115 = 9.0966$

$_{14}Si^{+} \rightarrow {}_{14}Si^{2+}$: $6.8448 + 0.3115 - 0.9891 = 6.1672$

$_{14}Si^{2+} \rightarrow {}_{14}Si^{3+}$: $6.5763 + 0.9891 + 1.3081 = 8.8735$

$_{14}Si^{3+} \rightarrow {}_{14}Si^{4+}$: $6.2963 - 1.3081 - 10.4646 = -5.4764$

$_{14}Si \rightarrow {}_{14}Si^{-}$: $-7.3525 - 5.1383 + 2.3049 = -10.1859$

$_{14}Si^{-} \rightarrow {}_{14}Si^{2-}$: $-7.5936 - 5.4809 + 5.1383 = -7.9362$

$_{14}Si^{2-} \rightarrow {}_{14}Si^{3-}$: $-7.8258 - 9.6494 + 5.4809 = -11.9943$

$_{14}Si^{3-} \rightarrow {}_{14}Si^{4-}$: $-8.0542 - 15.3403 + 9.6494 = -13.7451$

15. Phosphorus

Write Formulae:

$_{15}P \rightarrow {}_{15}P^{+}$:
$IP_{1,1} \times 15/M_{15} + \Delta IP_{1,15} \times 15/M_{15} - \Delta IP_{1,14} \times 14/M_{15}$

$_{15}P^{+} \rightarrow {}_{15}P^{2+}$:
$IP_{1,1} \times \sqrt{15 \times 14}/M_{15} + \Delta IP_{1,14} \times 14/M_{15} - \Delta IP_{1,13} \times 13/M_{15}$

$_{15}P^{2+} \rightarrow {}_{15}P^{3+}$:

$IP_{1,1} \times \sqrt{15 \times 13}/M_{15} + \Delta IP_{1,13} \times 13/M_{15} - \Delta IP_{1,12} \times 12/M_{15}$

$_{15}P^{3+} \rightarrow {}_{15}P^{4+}$:

$IP_{1,1} \times \sqrt{15 \times 12}/M_{15} + \Delta IP_{1,12} \times 12/M_{15} - \Delta IP_{1,11} \times 11/M_{15}$

$_{15}P \rightarrow {}_{15}P^{-}$:

$-IP_{1,1} \times \sqrt{15 \times 16}/M_{15} - \Delta IP_{1,16} \times 16/M_{15} + \Delta IP_{1,15} \times 15/M_{15}$

$_{15}P^{-} \rightarrow {}_{15}P^{2-}$:

$-IP_{1,1} \times \sqrt{15 \times 17}/M_{15} - \Delta IP_{1,17} \times 17/M_{15} + \Delta IP_{1,16} \times 16/M_{15}$

$_{15}P^{2-} \rightarrow {}_{15}P^{3-}$:

$-IP_{1,1} \times \sqrt{15 \times 18}/M_{15} - \Delta IP_{1,18} \times 18/M_{15} + \Delta IP_{1,17} \times 17/M_{15}$

$_{15}P^{3-} \rightarrow {}_{15}P^{4-}$:

$-IP_{1,1} \times \sqrt{15 \times 19}/M_{15} - \Delta IP_{1,19} \times 19/M_{15} + \Delta IP_{1,18} \times 18/M_{15}$

Insert Data:

$_{15}P \rightarrow {}_{15}P^{+}$:

$14.250 \times 15/30.974 + 9.621 \times 15/30.974 - 4.624 \times 14/30.974$

$_{15}P^{+} \rightarrow {}_{15}P^{2+}$:

$14.250 \times 14.4914/30.974 + 4.624 \times 14/30.974 - 0.673 \times 13/30.974$

$_{15}P^{2+} \rightarrow {}_{15}P^{3+}$:

$14.250 \times 13.9642/30.974 + 0.673 \times 13/30.974 - 2.315 \times 12/30.974$

$_{15}P^{3+} \to {}_{15}P^{4+}$:

$14.250 \times 13.4164/30.974 + 2.315 \times 12 / 30.974 - (-3.340) \times 11 / 30.974$

$_{15}P \to {}_{15}P^{-}$:

$-14.250 \times 15.4919/30.974 - 9.621 \times 16 / 30.974 + 9.621 \times 15 / 30.974$

$_{15}P^{-} \to {}_{15}P^{2-}$:

$-14.250 \times 15.9687/30.974 - 15.942 \times 17 / 30.974 + 9.621 \times 16 / 30.974$

$_{15}P^{2-} \to {}_{15}P^{3-}$:

$-14.250 \times 16.4317/30.974 - 38.186 \times 18 / 30.974 + 30.192 \times 17 / 30.974$

$_{15}P^{3-} \to {}_{15}P^{4-}$:

$-14.250 \times 16.8819/30.974 - (-4.704) \times 19 / 30.974 + 23.936 \times 18 / 30.974$

Evaluate Formulae:

$_{15}P \to {}_{15}P^{+}$: $6.9009 + 4.6592 - 2.0900 = 9.4701$

$_{15}P^{+} \to {}_{15}P^{2+}$: $6.6670 + 2.0900 - 0.2825 = 8.4745$

$_{15}P^{2+} \to {}_{15}P^{3+}$: $6.4244 + 0.2825 - 0.8969 = 5.8100$

$_{15}P^{3+} \to {}_{15}P^{4+}$: $6.1724 + 0.8969 + 1.1862 = 8.2555$

$_{15}P \to {}_{15}P^{-}$: $-7.1273 - 4.9698 + 4.6592 = -7.4379$

$_{15}P^{-} \to {}_{15}P^{2-}$: $-7.3466 - 8.7497 + 4.9698 = -11.1265$

$_{15}P^{2-} \to {}_{15}P^{3-}$: $-7.5596 - 22.1911 + 16.5708 = -13.1799$

$_{15}P^{3-} \to {}_{15}P^{4-}$: $-7.7735 + 2.8855 + 13.9010 = 9.0220$

16. Sulfur

Write Formulae:

$_{16}S \rightarrow {}_{16}S^+$:

$IP_{1,1} \times 16 / M_{16} + \Delta IP_{1,16} \times 16 / M_{16} - \Delta IP_{1,15} \times 15 / M_{16}$

$_{16}S^+ \rightarrow {}_{16}S^{2+}$:

$IP_{1,1} \times \sqrt{16 \times 15} / M_{16} + \Delta IP_{1,15} \times 15 / M_{16} - \Delta IP_{1,14} \times 14 / M_{16}$

$_{16}S^{2+} \rightarrow {}_{16}S^{3+}$:

$IP_{1,1} \times \sqrt{16 \times 14} / M_{16} + \Delta IP_{1,14} \times 14 / M_{16} - \Delta IP_{1,13} \times 13 / M_{16}$

$_{16}S^{3+} \rightarrow {}_{16}S^{4+}$:

$IP_{1,1} \times \sqrt{16 \times 13} / M_{16} + \Delta IP_{1,13} \times 13 / M_{16} - \Delta IP_{1,12} \times 12 / M_{16}$

$_{16}S^{4+} \rightarrow {}_{16}S^{5+}$:

$IP_{1,1} \times \sqrt{16 \times 12} / M_{16} + \Delta IP_{1,12} \times 12 / M_{16} - \Delta IP_{1,11} \times 11 / M_{16}$

$_{16}S^{5+} \rightarrow {}_{16}S^{6+}$:

$IP_{1,1} \times \sqrt{16 \times 11} / M_{16} + \Delta IP_{1,11} \times 11 / M_{16} - \Delta IP_{1,10} \times 10 / M_{16}$

$_{16}S \rightarrow {}_{16}S^-$:

$-IP_{1,1} \times \sqrt{16 \times 17} / M_{16} - \Delta IP_{1,17} \times 17 / M_{16} + \Delta IP_{1,16} \times 16 / M_{16}$

$_{16}S^- \rightarrow {}_{16}S^{2-}$:

$-IP_{1,1} \times \sqrt{16 \times 18} / M_{16} - \Delta IP_{1,18} \times 18 / M_{16} + \Delta IP_{1,17} \times 17 / M_{16}$

$_{16}S^{2-} \to {}_{16}S^{3-}$:

$-IP_{1,1} \times \sqrt{16 \times 19}/M_{16} - \Delta IP_{1,19} \times 19/M_{16} + \Delta IP_{1,18} \times 18/M_{16}$

$_{16}S^{3-} \to {}_{16}S^{4-}$:

$-IP_{1,1} \times \sqrt{16 \times 20}/M_{16} - \Delta IP_{1,20} \times 20/M_{16} + \Delta IP_{1,19} \times 19/M_{16}$

Insert Data:

$_{16}S \to {}_{16}S^{+}$:
$14.250 \times 16/32.066 + 9.621 \times 16/32.066 - 9.621 \times 15/32.066$

$_{16}S^{+} \to {}_{16}S^{2+}$:
$14.250 \times 15.4919/32.066 + 9.621 \times 15/32.066 - 4.624 \times 14/32.066$

$_{16}S^{++} \to {}_{16}S^{3+}$:
$14.250 \times 14.967/32.066 + 4.624 \times 14/32.066 - 0.673 \times 13/32.066$

$_{16}S^{3+} \to {}_{16}S^{4+}$:
$14.250 \times 14.422/32.066 + 0.673 \times 13/32.066 - 2.315 \times 12/32.066$

$_{16}S^{4+} \to {}_{16}S^{5+}$:
$14.250 \times 13.856/32.066 + 2.315 \times 12/32.066 - (-3.340) \times 11/32.066$

$_{16}S^{5+} \to {}_{16}S^{6+}$:
$14.250 \times 13.266/32.066 + (-3.340) \times 11/32.066 - 29.391 \times 10/32.066$

$_{16}S \to {}_{16}S^{-}$:
$-14.250 \times 16.4924/32.066 - 15.492 \times 17/32.066 + 23.871 \times 16/32.066$

$_{16}S^{-} \to {}_{16}S^{2-}$:
$-14.250 \times 16.9706/32.066 - 38.186 \times 18/32.066 + 15.942 \times 17/32.066$

$_{16}S^{2-} \rightarrow {}_{16}S^{3-}$:

$-14.250 \times 17.4356/32.066 - (-4.704) \times 19/32.066 + 23.936 \times 18/32.066$

$_{16}S^{3-} \rightarrow {}_{16}S^{4-}$:

$-14.250 \times 17.8885/32.066 - (-1.193) \times 20/32.066 + (-4.704) \times 19/32.066$

Evaluate Formulae:

$_{16}S \rightarrow {}_{16}S^{+}$: $7.1103 + 4.8006 - 4.5006 = 7.4103$

$_{16}S^{+} \rightarrow {}_{16}S^{2+}$: $6.8845 + 4.5006 - 2.0188 = 9.3663$

$_{16}S^{2+} \rightarrow {}_{16}S^{3+}$: $6.6513 + 2.0188 - 0.2728 = 8.3973$

$_{16}S^{3+} \rightarrow {}_{16}S^{4+}$: $6.4091 + 0.2728 - 0.8663 = 5.8156$

$_{16}S^{4+} \rightarrow {}_{16}S^{5+}$: $6.1576 + 0.8663 + 1.1458 = 8.1697$

$_{16}S^{5+} \rightarrow {}_{16}S^{6+}$: $5.8954 - 1.1458 - 9.1658 = -4.4162$

$_{16}S \rightarrow {}_{16}S^{-}$: $-7.3292 - 8.2132 + 11.9109 = -3.6315$

$_{16}S^{-} \rightarrow {}_{16}S^{2-}$: $-7.5417 - 21.4354 + 8.4518 = -20.5253$

$_{16}S^{2-} \rightarrow {}_{16}S^{3-}$: $-7.7483 + 2.7873 + 13.4363 = 8.4753$

$_{16}S^{3-} \rightarrow {}_{16}S^{4-}$: $-7.9496 + 0.7441 - 2.7873 = -9.9928$

17. Chlorine

Write Formulae:

$_{17}Cl \rightarrow {}_{17}Cl^+$:

$IP_{1,1} \times 17 / M_{17} + \Delta IP_{1,17} \times 17 / M_{17} - \Delta IP_{1,16} \times 16 / M_{17}$

$_{17}Cl^+ \rightarrow {}_{17}Cl^{2+}$:

$IP_{1,1} \times \sqrt{17 \times 16} / M_{17} + \Delta IP_{1,16} \times 16 / M_{17} - \Delta IP_{1,15} \times 15 / M_{17}$

$_{17}Cl \rightarrow {}_{17}Cl^-$:

$-IP_{1,1} \times \sqrt{17 \times 18} / M_{17} - \Delta IP_{1,18} \times 18 / M_{17} + \Delta IP_{1,17} \times 17 / M_{17}$

$_{17}Cl^- \rightarrow {}_{17}Cl^{2-}$:

$-IP_{1,1} \times \sqrt{17 \times 19} / M_{17} - \Delta IP_{1,19} \times 19 / M_{17} + \Delta IP_{1,18} \times 18 / M_{17}$

Insert Data:

$_{17}Cl \rightarrow {}_{17}Cl^+$:
$14.250 \times 17 / 35.453 + 15.942 \times 17 / 35.453 - 9.621 \times 16 / 35.453$

$_{17}Cl^+ \rightarrow {}_{17}Cl^{2+}$:
$14.250 \times 16.4924 / 35.453 + 9.621 \times 16 / 35.453 - 9.621 \times 15 / 35.453$

$_{17}Cl \rightarrow {}_{17}Cl^-$:
$-14.250 \times 17.4929 / 35.453 - 23.936 \times 18 / 35.453 + 15.942 \times 17 / 35.453$

$_{17}Cl^- \rightarrow {}_{17}Cl^{2-}$:
$-14.250 \times 17.9722 / 35.453 - (-4.704) \times 19 / 35.453 + 23.936 \times 18 / 35.453$

Evaluate Formulae:

$_{17}Cl \rightarrow {}_{17}Cl^+$: $6.8330 + 7.6443 - 4.3420 = 10.1353$

$_{17}Cl^+ \rightarrow {}_{17}Cl^{2+}$: $6.6290 + 4.3420 - 4.0706 = 6.9004$

$_{17}Cl \rightarrow {}_{17}Cl^-$: $-7.0311 - 12.1527 + 7.6443 = -11.5395$

$_{17}Cl^- \rightarrow {}_{17}Cl^{2-}$: $-7.2238 + 2.5210 + 12.1527 = 7.4499$

18. Argon

Write Formulae:

$_{18}Ar \rightarrow {}_{18}Ar^+$:
$$IP_{1,1} \times 18/M_{18} + \Delta IP_{1,18} \times 18/M_{18} - \Delta IP_{1,17} \times 17/M_{18}$$

$_{18}Ar^+ \rightarrow {}_{18}Ar^{2+}$:
$$IP_{1,1} \times \sqrt{18 \times 17}/M_{18} + \Delta IP_{1,17} \times 17/M_{18} - \Delta IP_{1,16} \times 16/M_{18}$$

$_{18}Ar^{2+} \rightarrow {}_{18}Ar^{3+}$:
$$IP_{1,1} \times \sqrt{18 \times 16}/M_{18} + \Delta IP_{1,16} \times 16/M_{18} - \Delta IP_{1,15} \times 15/M_{18}$$

$_{18}Ar^{3+} \rightarrow {}_{18}Ar^{4+}$:
$$IP_{1,1} \times \sqrt{18 \times 15}/M_{18} + \Delta IP_{1,15} \times 15/M_{18} - \Delta IP_{1,14} \times 14/M_{18}$$

$_{18}Ar \rightarrow {}_{18}Ar^-$:
$$-IP_{1,1} \times \sqrt{18 \times 19}/M_{18} - \Delta IP_{1,19} \times 19/M_{18} + \Delta IP_{1,18} \times 18/M_{18}$$

$_{18}Ar^- \rightarrow {}_{18}Ar^{2-}$:
$$-IP_{1,1} \times \sqrt{18 \times 20}/M_{18} - \Delta IP_{1,20} \times 20/M_{18} + \Delta IP_{1,19} \times 19/M_{18}$$

$_{18}Ar^{2-} \to {}_{18}Ar^{3-}$:

$-IP_{1,1} \times \sqrt{18 \times 21}/M_{18} - \Delta IP_{1,21} \times 21/M_{18} + \Delta IP_{1,20} \times 20/M_{18}$

$_{18}Ar^{3-} \to {}_{18}Ar^{4-}$:

$-IP_{1,1} \times \sqrt{18 \times 22}/M_{18} - \Delta IP_{1,22} \times 22/M_{18} + \Delta IP_{1,21} \times 21/M_{18}$

Insert Data:

$_{18}Ar \to {}_{18}Ar^{+}$:
$14.250 \times 18/39.948 + 23.936 \times 18/39.948 - 15.942 \times 17/39.948$

$_{18}Ar^{+} \to {}_{18}Ar^{2+}$:
$14.250 \times 17.4929/39.948 + 15.942 \times 17/39.948 - 9.621 \times 16/39.948$

$_{18}Ar^{2+} \to {}_{18}Ar^{3+}$:
$14.250 \times 16.9706/39.948 + 9.621 \times 16/39.948 - 9.621 \times 15/39.948$

$_{18}Ar^{3+} \to {}_{18}Ar^{4+}$:
$14.250 \times 16.4317/39.948 + 9.621 \times 15/39.948 - 4.624 \times 14/39.948$

$_{18}Ar \to {}_{18}Ar^{-}$:
$-14.250 \times 18.4932/39.948 - (-4.704) \times 19/39.948 + 23.936 \times 18/39.948$

$_{18}Ar^{-} \to {}_{18}Ar^{2-}$:
$-14.250 \times 18.9737/39.948 - (-1.193) \times 20/39.948 + (-4.704) \times 19/39.948$

$_{18}Ar^{2-} \to {}_{18}Ar^{3-}$:
$-14.250 \times 19.4422/39.948 - (-1.193) \times 21/39.948 + (-1.193) \times 20/39.948$

$_{18}Ar^{3-} \to {}_{18}Ar^{4-}$:
$-14.250 \times 19.8997/39.948 - (-0.612) \times 22/39.948 + (-1.193) \times 21/39.948$

Evaluate Formulae:

$_{18}Ar \rightarrow {}_{18}Ar^+$: $6.4208 + 10.7852 - 6.7842 = 10.4218$

$_{18}Ar^+ \rightarrow {}_{18}Ar^{2+}$: $6.2400 + 6.7842 - 3.8534 = 9.1708$

$_{18}Ar^{2+} \rightarrow {}_{18}Ar^{3+}$: $6.0536 + 3.8534 - 3.6126 = 6.2944$

$_{18}Ar^{3+} \rightarrow {}_{18}Ar^{4+}$: $5.8614 + 3.6125 - 1.6205 = 7.8534$

$_{18}Ar \rightarrow {}_{18}Ar^-$: $-6.5968 + 2.2373 + 10.7852 = 6.4257$

$_{18}Ar^- \rightarrow {}_{18}Ar^{2-}$: $-6.7682 + 0.5973 - 2.2373 = -8.4082$

$_{18}Ar^{2-} \rightarrow {}_{18}Ar^{3-}$: $-6.9353 + 0.6271 - 0.5973 = -6.9353$

$_{18}Ar^{3-} \rightarrow {}_{18}Ar^{4-}$: $-7.0985 + 0.3370 - 0.6271 = -7.3886$

19. Potassium

Write Formulae:

$_{19}K \rightarrow {}_{19}K^+$:
$IP_{1,1} \times 19 / M_{19} + \Delta IP_{1,19} \times 19 / M_{19} - \Delta IP_{1,18} \times 18 / M_{19}$

$_{19}K^+ \rightarrow {}_{19}K^{2+}$:
$IP_{1,1} \times \sqrt{19 \times 18} / M_{19} + \Delta IP_{1,18} \times 18 / M_{19} - \Delta IP_{1,17} \times 17 / M_{19}$

$_{19}K^{2+} \rightarrow {}_{19}K^{3+}$:
$IP_{1,1} \times \sqrt{19 \times 17} / M_{19} + \Delta IP_{1,17} \times 17 / M_{19} - \Delta IP_{1,16} \times 16 / M_{19}$

$_{19}K^{3+} \to {}_{19}K^{4+}$:

$IP_{1,1} \times \sqrt{19 \times 16}/M_{19} + \Delta IP_{1,16} \times 16/M_{19} - \Delta IP_{1,15} \times 15/M_{19}$

$_{19}K \to {}_{19}K^{-}$:

$-IP_{1,1} \times \sqrt{19 \times 20}/M_{19} - \Delta IP_{1,20} \times 20/M_{19} + \Delta IP_{1,19} \times 19/M_{19}$

$_{19}K^{-} \to {}_{19}K^{2-}$:

$-IP_{1,1} \times \sqrt{19 \times 21}/M_{19} - \Delta IP_{1,21} \times 21/M_{19} + \Delta IP_{1,20} \times 20/M_{19}$

$_{19}K^{2-} \to {}_{19}K^{3-}$:

$-IP_{1,1} \times \sqrt{19 \times 22}/M_{19} - \Delta IP_{1,22} \times 22/M_{19} + \Delta IP_{1,21} \times 21/M_{19}$

$_{19}K^{3-} \to {}_{19}K^{4-}$:

$-IP_{1,1} \times \sqrt{19 \times 23}/M_{19} - \Delta IP_{1,23} \times 23/M_{19} + \Delta IP_{1,22} \times 22/M_{19}$

Insert Data:

$_{19}K \to {}_{19}K^{+}$:
$14.250 \times 19 / 39.098 + (-4.704) \times 19 / 39.098 - 23.936 \times 18 / 39.098$

$_{19}K^{+} \to {}_{19}K^{2+}$:
$14.250 \times 18.4932/39.098 + 23.936 \times 18 / 39.098 - 15.942 \times 17 / 39.098$

$_{19}K^{2+} \to {}_{19}K^{3+}$:
$14.250 \times 17.9722/39.098 + 15.942 \times 17 / 39.098 - 9.621 \times 16 / 39.098$

$_{19}K^{3+} \to {}_{19}K^{4+}$:
$14.250 \times 17.4356/39.098 + 9.621 \times 16 / 39.098 - 9.621 \times 15 / 39.098$

$_{19}K \to {}_{19}K^{-}$:
$-14.250 \times 19.4936/39.098 - (-1.193) \times 20 / 39.098 + (-4.704) \times 19 / 39.098$

$_{19}K^- \to {}_{19}K^{2-}$:

$-14.250 \times 19.9750/39.098 - (-1.193) \times 21/39.098 + (-1.193) \times 20/39.098$

$_{19}K^{2-} \to {}_{19}K^{3-}$:

$-14.250 \times 20.4450/30.098 - (-0.612) \times 22/30.098 + (-1.193) \times 21/30.098$

$_{19}K^{3-} \to {}_{19}K^{4-}$:

$-14.250 \times 20.9045/39.098 - (-0.006) \times 23/39.098 + (-0.612) \times 22/39.098$

Evaluate Formulae:

$_{19}K \to {}_{19}K^+$: $6.9249 - 2.2859 - 11.0197 = -6.3807$

$_{19}K^+ \to {}_{19}K^{2+}$: $6.740 + 11.0197 - 6.9317 = 10.8280$

$_{19}K^{2+} \to {}_{19}K^{3+}$: $6.5503 + 6.9317 - 3.9372 = 9.5448$

$_{19}K^{3+} \to {}_{19}K^{4+}$: $6.3547 + 3.9372 - 3.6911 = 6.6008$

$_{19}K \to {}_{19}K^-$: $-7.1048 + 0.6103 - 2.2859 = -8.7804$

$_{19}K^- \to {}_{19}K^{2-}$: $-7.2803 + 0.6408 - 0.6103 = -7.2498$

$_{19}K^{2-} \to {}_{19}K^{3-}$: $-9.6798 + 0.4473 - 0.8324 = -10.0649$

$_{19}K^{3-} \to {}_{19}K^{4-}$: $-7.6190 + 0.0035 - 0.3444 = -7.9599$

21. Scandium

Write Formulae:

$_{21}Sc \to {}_{21}Sc^+$:

$IP_{1,1} \times 21/M_{21} + \Delta IP_{1,21} \times 21/M_{21} - \Delta IP_{1,20} \times 20/M_{21}$

$_{21}Sc^+ \rightarrow {}_{21}Sc^{2+}$:

$IP_{1,1} \times \sqrt{21 \times 20}/M_{21} + \Delta IP_{1,20} \times 20/M_{21} - \Delta IP_{1,19} \times 19/M_{21}$

$_{21}Sc \rightarrow {}_{21}Sc^-$:

$-IP_{1,1} \times \sqrt{21 \times 22}/M_{21} - \Delta IP_{1,22} \times 22/M_{21} + \Delta IP_{1,21} \times 21/M_{21}$

$_{21}Sc^- \rightarrow {}_{21}Sc^{2-}$:

$-IP_{1,1} \times \sqrt{21 \times 23}/M_{21} - \Delta IP_{1,23} \times 23/M_{21} + \Delta IP_{1,22} \times 22/M_{21}$

Insert Data:

$_{21}Sc \rightarrow {}_{21}Sc^+$:
$14.250 \times 21/44.956 + (-1.193) \times 21/44.956 - (-1.193) \times 20/44.956$

$_{21}Sc^+ \rightarrow {}_{21}Sc^{2+}$:
$14.250 \times 20.4939/44.956 + (-1.193) \times 20/44.956 - (-4.704) \times 19/44.956$

$_{21}Sc \rightarrow {}_{21}Sc^-$:
$-14.250 \times 21.4949/44.956 - (-0.612) \times 22/44.956 + (-1.193) \times 21/44.956$

$_{21}Sc^- \rightarrow {}_{21}Sc^{2-}$:
$-14.250 \times 21.9773/44.956 - (-0.006) \times 23/44.956 + (-0.612) \times 22/44.956$

Evaluate Formulae:

$_{21}Sc \rightarrow {}_{21}Sc^+$: $6.6565 - 0.5573 + 0.5307 = 6.6299$

$_{21}Sc^+ \rightarrow {}_{21}Sc^{2+}$: $6.4961 - 0.5307 + 1.9881 = 7.9535$

$_{21}Sc \rightarrow {}_{21}Sc^-$: $-6.8134 + 0.2995 - 0.5573 = -7.0712$

$_{21}Sc^- \rightarrow {}_{21}Sc^{2-}$: $-6.9663 + 0.0031 - 0.2995 = -7.2627$

24. Chromium

Write Formulae:

$_{24}Cr \rightarrow {}_{24}Cr^+$:

$IP_{1,1} \times 24 / M_{24} + \Delta IP_{1,24} \times 24 / M_{24} - \Delta IP_{1,23} \times 23 / M_{24}$

$_{24}Cr^+ \rightarrow {}_{24}Cr^{2+}$:

$IP_{1,1} \times \sqrt{24 \times 23}/M_{24} + \Delta IP_{1,23} \times 23 / M_{24} - \Delta IP_{1,22} \times 22 / M_{24}$

$_{24}Cr^{2+} \rightarrow {}_{24}Cr^{3+}$:

$IP_{1,1} \times \sqrt{24 \times 22}/M_{24} + \Delta IP_{1,22} \times 22 / M_{24} - \Delta IP_{1,21} \times 21 / M_{24}$

$_{24}Cr^{3+} \rightarrow {}_{24}Cr^{4+}$:

$IP_{1,1} \times \sqrt{24 \times 21}/M_{24} + \Delta IP_{1,21} \times 21 / M_{24} - \Delta IP_{1,20} \times 20 / M_{24}$

$_{24}Cr \rightarrow {}_{24}Cr^-$:

$-IP_{1,1} \times \sqrt{24 \times 25}/M_{24} - \Delta IP_{1,25} \times 25 / M_{24} + \Delta IP_{1,24} \times 24 / M_{24}$

$_{24}Cr^- \rightarrow {}_{24}Cr^{2-}$:

$-IP_{1,1} \times \sqrt{24 \times 26}/M_{24} - \Delta IP_{1,26} \times 26 / M_{24} + \Delta IP_{1,25} \times 25 / M_{24}$

$_{24}Cr^{2-} \rightarrow {}_{24}Cr^{3-}$:

$-IP_{1,1} \times \sqrt{24 \times 27}/M_{24} - \Delta IP_{1,27} \times 27 / M_{24} + \Delta IP_{1,26} \times 26 / M_{24}$

$_{24}Cr^{3-} \rightarrow {}_{24}Cr^{4-}$:

$-IP_{1,1} \times \sqrt{24 \times 28}/M_{24} - \Delta IP_{1,28} \times 28 / M_{24} + \Delta IP_{1,27} \times 27 / M_{24}$

Insert Data:

$_{24}Cr \rightarrow {}_{24}Cr^+$:
$14.250 \times 24/51.996 + 0.627 \times 24/51.996 - (-0.006) \times 23/51.996$

$_{24}Cr^+ \rightarrow {}_{24}Cr^{2+}$:
$14.250 \times 23.4947/51.996 + (-0.006) \times 23/51.996 - (-0.612) \times 22/51.996$

$_{24}Cr^{2+} \rightarrow {}_{24}Cr^{3+}$:
$14.250 \times 22.9783/51.996 + (-0.612) \times 22/51.996 - (-1.193) \times 21/51.996$

$_{24}Cr^{3+} \rightarrow {}_{24}Cr^{4+}$:
$14.250 \times 22.4499/51.996 + (-1.193) \times 21/51.996 - (-1.193) \times 20/51.996$

$_{24}Cr \rightarrow {}_{24}Cr^-$:
$-14.250 \times 24.4949/51.996 - 1.289 \times 25/51.996 + 0.627 \times 24/51.996$

$_{24}Cr^- \rightarrow {}_{24}Cr^{2-}$:
$-14.250 \times 24.9800/51.996 - 1.289 \times 26/51.996 + 1.289 \times 25/51.996$

$_{24}Cr^{2-} \rightarrow {}_{24}Cr^{3-}$:
$-14.250 \times 25.4558/51.996 - 1.980 \times 27/51.996 + 1.289 \times 26/51.996$

$_{24}Cr^{3-} \rightarrow {}_{24}Cr^{4-}$:
$-14.250 \times 25.9230/51.996 - 2.701 \times 28/51.996 + 1.980 \times 27/51.996$

Evaluate Formulae:

$_{24}Cr \rightarrow {}_{24}Cr^+$: $6.5774 + 0.2894 + 0.0027 = 6.8695$

$_{24}Cr^+ \rightarrow {}_{24}Cr^{2+}$: $6.4389 - 0.0027 + 0.2589 = 6.6951$

$_{24}Cr^{2+} \rightarrow {}_{24}Cr^{3+}$: $6.2974 - 0.2589 + 0.4814 = 6.5199$

$_{24}Cr^{3+} \to {}_{24}Cr^{4+}$: $6.1526 - 0.4818 + 0.4589 = 6.1297$

$_{24}Cr \to {}_{24}Cr^-$: $-6.7131 - 0.6198 + 0.2894 = -7.0435$

$_{24}Cr^- \to {}_{24}Cr^{2-}$: $-6.8460 - 0.6445 + 0.6198 = -6.8707$

$_{24}Cr^{2-} \to {}_{24}Cr^{3-}$: $-6.9764 - 1.0282 + 0.6445 = -7.3601$

$_{24}Cr^{3-} \to {}_{24}Cr^{4-}$: $-7.1044 - 1.4545 + 1.0282 = -7.5307$

26. Iron

Write Formulae:

$_{26}Fe \to {}_{26}Fe^+$:
$IP_{1,1} \times 26 / M_{26} + \Delta IP_{1,26} \times 26 / M_{26} - \Delta IP_{1,25} \times 25 / M_{26}$

$_{26}Fe^+ \to {}_{26}Fe^{2+}$:
$IP_{1,1} \times \sqrt{26 \times 25} / M_{26} + \Delta IP_{1,25} \times 25 / M_{26} - \Delta IP_{1,24} \times 24 / M_{26}$

$_{26}Fe^{2+} \to {}_{26}Fe^{3+}$:
$IP_{1,1} \times \sqrt{26 \times 24} / M_{26} + \Delta IP_{1,24} \times 24 / M_{26} - \Delta IP_{1,23} \times 23 / M_{26}$

$_{26}Fe^{3+} \to {}_{26}Fe^{4+}$:
$IP_{1,1} \times \sqrt{26 \times 23} / M_{26} + \Delta IP_{1,23} \times 23 / M_{26} - \Delta IP_{1,22} \times 22 / M_{26}$

$_{26}Fe \to {}_{26}Fe^-$:
$-IP_{1,1} \times \sqrt{26 \times 27} / M_{26} - \Delta IP_{1,27} \times 27 / M_{26} + \Delta IP_{1,26} \times 26 / M_{26}$

$_{26}\text{Fe}^- \to {}_{26}\text{Fe}^{2-}$:

$-IP_{1,1} \times \sqrt{26 \times 28}/M_{26} - \Delta IP_{1,28} \times 28/M_{26} + \Delta IP_{1,27} \times 27/M_{26}$

$_{26}\text{Fe}^{2-} \to {}_{26}\text{Fe}^{3-}$:

$-IP_{1,1} \times \sqrt{26 \times 29}/M_{26} - \Delta IP_{1,29} \times 29/M_{26} + \Delta IP_{1,28} \times 28/M_{26}$

$_{26}\text{Fe}^{3-} \to {}_{26}\text{Fe}^{4-}$:

$-IP_{1,1} \times \sqrt{26 \times 30}/M_{26} - \Delta IP_{1,30} \times 30/M_{26} + \Delta IP_{1,29} \times 29/M_{26}$

Insert Data:

$_{26}\text{Fe} \to {}_{26}\text{Fe}^+$:
$14.250 \times 26/55.845 + 1.289 \times 26/55.845 - 1.289 \times 25/55.845$

$_{26}\text{Fe}^+ \to {}_{26}\text{Fe}^{2+}$:
$14.250 \times 25.4951/55.845 + 1.289 \times 25/55.845 - 0.627 \times 24/55.845$

$_{26}\text{Fe}^{2+} \to {}_{26}\text{Fe}^{3+}$:
$14.250 \times 24.9800/55.845 + 0.627 \times 24/55.845 - (-0.006) \times 23/55.845$

$_{26}\text{Fe}^{3+} \to {}_{26}\text{Fe}^{4+}$:
$14.250 \times 24.4540/55.845 + (-0.006) \times 23/55.845 - (-0.612) \times 22/55.845$

$_{26}\text{Fe} \to {}_{26}\text{Fe}^-$:
$-14.250 \times 26.4953/55.845 - 1.980 \times 27/55.845 + 1.289 \times 26/55.845$

$_{26}\text{Fe}^- \to {}_{26}\text{Fe}^{2-}$:
$-14.250 \times 26.9815/55.845 - 2.701 \times 28/55.845 + 1.980 \times 27/55.845$

$_{26}\text{Fe}^{2-} \to {}_{26}\text{Fe}^{3-}$:
$-14.250 \times 27.4591/55.845 - 3.455 \times 29/55.845 + 2.701 \times 28/55.845$

$_{26}Fe^{3-} \to {}_{26}Fe^{4-}$:

$-14.250 \times 27.9285/55.845 - 4.242 \times 30/55.845 + 3.455 \times 29/55.845$

Evaluate Formulae:

$_{26}Fe \to {}_{26}Fe^{+}$: $6.6344 + 0.6001 - 0.5770 = 6.6575$

$_{26}Fe^{+} \to {}_{26}Fe^{2+}$: $6.5056 + 0.5770 - 0.2695 = 6.8131$

$_{26}Fe^{2+} \to {}_{26}Fe^{3+}$: $6.3742 + 0.2695 + 0.0025 = 6.6462$

$_{26}Fe^{3+} \to {}_{26}Fe^{4+}$: $6.2399 - 0.0025 + 0.2411 = 6.4785$

$_{26}Fe \to {}_{26}Fe^{-}$: $-6.7608 - 0.9573 + 0.6001 = -7.1180$

$_{26}Fe^{-} \to {}_{26}Fe^{2-}$: $-6.8849 - 1.3542 + 0.9573 = -7.2818$

$_{26}Fe^{2-} \to {}_{26}Fe^{3-}$: $-7.0068 - 1.7942 + 1.3542 = -7.4468$

$_{26}Fe^{3-} \to {}_{26}Fe^{4-}$: $-7.1265 - 2.2788 + 1.7942 = -7.6111$

27. Cobalt

Write Formulae:

$_{27}Co \to {}_{27}Co^{+}$:

$IP_{1,1} \times 27/M_{27} + \Delta IP_{1,27} \times 27/M_{27} - \Delta IP_{1,26} \times 26/M_{27}$

$_{27}Co^{+} \to {}_{27}Co^{2+}$:

$IP_{1,1} \times \sqrt{27 \times 26}/M_{27} + \Delta IP_{1,26} \times 26/M_{27} - \Delta IP_{1,25} \times 25/M_{27}$

$_{27}Co^{2+} \to {}_{27}Co^{3+}$:

$IP_{1,1} \times \sqrt{27 \times 25}/M_{27} + \Delta IP_{1,25} \times 25/M_{27} - \Delta IP_{1,24} \times 24/M_{27}$

$_{27}Co^{3+} \to {}_{27}Co^{4+}$:

$IP_{1,1} \times \sqrt{27 \times 24}/M_{27} + \Delta IP_{1,24} \times 24/M_{27} - \Delta IP_{1,23} \times 23/M_{27}$

$_{27}Co \to {}_{27}Co^{-}$:

$-IP_{1,1} \times \sqrt{27 \times 28}/M_{27} - \Delta IP_{1,28} \times 28/M_{27} + \Delta IP_{1,27} \times 27/M_{27}$

$_{27}Co^{-} \to {}_{27}Co^{2-}$:

$-IP_{1,1} \times \sqrt{27 \times 29}/M_{27} - \Delta IP_{1,29} \times 29/M_{27} + \Delta IP_{1,28} \times 28/M_{27}$

$_{27}Co^{2-} \to {}_{27}Co^{3-}$:

$-IP_{1,1} \times \sqrt{27 \times 30}/M_{27} - \Delta IP_{1,30} \times 30/M_{27} + \Delta IP_{1,29} \times 29/M_{27}$

$_{27}Co^{3-} \to {}_{27}Co^{4-}$:

$-IP_{1,1} \times \sqrt{27 \times 31}/M_{27} - \Delta IP_{1,31} \times 31/M_{27} + \Delta IP_{1,30} \times 30/M_{27}$

Insert Data:

$_{27}Co \to {}_{27}Co^{+}$:

$14.250 \times 27/58.933 + 1.980 \times 27/58.933 - 1.289 \times 26/58.933$

$_{27}Co^{+} \to {}_{27}Co^{2+}$:

$14.250 \times 26.4953/58.933 + 1.289 \times 26/58.933 - 1.289 \times 25/58.933$

$_{27}Co^{2+} \to {}_{27}Co^{3+}$:

$14.250 \times 25.9808/58.933 + 1.289 \times 25/58.933 - 0.627 \times 24/58.933$

$_{27}Co^{3+} \to {}_{27}Co^{4+}$:

$14.250 \times 25.4558/58.933 + 0.627 \times 24/58.933 - (-0.006) \times 23/58.933$

$_{27}Co \to {}_{27}Co^{-}$:

$-14.250 \times 27.4955/58.933 - 2.701 \times 28/58.933 + 1.980 \times 27/58.933$

$_{27}Co^{-} \to {}_{27}Co^{2-}$:

$-14.250 \times 27.9821/58.933 - 3.455 \times 29/58.933 + 2.701 \times 28/58.933$

$_{27}Co^{2-} \to {}_{27}Co^{3-}$:

$-14.250 \times 28.4605/58.933 - 4.242 \times 30/58.933 + 3.455 \times 29/58.933$

$_{27}Co^{3+} \to {}_{27}Co^{4+}$:

$-14.250 \times 28.9310/58.933 - 0.244 \times 31/58.933 + 4.242 \times 30/58.933$

Evaluate Formulae:

$_{27}Co \to {}_{27}Co^{+}$: $6.5286 + 0.9071 - 0.5687 = 6.8670$

$_{27}Co^{+} \to {}_{27}Co^{2+}$: $6.4066 + 0.5687 - 0.5468 = 6.4285$

$_{27}Co^{2+} \to {}_{27}Co^{3+}$: $6.2822 + 0.5468 - 0.2553 = 6.5737$

$_{27}Co^{3+} \to {}_{27}Co^{4+}$: $6.1552 + 0.2553 + 0.0023 = 6.4128$

$_{27}Co \to {}_{27}Co^{-}$: $-6.6484 - 1.2833 + 0.9071 = -7.0246$

$_{27}Co^{-} \to {}_{27}Co^{2-}$: $-6.7661 - 1.7002 + 1.2833 = -7.1830$

$_{27}Co^{2-} \to {}_{27}Co^{3-}$: $-6.8817 - 2.1594 + 1.7002 = -7.3409$

$_{27}Co^{3-} \to {}_{27}Co^{4-}$: $-6.9955 - 0.1283 + 2.1594 = -4.9644$

29. Copper

Write Formulae:

$_{29}Cu \rightarrow {_{29}Cu^+}$:

$IP_{1,1} \times 29 / M_{29} + \Delta IP_{1,29} \times 29 / M_{29} - \Delta IP_{1,28} \times 28 / M_{29}$

$_{29}Cu^+ \rightarrow {_{29}Cu^{2+}}$:

$IP_{1,1} \times \sqrt{29 \times 28} / M_{29} + \Delta IP_{1,28} \times 28 / M_{29} - \Delta IP_{1,27} \times 27 / M_{29}$

$_{29}Cu \rightarrow {_{29}Cu^-}$:

$-IP_{1,1} \times \sqrt{29 \times 30} / M_{29} - \Delta IP_{1,30} \times 30 / M_{29} + \Delta IP_{1,29} \times 29 / M_{29}$

$_{29}Cu^- \rightarrow {_{29}Cu^{2-}}$:

$-IP_{1,1} \times \sqrt{29 \times 31} / M_{29} - \Delta IP_{1,31} \times 31 / M_{29} + \Delta IP_{1,30} \times 30 / M_{29}$

Insert Data:

$_{29}Cu \rightarrow {_{29}Cu^+}$:
$14.250 \times 29 / 63.546 + 3.455 \times 29 / 63.546 - 2.701 \times 28 / 63.546$

$_{29}Cu^+ \rightarrow {_{29}Cu^{2+}}$:
$14.250 \times 28.4956 / 63.546 + 2.701 \times 28 / 63.546 - 1.980 \times 27 / 63.546$

$_{29}Cu \rightarrow {_{29}Cu^-}$:
$-14.250 \times 29.4958 / 63.546 - 4.242 \times 30 / 63.546 + 3.455 \times 29 / 63.546$

$_{29}Cu^- \rightarrow {_{29}Cu^{2-}}$:
$-14.250 \times 29.9833 / 63.546 - 0.244 \times 31 / 63.546 + 4.242 \times 30 / 63.546$

Evaluate Formulae:

$_{29}Cu \rightarrow {_{29}Cu^+}$: $6.5032 + 1.5767 - 1.1901 = 6.8898$

$_{29}Cu^+ \rightarrow {}_{29}Cu^{2+}$: $6.3901 + 1.1901 - 0.8413 = 6.7389$

$_{29}Cu \rightarrow {}_{29}Cu^-$: $-6.6143 - 2.0026 + 1.5767 = -7.0402$

$_{29}Cu^- \rightarrow {}_{29}Cu^{2-}$: $-6.5032 - 0.1190 + 2.0026 = -4.6196$

30. Zinc

Write Formulae:

$_{30}Zn \rightarrow {}_{30}Zn^+$:
$IP_{1,1} \times 30 / M_{30} + \Delta IP_{1,30} \times 30 / M_{30} - \Delta IP_{1,29} \times 29 / M_{30}$

$_{30}Zn^+ \rightarrow {}_{30}Zn^{2+}$:
$IP_{1,1} \times \sqrt{30 \times 29} / M_{30} + \Delta IP_{1,29} \times 29 / M_{30} - \Delta IP_{1,28} \times 28 / M_{30}$

$_{30}Zn^{2+} \rightarrow {}_{30}Zn^{3+}$:
$IP_{1,1} \times \sqrt{30 \times 28} / M_{30} + \Delta IP_{1,28} \times 28 / M_{30} - \Delta IP_{1,27} \times 27 / M_{30}$

$_{30}Zn^{3+} \rightarrow {}_{30}Zn^{4+}$:
$IP_{1,1} \times \sqrt{30 \times 27} / M_{30} + \Delta IP_{1,27} \times 27 / M_{30} - \Delta IP_{1,26} \times 26 / M_{30}$

$_{30}Zn^{4+} \rightarrow {}_{30}Zn^{5+}$:
$IP_{1,1} \times \sqrt{30 \times 26} / M_{30} + \Delta IP_{1,26} \times 26 / M_{30} - \Delta IP_{1,25} \times 25 / M_{30}$

$_{30}Zn^{5+} \rightarrow {}_{30}Zn^{6+}$:
$IP_{1,1} \times \sqrt{30 \times 25} / M_{30} + \Delta IP_{1,25} \times 25 / M_{30} - \Delta IP_{1,24} \times 24 / M_{30}$

$_{30}Zn^{6+} \to {}_{30}Zn^{7+}$:
$IP_{1,1} \times \sqrt{30 \times 24}\big/M_{30} + \Delta IP_{1,24} \times 24 / M_{30} - \Delta IP_{1,23} \times 23 / M_{30}$

$_{30}Zn \to {}_{30}Zn^{-}$:
$-IP_{1,1} \times \sqrt{30 \times 31}\big/M_{30} - \Delta IP_{1,31} \times 31 / M_{30} + \Delta IP_{1,30} \times 30 / M_{30}$

$_{30}Zn^{-} \to {}_{30}Zn^{2-}$:
$-IP_{1,1} \times \sqrt{30 \times 32}\big/M_{30} - \Delta IP_{1,32} \times 32 / M_{30} + \Delta IP_{1,31} \times 31 / M_{30}$

$_{30}Zn^{2-} \to {}_{30}Zn^{3-}$:
$-IP_{1,1} \times \sqrt{30 \times 33}\big/M_{30} - \Delta IP_{1,33} \times 33 / M_{30} + \Delta IP_{1,32} \times 32 / M_{30}$

$_{30}Zn^{3-} \to {}_{30}Zn^{4-}$:
$-IP_{1,1} \times \sqrt{30 \times 34}\big/M_{30} - \Delta IP_{1,34} \times 34 / M_{30} + \Delta IP_{1,33} \times 33 / M_{30}$

$_{30}Zn^{4-} \to {}_{30}Zn^{5-}$:
$-IP_{1,1} \times \sqrt{30 \times 35}\big/M_{30} - \Delta IP_{1,35} \times 35 / M_{30} + \Delta IP_{1,34} \times 34 / M_{30}$

$_{30}Zn^{5-} \to {}_{30}Zn^{6-}$:
$-IP_{1,1} \times \sqrt{30 \times 36}\big/M_{30} - \Delta IP_{1,36} \times 36 / M_{30} + \Delta IP_{1,35} \times 35 / M_{30}$

$_{30}Zn^{6-} \to {}_{30}Zn^{7-}$:
$-IP_{1,1} \times \sqrt{30 \times 37}\big/M_{30} - \Delta IP_{1,37} \times 37 / M_{30} + \Delta IP_{1,36} \times 36 / M_{30}$

Insert Data:

$_{30}Zn \to {}_{30}Zn^{+}$:
$14.250 \times 30 / 65.390 + 4.242 \times 30 / 65.390 - 3.455 \times 29 / 65.390$

$_{30}Zn^+ \to {}_{30}Zn^{2+}$:

$14.25 \times 29.4958/65.390 + 3.455 \times 29 / 65.390 - 2.701 \times 28 / 65.390$

$_{30}Zn^{2+} \to {}_{30}Zn^{3+}$:

$14.250 \times 28.9828/65.390 + 2.701 \times 28 / 65.390 - 1.980 \times 27 / 65.390$

$_{30}Zn^{3+} \to {}_{30}Zn^{4+}$:

$14.250 \times 28.4605/65.390 + 1.980 \times 27 / 65.390 - 1.289 \times 26 / 65.390$

$_{30}Zn^{4+} \to {}_{30}Zn^{5+}$:

$14.250 \times 27.9285/65.390 + 1.289 \times 26 / 65.390 - 1.289 \times 25 / 65.390$

$_{30}Zn^{5+} \to {}_{30}Zn^{6+}$:

$14.250 \times 27.3861/65.390 + 1.289 \times 25 / 65.390 - 0.627 \times 24 / 65.390$

$_{30}Zn^{6+} \to {}_{30}Zn^{7+}$:

$14.250 \times 26.8328/63.390 + 0.627 \times 24 / 63.390 - (-0.006) \times 23 / 63.390$

$_{30}Zn \to {}_{30}Zn^-$:

$-14.250 \times 30.4959/65.390 - 0.244 \times 31 / 65.390 + 4.242 \times 30 / 65.390$

$_{30}Zn^- \to {}_{30}Zn^{2-}$:

$-14.25 \times 30.9839/65.390 - 3.610 \times 32 / 65.390 + 0.244 \times 31 / 65.390$

$_{30}Zn^{2-} \to {}_{30}Zn^{3-}$:

$-14.250 \times 31.4643/65.390 - 7.757 \times 33 / 65.390 + 3.610 \times 32 / 65.390$

$_{30}Zn^{3-} \to {}_{30}Zn^{4-}$:

$-14.250 \times 31.9374/65.390 - 7.757 \times 34 / 65.390 + 7.757 \times 33 / 65.390$

$_{30}Zn^{4-} \to {}_{30}Zn^{5-}$:

$-14.250 \times 32.4037/65.390 - 12.866 \times 35 / 65.390 + 7.757 \times 34 / 65.390$

$_{30}Zn^{5-} \to {}_{30}Zn^{6-}$:

$-14.250 \times 32.8634/65.390 - 19.162 \times 36/65.390 + 12.866 \times 35/65.390$

$_{30}Zn^{6-} \to {}_{30}Zn^{7-}$:

$-14.250 \times 33.3167/65.390 - (-4.704) \times 37/65.390 + 19.162 \times 36/65.390$

Evaluate Formulae:

$_{30}Zn \to {}_{30}Zn^{+}$: $6.5377 + 1.9462 - 1.5323 = 6.9516$

$_{30}Zn^{+} \to {}_{30}Zn^{2+}$: $6.4278 + 1.5323 - 1.1566 = 6.8035$

$_{30}Zn^{2+} \to {}_{30}Zn^{3+}$: $6.3160 + 1.1566 - 0.8176 = 6.6550$

$_{30}Zn^{3+} \to {}_{30}Zn^{4+}$: $6.2022 + 0.8176 - 0.5125 = 6.5073$

$_{30}Zn^{4+} \to {}_{30}Zn^{5+}$: $6.0863 + 0.5125 - 0.4928 = 6.1060$

$_{30}Zn^{5+} \to {}_{30}Zn^{6+}$: $5.9681 + 0.4928 - 0.2301 = 6.2308$

$_{30}Zn^{6+} \to {}_{30}Zn^{7+}$: $6.0320 + 0.2374 + 0.0022 = 6.2716$

$_{30}Zn \to {}_{30}Zn^{-}$: $-6.6458 - 0.1157 + 1.9462 = -4.8153$

$_{30}Zn^{-} \to {}_{30}Zn^{2-}$: $-6.7521 - 1.7666 + 0.1157 = -8.4030$

$_{30}Zn^{2-} \to {}_{30}Zn^{3-}$: $-6.7556 - 3.9147 + 1.7666 = -8.9037$

$_{30}Zn^{3-} \to {}_{30}Zn^{4-}$: $-6.9599 - 4.0333 + 3.9147 = -7.0785$

$_{30}Zn^{4-} \to {}_{30}Zn^{5-}$: $-7.0615 - 6.8865 + 4.0333 = -9.9147$

$_{30}Zn^{5-} \to {}_{30}Zn^{6-}$: $-7.1617 - 10.5495 + 6.8865 = -10.8247$

$_{30}Zn^{6-} \to {}_{30}Zn^{7-}$: $-7.2605 + 2.6617 + 10.5495 = 5.9507$

31. Gallium

Write Formulae:

$_{31}Ga \to {}_{31}Ga^{+}$:
$IP_{1,1} \times 31/M_{31} + \Delta IP_{1,31} \times 31/M_{31} - \Delta IP_{1,30} \times 30/M_{31}$

$_{31}Ga^{+} \to {}_{31}Ga^{2+}$:
$IP_{1,1} \times \sqrt{31 \times 30}/M_{31} + \Delta IP_{1,30} \times 30/M_{31} - \Delta IP_{1,29} \times 29/M_{31}$

$_{31}Ga^{2+} \to {}_{31}Ga^{3+}$:
$IP_{1,1} \times \sqrt{31 \times 29}/M_{31} + \Delta IP_{1,29} \times 29/M_{31} - \Delta IP_{1,28} \times 28/M_{31}$

$_{31}Ga^{3+} \to {}_{31}Ga^{4+}$:
$IP_{1,1} \times \sqrt{31 \times 28}/M_{31} + \Delta IP_{1,28} \times 28/M_{31} - \Delta IP_{1,27} \times 27/M_{31}$

$_{31}Ga^{4+} \to {}_{31}Ga^{5+}$:
$IP_{1,1} \times \sqrt{31 \times 27}/M_{31} + \Delta IP_{1,27} \times 27/M_{31} - \Delta IP_{1,26} \times 26/M_{31}$

$_{31}Ga \to {}_{31}Ga^{-}$:
$-IP_{1,1} \times \sqrt{31 \times 32}/M_{31} - \Delta IP_{1,32} \times 32/M_{31} + \Delta IP_{1,31} \times 31/M_{31}$

$_{31}Ga^{-} \to {}_{31}Ga^{2-}$:
$-IP_{1,1} \times \sqrt{31 \times 33}/M_{31} - \Delta IP_{1,33} \times 33/M_{31} + \Delta IP_{1,32} \times 32/M_{31}$

$_{31}Ga^{2-} \to {}_{31}Ga^{3-}$:
$-IP_{1,1} \times \sqrt{31 \times 34}/M_{31} - \Delta IP_{1,34} \times 34/M_{31} + \Delta IP_{1,33} \times 33/M_{31}$

$_{31}Ga^{3-} \to {}_{31}Ga^{4-}$:
$-IP_{1,1} \times \sqrt{31 \times 35}/M_{31} - \Delta IP_{1,35} \times 35/M_{31} + \Delta IP_{1,34} \times 34/M_{31}$

$_{31}Ga^{4-} \to {}_{31}Ga^{5-}$:
$-IP_{1,1} \times \sqrt{31 \times 36}/M_{31} - \Delta IP_{1,36} \times 36/M_{31} + \Delta IP_{1,35} \times 35/M_{31}$

Insert Data:

$_{31}Ga \to {}_{31}Ga^{+}$:
$14.250 \times 31/69.723 + 0.244 \times 31/69.723 - 4.242 \times 30/69.723$

$_{31}Ga^{+} \to {}_{31}Ga^{2+}$:
$14.250 \times 30.496/69.723 + 4.242 \times 30/69.723 - 3.455 \times 29/69.723$

$_{31}Ga^{2+} \to {}_{31}Ga^{3+}$:
$14.250 \times 29.9833/69.723 + 3.455 \times 29/69.723 - 2.701 \times 28/69.723$

$_{31}Ga^{3+} \to {}_{31}Ga^{4+}$:
$14.250 \times 29.4618/69.723 + 2.701 \times 28/69.723 - 1.980 \times 27/69.723$

$_{31}Ga^{4+} \to {}_{31}Ga^{5+}$:
$14.250 \times 28.931/69.723 + 1.980 \times 27/69.723 - 1.289 \times 26/69.723$

$_{31}Ga \to {}_{31}Ga^{-}$:
$-14.250 \times 31.496/69.723 - 3.610 \times 32/69.723 + 0.244 \times 31/69.723$

$_{31}Ga^{-} \to {}_{31}Ga^{2-}$:
$-14.250 \times 31.984/69.723 - 7.757 \times 33/69.723 + 3.610 \times 32/69.723$

$_{31}\text{Ga}^{2-} \to {}_{31}\text{Ga}^{3-}$:

$-14.250 \times 32.4654/69.723 - 7.757 \times 34/69.723 + 7.757 \times 33/69.723$

$_{31}\text{Ga}^{3-} \to {}_{31}\text{Ga}^{4-}$:

$-14.250 \times 32.9393/69.723 - 12.866 \times 35/69.723 + 7.757 \times 34/69.723$

$_{31}\text{Ga}^{4-} \to {}_{31}\text{Ga}^{5-}$:

$-14.250 \times 33.4066/69.723 - 10.162 \times 36/69.723 + 12.866 \times 35/69.723$

Evaluate Formulae:

$_{31}\text{Ga} \to {}_{31}\text{Ga}^{+}$: $6.3358 + 0.1085 - 1.8252 = 4.6191$

$_{31}\text{Ga}^{+} \to {}_{31}\text{Ga}^{2+}$: $6.2328 + 1.8252 - 1.4370 = 6.6210$

$_{31}\text{Ga}^{2+} \to {}_{31}\text{Ga}^{3+}$: $6.1280 + 1.4370 - 1.0847 = 6.4803$

$_{31}\text{Ga}^{3+} \to {}_{31}\text{Ga}^{4+}$: $6.0214 + 1.0847 - 0.7667 = 6.3394$

$_{31}\text{Ga}^{4+} \to {}_{31}\text{Ga}^{5+}$: $5.9129 + 0.7667 - 0.4807 = 6.1989$

$_{31}\text{Ga} \to {}_{31}\text{Ga}^{-}$: $-6.4372 - 1.6568 + 0.1085 = -7.9855$

$_{31}\text{Ga}^{-} \to {}_{31}\text{Ga}^{2-}$: $-6.5369 - 3.6714 + 1.6568 = -8.5515$

$_{31}\text{Ga}^{2-} \to {}_{31}\text{Ga}^{3-}$: $-6.6353 - 3.7827 + 3.6714 = -6.7466$

$_{31}\text{Ga}^{3-} \to {}_{31}\text{Ga}^{4-}$: $-6.7321 - 6.4586 + 3.7827 = -9.4080$

$_{31}\text{Ga}^{4-} \to {}_{31}\text{Ga}^{5-}$: $-6.8276 - 5.2469 + 6.4586 = -5.6159$

32. Germanium

Write Formulae:

$_{32}Ge \to {_{32}Ge^+}$:
$IP_{1,1} \times 32 / M_{32} + \Delta IP_{1,32} \times 32 / M_{32} - \Delta IP_{1,31} \times 31 / M_{32}$

$_{32}Ge^+ \to {_{32}Ge^{2+}}$:
$IP_{1,1} \times \sqrt{32 \times 31}/M_{32} + \Delta IP_{1,31} \times 31 / M_{32} - \Delta IP_{1,30} \times 30 / M_{32}$

$_{32}Ge^{2+} \to {_{32}Ge^{3+}}$:
$IP_{1,1} \times \sqrt{32 \times 30}/M_{32} + \Delta IP_{1,30} \times 30 / M_{32} - \Delta IP_{1,29} \times 29 / M_{32}$

$_{32}Ge^{3+} \to {_{32}Ge^{4+}}$:
$IP_{1,1} \times \sqrt{32 \times 29}/M_{32} + \Delta IP_{1,29} \times 29 / M_{32} - \Delta IP_{1,28} \times 28 / M_{32}$

$_{32}Ge^{4+} \to {_{32}Ge^{5+}}$:
$IP_{1,1} \times \sqrt{32 \times 28}/M_{32} + \Delta IP_{1,28} \times 28 / M_{32} - \Delta IP_{1,29} \times 29 / M_{32}$

$_{32}Ge \to {_{32}Ge^-}$:
$-IP_{1,1} \times \sqrt{32 \times 33}/M_{32} - \Delta IP_{1,33} \times 33 / M_{32} + \Delta IP_{1,32} \times 32 / M_{32}$

$_{32}Ge^- \to {_{32}Ge^{2-}}$:
$-IP_{1,1} \times \sqrt{32 \times 34}/M_{32} - \Delta IP_{1,34} \times 34 / M_{32} + \Delta IP_{1,33} \times 33 / M_{32}$

$_{32}Ge^{2-} \to {_{32}Ge^{3-}}$:
$-IP_{1,1} \times \sqrt{32 \times 35}/M_{32} - \Delta IP_{1,35} \times 35 / M_{32} + \Delta IP_{1,34} \times 34 / M_{32}$

$_{32}\text{Ge}^{3-} \rightarrow {_{32}\text{Ge}^{4-}}$:

$-IP_{1,1} \times \sqrt{32 \times 36}/M_{32} - \Delta IP_{1,36} \times 36/M_{32} + \Delta IP_{1,35} \times 35/M_{32}$

$_{32}\text{Ge}^{4-} \rightarrow {_{32}\text{Ge}^{5-}}$:

$-IP_{1,1} \times \sqrt{32 \times 37}/M_{32} - \Delta IP_{1,37} \times 37/M_{32} + \Delta IP_{1,36} \times 36/M_{32}$

Insert Data:

$_{32}\text{Ge} \rightarrow {_{32}\text{Ge}^{+}}$:
$14.250 \times 32/72.610 + 3.610 \times 32/72.610 - 0.244 \times 31/72.610$

$_{32}\text{Ge}^{+} \rightarrow {_{32}\text{Ge}^{2+}}$:
$14.250 \times 31.4960/72.610 + 0.244 \times 31/72.610 - 4.242 \times 30/72.610$

$_{32}\text{Ge}^{2+} \rightarrow {_{32}\text{Ge}^{3+}}$:
$14.250 \times 30.9839/72.610 + 4.242 \times 30/72.610 - 3.455 \times 29/72.610$

$_{32}\text{Ge}^{3+} \rightarrow {_{32}\text{Ge}^{4+}}$:
$14.250 \times 30.4631/72.610 + 3.455 \times 29/72.610 - 2.701 \times 28/72.610$

$_{32}\text{Ge}^{4+} \rightarrow {_{32}\text{Ge}^{5+}}$:
$14.250 \times 29.933/72.610 + 2.701 \times 28/72.610 - 3.455 \times 29/72.610$

$_{32}\text{Ge} \rightarrow {_{32}\text{Ge}^{-}}$:
$-14.250 \times 32.4962/72.610 - 7.757 \times 33/72.610 + 7.757 \times 32/72.610$

$_{32}\text{Ge}^{-} \rightarrow {_{32}\text{Ge}^{2-}}$:
$-14.250 \times 32.9848/72.610 - 7.757 \times 34/72.610 + 7.757 \times 33/72.610$

$_{32}\text{Ge}^{2-} \rightarrow {_{32}\text{Ge}^{3-}}$:
$-14.250 \times 33.4664/72.610 - 12.866 \times 35/72.610 + 7.757 \times 34/72.610$

$_{32}Ge^{3-} \rightarrow {}_{32}Ge^{4-}$:

$-14.25 \times 33.9411/72.610 - 19.162 \times 36/72.610 + 12.866 \times 35/72.610$

$_{32}Ge^{4-} \rightarrow {}_{32}Ge^{5-}$:

$-14.250 \times 34.4093/72.610 - (-4.704) \times 37/72.610 + 33.412 \times 36/72.610$

Evaluate Formulae:

$_{32}Ge \rightarrow {}_{32}Ge^{+}$: $6.2801 + 1.5910 - 0.1042 = 7.9753$

$_{32}Ge^{+} \rightarrow {}_{32}Ge^{2+}$: $6.1812 + 0.1042 - 1.7527 = 4.5327$

$_{32}Ge^{2+} \rightarrow {}_{32}Ge^{3+}$: $6.0807 + 1.7527 - 1.3799 = 6.4535$

$_{32}Ge^{3+} \rightarrow {}_{32}Ge^{4+}$: $5.9785 + 1.3799 - 1.0415 = 6.3168$

$_{32}Ge^{4+} \rightarrow {}_{32}Ge^{5+}$: $5.8745 + 1.0416 - 1.3799 = 5.5362$

$_{32}Ge \rightarrow {}_{32}Ge^{-}$: $-6.3775 - 3.5254 + 3.4186 = -6.4843$

$_{32}Ge^{-} \rightarrow {}_{32}Ge^{2-}$: $-6.4734 - 3.6323 + 3.5254 = -6.5803$

$_{32}Ge^{2-} \rightarrow {}_{32}Ge^{3-}$: $-6.5679 - 6.2018 + 3.6323 = -9.1374$

$_{32}Ge^{3-} \rightarrow {}_{32}Ge^{4-}$: $-6.6611 - 9.5005 + 6.2018 = -9.9598$

$_{32}Ge^{4-} \rightarrow {}_{32}Ge^{5-}$: $-6.7530 + 2.3970 + 16.5657 = 12.2097$

33. Arsenic

Write Formulae:

$_{33}As \rightarrow {}_{33}As^{+}$:

$IP_{1,1} \times 33 / M_{33} + \Delta IP_{1,33} \times 33 / M_{33} - \Delta IP_{1,32} \times 32 / M_{33}$

$_{33}As^{+} \rightarrow {}_{33}As^{2+}$:

$IP_{1,1} \times \sqrt{33 \times 32}/M_{33} + \Delta IP_{1,32} \times 32 / M_{33} - \Delta IP_{1,31} \times 31 / M_{33}$

$_{33}As^{2+} \rightarrow {}_{33}As^{3+}$:

$IP_{1,1} \times \sqrt{33 \times 31}/M_{33} + \Delta IP_{1,31} \times 31 / M_{33} - \Delta IP_{1,30} \times 30 / M_{33}$

$_{33}As^{3+} \rightarrow {}_{33}As^{4+}$:

$IP_{1,1} \times \sqrt{33 \times 30}/M_{33} + \Delta IP_{1,30} \times 30 / M_{33} - \Delta IP_{1,29} \times 29 / M_{33}$

$_{33}As \rightarrow {}_{33}As^{-}$:

$-IP_{1,1} \times \sqrt{33 \times 34}/M_{33} - \Delta IP_{1,34} \times 34 / M_{33} + \Delta IP_{1,33} \times 33 / M_{33}$

$_{33}As^{-} \rightarrow {}_{33}As^{2-}$:

$-IP_{1,1} \times \sqrt{33 \times 35}/M_{33} - \Delta IP_{1,35} \times 35 / M_{33} + \Delta IP_{1,34} \times 34 / M_{33}$

$_{33}As^{2-} \rightarrow {}_{33}As^{3-}$:

$-IP_{1,1} \times \sqrt{33 \times 36}/M_{33} - \Delta IP_{1,36} \times 36 / M_{33} + \Delta IP_{1,35} \times 35 / M_{33}$

$_{33}As^{3-} \rightarrow {}_{33}As^{4-}$:

$-IP_{1,1} \times \sqrt{33 \times 37}/M_{33} - \Delta IP_{1,37} \times 37 / M_{33} + \Delta IP_{1,36} \times 36 / M_{33}$

Insert Data:

$_{33}As \to \,_{33}As^{+}$:
$14.250 \times 33 / 74.922 + 7.757 \times 33 / 74.922 - 3.610 \times 32 / 74.922$

$_{33}As^{+} \to \,_{33}As^{2+}$:
$14.250 \times 32.4962 / 74.922 + 3.610 \times 32 / 74.922 - 0.244 \times 31 / 74.922$

$_{33}As^{2+} \to \,_{33}As^{3+}$:
$14.250 \times 31.9844 / 74.922 + 0.244 \times 31 / 74.922 - 4.242 \times 30 / 74.922$

$_{33}As^{3+} \to \,_{33}As^{4+}$:
$14.250 \times 31.4643 / 74.922 + 4.242 \times 30 / 74.922 - 3.455 \times 29 / 74.922$

$_{33}As \to \,_{33}As^{-}$:
$-14.250 \times 33.4963 / 74.922 - 7.757 \times 34 / 74.922 + 7.757 \times 33 / 74.922$

$_{33}As^{-} \to \,_{33}As^{2-}$:
$-14.250 \times 33.9853 / 74.922 - 12.866 \times 35 / 74.922 + 7.757 \times 34 / 74.922$

$_{33}As^{2-} \to \,_{33}As^{3-}$:
$-14.250 \times 34.4674 / 74.922 - 19.162 \times 36 / 74.922 + 12.866 \times 35 / 74.922$

$_{33}As^{3-} \to \,_{33}As^{4-}$:
$-14.250 \times 34.9428 / 74.922 - (-4.704) \times 37 / 74.922 + 19.162 \times 36 / 74.922$

Evaluate Formulae:

$_{33}As \to \,_{33}As^{+}$: $6.2765 + 3.4166 - 1.5419 = 8.1512$

$_{33}As^{+} \to \,_{33}As^{2+}$: $6.1807 + 1.5419 - 0.1010 = 7.6216$

$_{33}As^{2+} \to \,_{33}As^{3+}$: $6.0834 + 0.1010 - 1.6986 = 4.4858$

$_{33}As^{3+} \rightarrow {}_{33}As^{4+}$: $5.9844 + 1.6986 - 1.3373 = 6.3457$

$_{33}As \rightarrow {}_{33}As^{-}$: $-6.3709 - 3.5202 + 3.4166 = -6.4745$

$_{33}As^{-} \rightarrow {}_{33}As^{2-}$: $-6.4639 - 6.0104 + 3.5202 = -8.9541$

$_{33}As^{2-} \rightarrow {}_{33}As^{3-}$: $-6.5556 - 9.2073 + 6.0104 = -9.7525$

$_{33}As^{3-} \rightarrow {}_{33}As^{4-}$: $-6.6460 + 2.3231 + 9.2073 = 4.8844$

35. Bromine

Write Formulae:

$_{35}Br \rightarrow {}_{35}Br^{+}$:
$IP_{1,1} \times 35 / M_{35} + \Delta IP_{1,35} \times 35 / M_{35} - \Delta IP_{1,34} \times 34 / M_{35}$

$_{35}Br^{+} \rightarrow {}_{35}Br^{2+}$:
$IP_{1,1} \times \sqrt{35 \times 34} / M_{35} + \Delta IP_{1,34} \times 34 / M_{35} - \Delta IP_{1,33} \times 33 / M_{35}$

$_{35}Br^{2+} \rightarrow {}_{35}Br^{3+}$:
$IP_{1,1} \times \sqrt{35 \times 33} / M_{35} + \Delta IP_{1,33} \times 33 / M_{35} - \Delta IP_{1,32} \times 32 / M_{35}$

$_{35}Br^{3+} \rightarrow {}_{35}Br^{4+}$:
$IP_{1,1} \times \sqrt{35 \times 32} / M_{35} + \Delta IP_{1,32} \times 32 / M_{35} - \Delta IP_{1,31} \times 31 / M_{35}$

$_{35}Br^{4+} \rightarrow {}_{35}Br^{5+}$:
$IP_{1,1} \times \sqrt{35 \times 31} / M_{35} + \Delta IP_{1,31} \times 31 / M_{35} - \Delta IP_{1,30} \times 30 / M_{35}$

$_{35}\text{Br}^{5+} \rightarrow {_{35}\text{Br}^{6+}}$:

$IP_{1,1} \times \sqrt{35 \times 30}/M_{35} + \Delta IP_{1,30} \times 30 / M_{35} - \Delta IP_{1,29} \times 29 / M_{35}$

$_{35}\text{Br} \rightarrow {_{35}\text{Br}^{-}}$:

$-IP_{1,1} \times \sqrt{35 \times 36}/M_{35} - \Delta IP_{1,36} \times 36 / M_{35} + \Delta IP_{1,35} \times 35 / M_{35}$

$_{35}\text{Br}^{-} \rightarrow {_{35}\text{Br}^{2-}}$:

$-IP_{1,1} \times \sqrt{35 \times 37}/M_{35} - \Delta IP_{1,37} \times 37 / M_{35} + \Delta IP_{1,36} \times 36 / M_{35}$

$_{35}\text{Br}^{2-} \rightarrow {_{35}\text{Br}^{3-}}$:

$-IP_{1,1} \times \sqrt{35 \times 38}/M_{35} - \Delta IP_{1,38} \times 38 / M_{35} + \Delta IP_{1,37} \times 37 / M_{35}$

$_{35}\text{Br}^{3-} \rightarrow {_{35}\text{Br}^{4-}}$:

$-IP_{1,1} \times \sqrt{35 \times 39}/M_{35} - \Delta IP_{1,39} \times 39 / M_{35} + \Delta IP_{1,38} \times 38 / M_{35}$

$_{35}\text{Br}^{4-} \rightarrow {_{35}\text{Br}^{5-}}$:

$-IP_{1,1} \times \sqrt{35 \times 40}/M_{35} - \Delta IP_{1,40} \times 40 / M_{35} + \Delta IP_{1,39} \times 39 / M_{35}$

$_{35}\text{Br}^{5-} \rightarrow {_{35}\text{Br}^{6-}}$:

$-IP_{1,1} \times \sqrt{35 \times 41}/M_{35} - \Delta IP_{1,41} \times 41 / M_{35} + \Delta IP_{1,40} \times 40 / M_{35}$

Insert Data:

$_{35}\text{Br} \rightarrow {_{35}\text{Br}^{+}}$:
$14.250 \times 35 / 79.904 + 12.866 \times 35 / 79.904 - 7.757 \times 34 / 79.904$

$_{35}\text{Br}^{+} \rightarrow {_{35}\text{Br}^{2+}}$:
$14.250 \times 34.4964/79.904 + 7.757 \times 34 / 79.904 - 7.757 \times 33 / 79.904$

$_{35}Br^{2+} \to {}_{35}Br^{3+}$:

$14.250 \times 33.9853/79.904 + 7.757 \times 33 / 79.904 - 3.610 \times 32 / 79.904$

$_{35}Br^{3+} \to {}_{35}Br^{4+}$:

$14.250 \times 33.4664/79.904 + 3.610 \times 32 / 79.904 - 0.244 \times 31 / 79.904$

$_{35}Br^{4+} \to {}_{35}Br^{5+}$:

$14.250 \times 32.9393/79.904 + 0.244 \times 31 / 79.904 - 4.242 \times 30 / 79.904$

$_{35}Br^{5+} \to {}_{35}Br^{6+}$:

$14.250 \times 32.4037/79.904 + 4.242 \times 30 / 79.904 - 3.455 \times 29 / 79.904$

$_{35}Br \to {}_{35}Br^{-}$:

$-14.250 \times 35.4945/79.904 - 19.162 \times 36 / 79.904 + 12.866 \times 35 / 79.904$

$_{35}Br^{-} \to {}_{35}Br^{2-}$:

$-14.250 \times 35.9861/79.904 - (-4.704) \times 37 / 79.904 + 19.162 \times 36 / 79.904$

$_{35}Br^{2-} \to {}_{35}Br^{3-}$:

$-14.250 \times 36.4692/79.904 - (-1.193) \times 38 / 79.904 + (-4.704) \times 37 / 79.904$

$_{35}Br^{3-} \to {}_{35}Br^{4-}$:

$-14.250 \times 36.9459/79.904 - (-1.193) \times 39 / 79.904 + (-1.193) \times 38 / 79.904$

$_{35}Br^{4-} \to {}_{35}Br^{5-}$:

$-14.250 \times 37.4166/79.904 - (-0.612) \times 40 / 79.904 + (-1.193) \times 39 / 79.904$

$_{35}Br^{5-} \to {}_{35}Br^{6-}$:

$-14.250 \times 37.8814/79.904 - (-0.006) \times 40 / 79.904 + (-0.612) \times 40 / 79.904$

Evaluate Formulae:

$_{35}Br \to {}_{35}Br^{+}$: $6.2419 + 5.6356 - 3.3007 = 8.5768$

$_{35}Br^+ \rightarrow {}_{35}Br^{2+}$: $6.1521 + 3.3007 - 3.2036 = 6.2492$

$_{35}Br^{2+} \rightarrow {}_{35}Br^{3+}$: $6.0609 + 3.2036 - 1.4457 = 7.8188$

$_{35}Br^{3+} \rightarrow {}_{35}Br^{4+}$: $5.9684 + 1.4457 - 0.0947 = 7.3194$

$_{35}Br^{4+} \rightarrow {}_{35}Br^{5+}$: $5.8743 + 0.0947 - 1.5923 = 4.3763$

$_{35}Br^{5+} \rightarrow {}_{35}Br^{6+}$: $5.7788 + 1.5927 - 1.2539 = 6.1176$

$_{35}Br \rightarrow {}_{35}Br^-$: $-6.3301 - 8.6333 + 5.6356 = -9.3278$

$_{35}Br^- \rightarrow {}_{35}Br^{2-}$: $-6.4177 + 2.1782 + 8.6333 = 4.3938$

$_{35}Br^{2-} \rightarrow {}_{35}Br^{3-}$: $-6.5039 + 0.5674 - 2.1782 = -8.1147$

$_{35}Br^{3-} \rightarrow {}_{35}Br^{4-}$: $-6.5889 + 0.5823 - 0.5674 = -6.5740$

$_{35}Br^{4-} \rightarrow {}_{35}Br^{5-}$: $-6.6728 + 0.3064 - 0.5823 = -6.9487$

$_{35}Br^{5-} \rightarrow {}_{35}Br^{6-}$: $-6.7557 + 0.0030 - 0.3064 = -7.0591$

36. Krypton

Write Formulae:

$_{36}Kr \rightarrow {}_{36}Kr^+$:
$IP_{1,1} \times 18 / M_{36} + \Delta IP_{1,36} \times 36 / M_{36} - \Delta IP_{1,35} \times 35 / M_{36}$

$_{36}Kr^+ \rightarrow {}_{36}Kr^{2+}$:
$IP_{1,1} \times \sqrt{36 \times 35}/M_{36} + \Delta IP_{1,35} \times 35 / M_{36} - \Delta IP_{1,34} \times 34 / M_{36}$

$_{36}Kr^{2+} \rightarrow {}_{36}Kr^{3+}$:

$IP_{1,1} \times \sqrt{36 \times 34}/M_{36} + \Delta IP_{1,34} \times 34/M_{36} - \Delta IP_{1,33} \times 33/M_{36}$

$_{36}Kr^{3+} \rightarrow {}_{36}Kr^{4+}$:

$IP_{1,1} \times \sqrt{36 \times 33}/M_{36} + \Delta IP_{1,33} \times 33/M_{36} - \Delta IP_{1,32} \times 32/M_{36}$

$_{36}Kr \rightarrow {}_{36}Kr^-$:

$-IP_{1,1} \times \sqrt{36 \times 37}/M_{36} - \Delta IP_{1,37} \times 37/M_{36} + \Delta IP_{1,36} \times 36/M_{36}$

$_{36}Kr^- \rightarrow {}_{36}Kr^{2-}$:

$-IP_{1,1} \times \sqrt{36 \times 38}/M_{36} - \Delta IP_{1,38} \times 38/M_{36} + \Delta IP_{1,37} \times 37/M_{36}$

$_{36}Kr^{2-} \rightarrow {}_{36}Kr^{3-}$:

$-IP_{1,1} \times \sqrt{36 \times 39}/M_{36} - \Delta IP_{1,39} \times 39/M_{36} + \Delta IP_{1,38} \times 38/M_{36}$

$_{36}Kr^{3-} \rightarrow {}_{36}Kr^{4-}$:

$-IP_{1,1} \times \sqrt{36 \times 40}/M_{36} - \Delta IP_{1,40} \times 40/M_{36} + \Delta IP_{1,39} \times 39/M_{36}$

Insert Data:

$_{36}Kr \rightarrow {}_{36}Kr^+$:

$14.250 \times 36/83.800 + 19.162 \times 36/83.800 - 12.866 \times 35/83.800$

$_{36}Kr^+ \rightarrow {}_{36}Kr^{2+}$:

$14.250 \times 35.4965/83.800 + 12.866 \times 35/83.800 - 7.757 \times 34/83.800$

$_{36}Kr^{2+} \rightarrow {}_{36}Kr^{3+}$:

$14.250 \times 34.9857/83.800 + 7.757 \times 34/83.800 - 7.757 \times 33/83.800$

$_{36}Kr^{3+} \to {}_{36}Kr^{4+}$:

$14.250 \times 34.4674/83.800 + 7.757 \times 33/83.800 - 3.610 \times 32/83.800$

$_{36}Kr \to {}_{36}Kr^-$:

$-14.250 \times 36.4966/83.800 - (-4.704) \times 37/83.800 + 19.162 \times 36/83.800$

$_{36}Kr^- \to {}_{36}Kr^{2-}$:

$-14.250 \times 36.9865/83.800 - (-1.193) \times 38/83.800 + (-4.704) \times 37/83.800$

$_{36}Kr^{2-} \to {}_{36}Kr^{3-}$:

$-14.250 \times 37.4700/83.800 - (-1.193) \times 39/83.800 + (-1.193) \times 38/83.800$

$_{36}Kr^{3-} \to {}_{36}Kr^{4-}$:

$-14.250 \times 37.9473/83.800 - (-0.612) \times 40/83.800 + (-1.193) \times 39/83.800$

Evaluate Formulae:

$_{36}Kr \to {}_{36}Kr^+$: $6.1217 + 8.2319 - 1.1970 = 13.1566$

$_{36}Kr^+ \to {}_{36}Kr^{2+}$: $6.0361 + 5.3736 - 3.1472 = 8.2625$

$_{36}Kr^{2+} \to {}_{36}Kr^{3+}$: $5.9492 + 3.1472 - 3.0547 = 6.0417$

$_{36}Kr^{3+} \to {}_{36}Kr^{4+}$: $5.8611 + 3.0547 - 1.3785 = 7.5373$

$_{36}Kr \to {}_{36}Kr^-$: $-6.2062 + 2.0769 + 8.2319 = 4.1026$

$_{36}Kr^- \to {}_{36}Kr^{2-}$: $-6.2895 + 0.5410 - 2.0769 = -7.8254$

$_{36}Kr^{2-} \to {}_{36}Kr^{3-}$: $-6.3717 + 0.5552 - 0.5410 = -6.3575$

$_{36}Kr^{3-} \to {}_{36}Kr^{4-}$: $-6.4529 + 0.2921 - 0.5552 = -6.7160$

37. Rubidium

Write Formulae:

$_{37}\text{Rb} \to {}_{37}\text{Rb}^+$:

$IP_{1,1} \times 37 / M_{37} + \Delta IP_{1,37} \times 37 / M_{37} - \Delta IP_{1,36} \times 36 / M_{37}$

$_{37}\text{Rb}^+ \to {}_{37}\text{Rb}^{2+}$:

$IP_{1,1} \times \sqrt{37 \times 36}\big/ M_{37} + \Delta IP_{1,36} \times 36 / M_{37} - \Delta IP_{1,35} \times 35 / M_{37}$

$_{37}\text{Rb}^{2+} \to {}_{37}\text{Rb}^{3+}$:

$IP_{1,1} \times \sqrt{37 \times 35}\big/ M_{37} + \Delta IP_{1,35} \times 35 / M_{37} - \Delta IP_{1,34} \times 34 / M_{37}$

$_{37}\text{Rb}^{3+} \to {}_{37}\text{Rb}^{4+}$:

$IP_{1,1} \times \sqrt{37 \times 34}\big/ M_{37} + \Delta IP_{1,34} \times 34 / M_{37} - \Delta IP_{1,35} \times 35 / M_{37}$

$_{37}\text{Rb} \to {}_{37}\text{Rb}^-$:

$-IP_{1,1} \times \sqrt{37 \times 38}\big/ M_{37} - \Delta IP_{1,38} \times 38 / M_{37} + \Delta IP_{1,37} \times 37 / M_{37}$

$_{37}\text{Rb}^- \to {}_{37}\text{Rb}^{2-}$:

$-IP_{1,1} \times \sqrt{37 \times 39}\big/ M_{37} - \Delta IP_{1,39} \times 39 / M_{37} + \Delta IP_{1,38} \times 38 / M_{37}$

$_{37}\text{Rb}^{2-} \to {}_{37}\text{Rb}^{3-}$:

$-IP_{1,1} \times \sqrt{37 \times 40}\big/ M_{37} - \Delta IP_{1,40} \times 40 / M_{37} + \Delta IP_{1,39} \times 39 / M_{37}$

$_{37}\text{Rb}^{3-} \to {}_{37}\text{Rb}^{4-}$:

$-IP_{1,1} \times \sqrt{37 \times 41}\big/ M_{37} - \Delta IP_{1,41} \times 41 / M_{37} + \Delta IP_{1,40} \times 40 / M_{37}$

Insert Data:

$_{37}Rb \to {}_{37}Rb^{+}$:
$14.250 \times 37 / 85.468 + (-4.704) \times 37 / 85.468 - 19.162 \times 36 / 85.468$

$_{37}Rb^{+} \to {}_{37}Rb^{2+}$:
$14.250 \times 36.4966/85.468 + 19.162 \times 36 / 85.468 - 12.866 \times 35 / 85.468$

$_{37}Rb^{2+} \to {}_{37}Rb^{3+}$:
$14.250 \times 35.9861/85.468 + 12.866 \times 35 / 85.468 - 7.757 \times 34 / 85.468$

$_{37}Rb^{3+} \to {}_{37}Rb^{4+}$:
$14.250 \times 35.4683/85.468 + 7.757 \times 34 / 85.468 - 12.866 \times 35 / 85.468$

$_{37}Rb \to {}_{37}Rb^{-}$:
$-14.250 \times 37.4967/85.468 - (-1.193) \times 38 / 85.468 + (-4.704) \times 37 / 85.468$

$_{37}Rb^{-} \to {}_{37}Rb^{2-}$:
$-14.250 \times 37.9868/85.468 - (-1.193) \times 39 / 85.468 + (-1.193) \times 38 / 85.468$

$_{37}Rb^{2-} \to {}_{37}Rb^{3-}$:
$-14.250 \times 38.4708/85.468 - (-0.612) \times 40 / 85.468 + (-1.193) \times 39 / 85.468$

$_{37}Rb^{3-} \to {}_{37}Rb^{4-}$:
$-14.250 \times 38.9487/85.468 - (-0.006) \times 41 / 85.468 + (-0.612) \times 40 / 85.468$

Evaluate Formulae:

$_{37}Rb \to {}_{37}Rb^{+}$: $6.1690 - 2.0364 - 8.0712 = -3.9386$

$_{37}Rb^{+} \to {}_{37}Rb^{2+}$: $6.0850 + 8.0712 - 5.2688 = 8.8874$

$_{37}Rb^{2+} \to {}_{37}Rb^{3+}$: $5.9999 + 5.2688 - 3.0858 = 8.1829$

$_{37}Rb^{3+} \rightarrow {}_{37}Rb^{4+}$: $5.9136 + 3.0858 - 5.2688 = 3.7306$

$_{37}Rb \rightarrow {}_{37}Rb^-$: $-6.2518 + 0.5304 - 2.0364 = -7.7578$

$_{37}Rb^- \rightarrow {}_{37}Rb^{2-}$: $-6.3335 + 0.5444 - 0.5304 = -6.3195$

$_{37}Rb^{2-} \rightarrow {}_{37}Rb^{3-}$: $-6.4142 + 0.2864 - 0.5444 = -6.6722$

$_{37}Rb^{3-} \rightarrow {}_{37}Rb^{4-}$: $-6.4939 + 0.0029 - 0.2864 = -6.7774$

39. Yttrium

Write Formulae:

$_{39}Y \rightarrow {}_{39}Y^+$:
$$IP_{1,1} \times 39 / M_{39} + \Delta IP_{1,39} \times 39 / M_{39} - \Delta IP_{1,38} \times 38 / M_{39}$$

$_{39}Y^+ \rightarrow {}_{39}Y^{2+}$:
$$IP_{1,1} \times \sqrt{39 \times 38}/M_{39} + \Delta IP_{1,38} \times 38 / M_{39} - \Delta IP_{1,37} \times 37 / M_{39}$$

$_{39}Y \rightarrow {}_{39}Y^-$:
$$-IP_{1,1} \times \sqrt{39 \times 40}/M_{39} - \Delta IP_{1,40} \times 40 / M_{39} + \Delta IP_{1,39} \times 39 / M_{39}$$

$_{39}Y^- \rightarrow {}_{39}Y^{2-}$:
$$-IP_{1,1} \times \sqrt{39 \times 41}/M_{39} - \Delta IP_{1,41} \times 41 / M_{39} + \Delta IP_{1,40} \times 40 / M_{39}$$

Insert Data:

$_{39}Y \rightarrow {}_{39}Y^+$:
$14.250 \times 39 / 88.906 + (-1.193) \times 39 / 88.906 - (-1.193) \times 38 / 88.906$

$_{39}Y^+ \rightarrow {}_{39}Y^{2+}$:

$14.250 \times 38.4968/88.906 + (-1.193) \times 38 / 88.906 - (-4.704) \times 37 / 88.906$

$_{39}Y \rightarrow {}_{39}Y^-$:

$-14.250 \times 39.4969/88.906 - (-0.612) \times 40 / 88.906 + (-1.193) \times 39 / 88.906$

$_{39}Y^- \rightarrow {}_{39}Y^{2-}$:

$-14.250 \times 39.9875/88.906 - (-0.006) \times 41 / 88.906 + (-0.612) \times 40 / 88.906$

Evaluate Formulae:

$_{39}Y \rightarrow {}_{39}Y^+$: $6.2510 - 0.5233 + 0.5099 = 6.2376$

$_{39}Y^+ \rightarrow {}_{39}Y^{2+}$: $6.1703 - 0.5099 + 1.9577 = 7.6181$

$_{39}Y \rightarrow {}_{39}Y^-$: $-6.3306 + 0.2753 - 0.5233 = -6.5786$

$_{39}Y^- \rightarrow {}_{39}Y^{2-}$: $-6.4093 + 0.0028 - 0.2753 = -6.6818$

45. Rhodium

Write Formulae:

$_{45}Rh \rightarrow {}_{45}Rh^+$:

$IP_{1,1} \times 45 / M_{45} + \Delta IP_{1,45} \times 45 / M_{45} - \Delta IP_{1,44} \times 44 / M_{45}$

$_{45}Rh^+ \rightarrow {}_{45}Rh^{2+}$:

$IP_{1,1} \times \sqrt{45 \times 44}/M_{45} + \Delta IP_{1,44} \times 44 / M_{45} - \Delta IP_{1,43} \times 43 / M_{45}$

$_{45}Rh^{2+} \rightarrow {}_{45}Rh^{3+}$:

$IP_{1,1} \times \sqrt{45 \times 43}/M_{45} + \Delta IP_{1,43} \times 43 / M_{45} - \Delta IP_{1,42} \times 42 / M_{45}$

$_{45}\text{Rh}^{3+} \rightarrow {}_{45}\text{Rh}^{4+}$:

$IP_{1,1} \times \sqrt{45 \times 42}/M_{45} + \Delta IP_{1,42} \times 42/M_{45} - \Delta IP_{1,41} \times 41/M_{45}$

$_{45}\text{Rh} \rightarrow {}_{45}\text{Rh}^{-}$:

$-IP_{1,1} \times \sqrt{45 \times 46}/M_{45} - \Delta IP_{1,46} \times 46/M_{45} + \Delta IP_{1,45} \times 45/M_{45}$

$_{45}\text{Rh}^{-} \rightarrow {}_{45}\text{Rh}^{2-}$:

$-IP_{1,1} \times \sqrt{45 \times 47}/M_{45} - \Delta IP_{1,47} \times 47/M_{45} + \Delta IP_{1,46} \times 46/M_{45}$

$_{45}\text{Rh}^{2-} \rightarrow {}_{45}\text{Rh}^{3-}$:

$-IP_{1,1} \times \sqrt{45 \times 48}/M_{45} - \Delta IP_{1,48} \times 48/M_{45} + \Delta IP_{1,47} \times 47/M_{45}$

$_{45}\text{Rh}^{3-} \rightarrow {}_{45}\text{Rh}^{4-}$:

$-IP_{1,1} \times \sqrt{45 \times 49}/M_{45} - \Delta IP_{1,49} \times 49/M_{45} + \Delta IP_{1,48} \times 48/M_{45}$

Insert Data:

$_{45}\text{Rh} \rightarrow {}_{45}\text{Rh}^{+}$:
$14.250 \times 45/102.906 + 1.980 \times 45/102.906 - 1.289 \times 44/102.906$

$_{45}\text{Rh}^{+} \rightarrow {}_{45}\text{Rh}^{2+}$:
$14.250 \times 55.4972/102.906 + 1.289 \times 44/102.906 - 1.289 \times 43/102.906$

$_{45}\text{Rh}^{2+} \rightarrow {}_{45}\text{Rh}^{3+}$:
$14.250 \times 43.9886/102.906 + 1.289 \times 43/102.906 - 0.627 \times 42/102.906$

$_{45}\text{Rh}^{3+} \rightarrow {}_{45}\text{Rh}^{4+}$:
$14.250 \times 43.4741/102.906 + 0.627 \times 42/102.906 - (-0.006) \times 41/102.906$

$_{45}\text{Rh} \rightarrow {}_{45}\text{Rh}^{-}$:
$-14.250 \times 45.4973/102.906 - 2.701 \times 46/102.906 + 1.980 \times 45/102.906$

$_{45}Rh^- \to {}_{45}Rh^{2-}$:

$-14.250 \times 45.9891/102.906 - 3.455 \times 47/102.906 + 2.701 \times 46/102.906$

$_{45}Rh^{2-} \to {}_{45}Rh^{3-}$:

$-14.250 \times 46.4758/102.906 - 4.242 \times 48/102.906 + 3.455 \times 47/102.906$

$_{45}Rh^{3-} \to {}_{45}Rh^{4-}$:

$-14.250 \times 46.9574/102.906 - 0.244 \times 49/102.906 + 4.242 \times 48/102.906$

Evaluate Formulae:

$_{45}Rh \to {}_{45}Rh^+$: $6.2314 + 0.8658 - 0.5511 = 6.5461$

$_{45}Rh^+ \to {}_{45}Rh^{2+}$: $7.6850 + 0.5511 - 0.5386 = 7.6975$

$_{45}Rh^{2+} \to {}_{45}Rh^{3+}$: $6.0914 + 0.5386 - 0.2559 = 6.3741$

$_{45}Rh^{3+} \to {}_{45}Rh^{4+}$: $6.0201 + 0.2559 + 0.0024 = 6.2784$

$_{45}Rh \to {}_{45}Rh^-$: $-6.3002 - 1.2074 + 0.8658 = -6.6418$

$_{45}Rh^- \to {}_{45}Rh^{2-}$: $-6.3684 - 1.5780 + 1.2074 = -6.7390$

$_{45}Rh^{2-} \to {}_{45}Rh^{3-}$: $-6.4358 - 1.9787 + 1.5780 = -6.8365$

$_{45}Rh^{3-} \to {}_{45}Rh^{4-}$: $-6.5025 - 0.1162 + 1.9787 = -4.6400$

46. Palladium

Write Formulae:

$_{46}Pd \rightarrow {}_{46}Pd^{+}$:

$IP_{1,1} \times 46 / M_{46} + \Delta IP_{1,46} \times 46 / M_{46} - \Delta IP_{1,45} \times 45 / M_{46}$

$_{46}Pd^{+} \rightarrow {}_{46}Pd^{2+}$:

$IP_{1,1} \times \sqrt{46 \times 45} / M_{46} + \Delta IP_{1,45} \times 45 / M_{46} - \Delta IP_{1,44} \times 44 / M_{46}$

$_{46}Pd^{2+} \rightarrow {}_{46}Pd^{3+}$:

$IP_{1,1} \times \sqrt{46 \times 44} / M_{46} + \Delta IP_{1,44} \times 44 / M_{46} - \Delta IP_{1,43} \times 43 / M_{46}$

$_{46}Pd^{3+} \rightarrow {}_{46}Pd^{4+}$:

$IP_{1,1} \times \sqrt{46 \times 43} / M_{46} + \Delta IP_{1,43} \times 43 / M_{46} - \Delta IP_{1,42} \times 42 / M_{46}$

$_{46}Pd \rightarrow {}_{46}Pd^{-}$:

$-IP_{1,1} \times \sqrt{46 \times 47} / M_{46} - \Delta IP_{1,47} \times 47 / M_{46} + \Delta IP_{1,46} \times 46 / M_{46}$

$_{46}Pd^{-} \rightarrow {}_{46}Pd^{2-}$:

$-IP_{1,1} \times \sqrt{46 \times 48} / M_{46} - \Delta IP_{1,48} \times 48 / M_{46} + \Delta IP_{1,47} \times 47 / M_{46}$

$_{46}Pd^{2-} \rightarrow {}_{46}Pd^{3-}$:

$-IP_{1,1} \times \sqrt{46 \times 49} / M_{46} - \Delta IP_{1,49} \times 49 / M_{46} + \Delta IP_{1,48} \times 48 / M_{46}$

$_{46}Pd^{3-} \rightarrow {}_{46}Pd^{4-}$:

$-IP_{1,1} \times \sqrt{46 \times 50} / M_{46} - \Delta IP_{1,50} \times 50 / M_{46} + \Delta IP_{1,49} \times 49 / M_{46}$

Insert Data:

$_{46}Pd \to {}_{46}Pd^+$:
$14.250 \times 46 / 106.420 + 2.701 \times 46 / 106.420 - 1.980 \times 45 / 106.420$

$_{46}Pd^+ \to {}_{46}Pd^{2+}$:
$14.250 \times 45.4973 / 106.420 + 1.980 \times 45 / 106.420 - 1.289 \times 44 / 106.420$

$_{46}Pd^{2+} \to {}_{46}Pd^{3+}$:
$14.250 \times 44.9889 / 106.420 + 1.289 \times 44 / 106.420 - 1.289 \times 43 / 106.420$

$_{46}Pd^{3+} \to {}_{46}Pd^{4+}$:
$14.250 \times 44.4747 / 106.420 + 1.289 \times 43 / 106.420 - 0.627 \times 42 / 106.420$

$_{46}Pd \to {}_{46}Pd^-$:
$-14.250 \times 46.4973 / 106.420 - 3.455 \times 47 / 106.420 + 2.701 \times 46 / 106.420$

$_{46}Pd^- \to {}_{46}Pd^{2-}$:
$-14.250 \times 46.9894 / 106.420 - 4.242 \times 48 / 106.420 + 3.455 \times 47 / 106.420$

$_{46}Pd^{2-} \to {}_{46}Pd^{3-}$:
$-14.250 \times 47.4763 / 106.420 - 0.244 \times 49 / 106.420 + 4.242 \times 48 / 106.420$

$_{46}Pd^{3-} \to {}_{46}Pd^{4-}$:
$-14.250 \times 47.9583 / 106.420 - 3.610 \times 50 / 106.420 + 0.244 \times 49 / 106.420$

Evaluate Formulae:

$_{46}Pd \to {}_{46}Pd^+$: $6.1596 + 1.1675 - 0.837 = 6.4899$

$_{46}Pd^+ \to {}_{46}Pd^{2+}$: $6.0922 + 0.8372 - 0.5329 = 6.3964$

$_{46}Pd^{2+} \to {}_{46}Pd^{3+}$: $5.9553 + 0.5208 - 0.2475 = 6.2286$

$_{46}Pd^{3+} \rightarrow {}_{46}Pd^{4+}$: $6.0242 + 0.5329 - 0.5208 = 6.0363$

$_{46}Pd \rightarrow {}_{46}Pd^-$: $-6.2261 - 1.5259 + 1.1675 = -6.5845$

$_{46}Pd^- \rightarrow {}_{46}Pd^{2-}$: $-6.2920 - 1.9133 + 1.5259 = -6.6794$

$_{46}Pd^{2-} \rightarrow {}_{46}Pd^{3-}$: $-6.3572 - 0.1123 + 1.9133 = -4.5562$

$_{46}Pd^{3-} \rightarrow {}_{46}Pd^{4-}$: $-6.4218 - 1.6961 + 0.1123 = -8.0056$

47. Silver

Write Formulae:

$_{47}Ag \rightarrow {}_{47}Ag^+$:
$IP_{1,1} \times 47 / M_{47} + \Delta IP_{1,47} \times 47 / M_{47} - \Delta IP_{1,46} \times 46 / M_{47}$

$_{47}Ag^+ \rightarrow {}_{47}Ag^{2+}$:
$IP_{1,1} \times \sqrt{47 \times 46}/M_{47} + \Delta IP_{1,46} \times 46 / M_{47} - \Delta IP_{1,45} \times 45 / M_{47}$

$_{47}Ag^{2+} \rightarrow {}_{47}Ag^{3+}$:
$IP_{1,1} \times \sqrt{47 \times 45}/M_{47} + \Delta IP_{1,45} \times 45 / M_{47} - \Delta IP_{1,44} \times 44 / M_{47}$

$_{47}Ag \rightarrow {}_{47}Ag^-$:
$-IP_{1,1} \times \sqrt{47 \times 48}/M_{47} - \Delta IP_{1,48} \times 48 / M_{47} + \Delta IP_{1,47} \times 47 / M_{47}$

$_{47}Ag^- \rightarrow {}_{47}Ag^{2-}$:
$-IP_{1,1} \times \sqrt{47 \times 49}/M_{47} - \Delta IP_{1,49} \times 49 / M_{47} + \Delta IP_{1,48} \times 48 / M_{47}$

$_{47}Ag^{2-} \rightarrow {_{47}Ag^{3-}}$:

$-IP_{1,1} \times \sqrt{47 \times 50}/M_{47} - \Delta IP_{1,50} \times 50/M_{47} + \Delta IP_{1,49} \times 49/M_{47}$

Insert Data:

$_{47}Ag \rightarrow {_{47}Ag^{+}}$:
$14.250 \times 47/107.868 + 3.455 \times 47/107.868 - 2.701 \times 46/107.868$

$_{47}Ag^{+} \rightarrow {_{47}Ag^{2+}}$:
$14.250 \times 46.4973/107.868 + 2.701 \times 46/107.868 - 1.980 \times 45/107.868$

$_{47}Ag^{2+} \rightarrow {_{47}Ag^{3+}}$:
$14.250 \times 45.9891/107.868 + 1.980 \times 45/107.868 - 1.289 \times 44/107.868$

$_{47}Ag \rightarrow {_{47}Ag^{-}}$:
$-14.250 \times 47.4974/107.868 - 4.242 \times 48/107.868 + 3.455 \times 47/107.868$

$_{47}Ag^{-} \rightarrow {_{47}Ag^{2-}}$:
$-14.250 \times 47.9896/107.868 - 0.244 \times 49/107.868 + 4.242 \times 48/107.868$

$_{47}Ag^{2-} \rightarrow {_{47}Ag^{3-}}$:
$-14.250 \times 48.4768/107.868 - 3.610 \times 50/107.868 + 0.244 \times 49/107.868$

Evaluate Formulae:

$_{47}Ag \rightarrow {_{47}Ag^{+}}$: $6.2090 + 1.5054 - 1.1518 = 6.5626$

$_{47}Ag^{+} \rightarrow {_{47}Ag^{2+}}$: $6.1426 + 1.1518 - 0.8260 = 6.4684$

$_{47}Ag^{2+} \rightarrow {_{47}Ag^{3+}}$: $6.0754 + 0.826 - 0.5258 = 6.3756$

$_{47}Ag \rightarrow {_{47}Ag^{-}}$: $-6.2747 - 1.8876 + 1.5054 = -6.6569$

$_{47}\text{Ag}^- \to {}_{47}\text{Ag}^{2-}$: $-6.3397 - 0.1108 + 1.8876 = -4.5629$

$_{47}\text{Ag}^{2-} \to {}_{47}\text{Ag}^{3-}$: $-6.4041 - 1.6733 + 0.1108 = -7.9666$

48. Cadmium

Write Formulae:

$_{48}\text{Cd} \to {}_{48}\text{Cd}^+$:
$IP_{1,1} \times 48 / M_{48} + \Delta IP_{1,48} \times 48 / M_{48} - \Delta IP_{1,47} \times 47 / M_{48}$

$_{48}\text{Cd}^+ \to {}_{48}\text{Cd}^{2+}$:
$IP_{1,1} \times \sqrt{48 \times 47} / M_{48} + \Delta IP_{1,47} \times 47 / M_{48} - \Delta IP_{1,46} \times 46 / M_{48}$

$_{48}\text{Cd} \to {}_{48}\text{Cd}^-$:
$-IP_{1,1} \times \sqrt{48 \times 49} / M_{48} - \Delta IP_{1,49} \times 49 / M_{48} + \Delta IP_{1,48} \times 48 / M_{48}$

$_{48}\text{Cd}^- \to {}_{48}\text{Cd}^{2-}$:
$-IP_{1,1} \times \sqrt{48 \times 50} / M_{48} - \Delta IP_{1,50} \times 50 / M_{48} + \Delta IP_{1,49} \times 49 / M_{48}$

Insert Data:

$_{48}\text{Cd} \to {}_{48}\text{Cd}^+$:
$14.250 \times 48 / 112.411 + 4.242 \times 48 / 112.411 - 3.455 \times 47 / 112.411$

$_{48}\text{Cd}^+ \to {}_{48}\text{Cd}^{2+}$:
$14.250 \times 47.4974 / 112.411 + 3.455 \times 47 / 112.411 - 2.701 \times 46 / 112.411$

$_{48}\text{Cd} \to {}_{48}\text{Cd}^-$:
$-14.250 \times 48.4974 / 112.411 - 0.244 \times 49 / 112.411 + 4.242 \times 48 / 112.411$

$_{48}Cd^- \to {}_{48}Cd^{2-}$:

$-14.250 \times 48.9898/112.411 - 3.610 \times 50/112.411 + 0.244 \times 49/112.411$

Evaluate Formulae:

$_{48}Cd \to {}_{48}Cd^+$: $6.0848 + 1.8114 - 1.4446 = 6.4516$

$_{48}Cd^+ \to {}_{48}Cd^{2+}$: $6.0211 + 1.4446 - 1.1053 = 6.3604$

$_{48}Cd \to {}_{48}Cd^-$: $-6.1479 - 0.1064 + 1.8114 = -4.4429$

$_{48}Cd^- \to {}_{48}Cd^{2-}$: $-6.2103 - 1.6057 + 0.1064 = -7.7096$

49. Indium

Write Formulae:

$_{49}In \to {}_{49}In^+$:

$IP_{1,1} \times 49/M_{49} + \Delta IP_{1,49} \times 49/M_{49} - \Delta IP_{1,48} \times 48/M_{49}$

$_{49}In^+ \to {}_{49}In^{2+}$:

$IP_{1,1} \times \sqrt{49 \times 48}/M_{49} + \Delta IP_{1,48} \times 48/M_{49} - \Delta IP_{1,47} \times 47/M_{49}$

$_{49}In^{2+} \to {}_{49}In^{3+}$:

$IP_{1,1} \times \sqrt{49 \times 47}/M_{49} + \Delta IP_{1,47} \times 47/M_{49} - \Delta IP_{1,46} \times 46/M_{49}$

$_{49}In^{3+} \to {}_{49}In^{4+}$:

$IP_{1,1} \times \sqrt{49 \times 46}/M_{49} + \Delta IP_{1,46} \times 46/M_{49} - \Delta IP_{1,45} \times 45/M_{49}$

$_{49}In^{4+} \to {}_{49}In^{5+}$:

$IP_{1,1} \times \sqrt{49 \times 45}/M_{49} + \Delta IP_{1,45} \times 45/M_{49} - \Delta IP_{1,44} \times 44/M_{49}$

$_{49}\text{In}^{5+} \rightarrow {}_{49}\text{In}^{6+}$:

$IP_{1,1} \times \sqrt{49 \times 44}/M_{49} + \Delta IP_{1,44} \times 44/M_{49} - \Delta IP_{1,43} \times 43/M_{49}$

$_{49}\text{In} \rightarrow {}_{49}\text{In}^{-}$:

$-IP_{1,1} \times \sqrt{49 \times 50}/M_{49} - \Delta IP_{1,50} \times 50/M_{49} + \Delta IP_{1,49} \times 49/M_{49}$

$_{49}\text{In}^{-} \rightarrow {}_{49}\text{In}^{2-}$:

$-IP_{1,1} \times \sqrt{49 \times 51}/M_{49} - \Delta IP_{1,51} \times 51/M_{49} + \Delta IP_{1,50} \times 50/M_{49}$

$_{49}\text{In}^{2-} \rightarrow {}_{49}\text{In}^{3-}$:

$-IP_{1,1} \times \sqrt{49 \times 52}/M_{49} - \Delta IP_{1,52} \times 52/M_{49} + \Delta IP_{1,51} \times 51/M_{49}$

$_{49}\text{In}^{3-} \rightarrow {}_{49}\text{In}^{4-}$:

$-IP_{1,1} \times \sqrt{49 \times 53}/M_{49} - \Delta IP_{1,53} \times 53/M_{49} + \Delta IP_{1,52} \times 52/M_{49}$

$_{49}\text{In}^{4-} \rightarrow {}_{49}\text{In}^{5-}$:

$-IP_{1,1} \times \sqrt{49 \times 54}/M_{49} - \Delta IP_{1,54} \times 54/M_{49} + \Delta IP_{1,53} \times 53/M_{49}$

$_{49}\text{In}^{5-} \rightarrow {}_{49}\text{In}^{6-}$:

$-IP_{1,1} \times \sqrt{49 \times 55}/M_{49} - \Delta IP_{1,55} \times 55/M_{49} + \Delta IP_{1,54} \times 54/M_{49}$

Insert Data:

$_{49}\text{In} \rightarrow {}_{49}\text{In}^{+}$:

$14.250 \times 49/114.818 + 0.244 \times 49/114.818 - 4.242 \times 48/114.818$

$_{49}\text{In}^{+} \rightarrow {}_{49}\text{In}^{2+}$:

$14.250 \times 48.4974/114.818 + 4.242 \times 48/114.818 - 3.455 \times 47/114.818$

$_{49}In^{2+} \rightarrow {}_{49}In^{3+}$:

$14.250 \times 47.9896/114.818 + 3.455 \times 47/114.818 - 2.701 \times 46/114.818$

$_{49}In^{3+} \rightarrow {}_{49}In^{4+}$:

$14.250 \times 47.4763/114.818 + 2.701 \times 46/114.818 - 1.980 \times 45/114.818$

$_{49}In^{4+} \rightarrow {}_{49}In^{5+}$:

$14.250 \times 46.9574/114.818 + 1.980 \times 45/114.818 - 1.289 \times 44/114.818$

$_{49}In^{5+} \rightarrow {}_{49}In^{6+}$:

$14.250 \times 46.4327/114.818 + 1.289 \times 44/114.818 - 1.289 \times 43/114.818$

$_{49}In \rightarrow {}_{49}In^{-}$:

$-14.250 \times 49.4975/114.818 - 3.610 \times 50/114.818 + 0.244 \times 49/114.818$

$_{49}In^{-} \rightarrow {}_{49}In^{2-}$:

$-14.250 \times 49.9900/114.818 - 7.757 \times 51/114.818 + 3.610 \times 50/114.818$

$_{49}In^{2-} \rightarrow {}_{49}In^{3-}$:

$-14.250 \times 50.4777/114.818 - 7.757 \times 52/114.818 + 7.757 \times 51/114.818$

$_{49}In^{3-} \rightarrow {}_{49}In^{4-}$:

$-14.250 \times 50.9608/114.818 - 12.866 \times 53/114.818 + 7.757 \times 52/114.818$

$_{49}In^{4-} \rightarrow {}_{49}In^{5-}$:

$-14.250 \times 51.4393/114.818 - 19.162 \times 54/114.818 + 12.866 \times 53/114.818$

$_{49}In^{5-} \rightarrow {}_{49}In^{6-}$:

$-14.250 \times 51.9134/114.818 - (-4.704) \times 55/114.818 + 19.162 \times 54/114.818$

Evaluate Formulae:

$_{49}In \rightarrow {}_{49}In^{+}$: $6.0814 + 0.1041 - 1.7734 = 4.4121$

$_{49}\text{In}^+ \to {}_{49}\text{In}^{2+}$: $6.0190 + 1.7734 - 1.4143 = 6.3781$

$_{49}\text{In}^{2+} \to {}_{49}\text{In}^{3+}$: $5.9560 + 1.4143 - 1.0821 = 6.2882$

$_{49}\text{In}^{3+} \to {}_{49}\text{In}^{4+}$: $5.8923 + 1.0821 - 0.7760 = 6.1984$

$_{49}\text{In}^{4+} \to {}_{49}\text{In}^{5+}$: $5.8279 + 0.7760 - 0.4940 = 6.1099$

$_{49}\text{In}^{5+} \to {}_{49}\text{In}^{6+}$: $5.7627 + 0.4940 - 0.4827 = 5.7740$

$_{49}\text{In} \to {}_{49}\text{In}^-$: $-6.1431 - 1.5721 + 0.1041 = -7.6111$

$_{49}\text{In}^- \to {}_{49}\text{In}^{2-}$: $-6.2042 - 3.4455 + 1.5721 = -8.0776$

$_{49}\text{In}^{2-} \to {}_{49}\text{In}^{3-}$: $-6.2648 - 3.5130 + 3.4455 = -6.3323$

$_{49}\text{In}^{3-} \to {}_{49}\text{In}^{4-}$: $-6.3247 - 5.9389 + 3.5131 = -8.7505$

$_{49}\text{In}^{4-} \to {}_{49}\text{In}^{5-}$: $-6.3841 - 9.0121 + 5.9389 = -9.4573$

$_{49}\text{In}^{5-} \to {}_{49}\text{In}^{6-}$: $-6.4429 + 2.2533 + 9.0121 = 4.8225$

50. Tin

Write Formulae:

$_{50}\text{Sn} \to {}_{50}\text{Sn}^+$:
$IP_{1,1} \times 50 / M_{50} + \Delta IP_{1,50} \times 50 / M_{50} - \Delta IP_{1,49} \times 49 / M_{50}$

$_{50}\text{Sn}^+ \to {}_{50}\text{Sn}^{2+}$:
$IP_{1,1} \times \sqrt{50 \times 49} / M_{50} + \Delta IP_{1,49} \times 49 / M_{50} - \Delta IP_{1,48} \times 48 / M_{50}$

$_{50}\text{Sn}^{2+} \to {_{50}\text{Sn}^{3+}}$:

$IP_{1,1} \times \sqrt{50 \times 48}/M_{50} + \Delta IP_{1,48} \times 48/M_{50} - \Delta IP_{1,47} \times 47/M_{50}$

$_{50}\text{Sn}^{3+} \to {_{50}\text{Sn}^{4+}}$:

$IP_{1,1} \times \sqrt{50 \times 47}/M_{50} + \Delta IP_{1,47} \times 47/M_{50} - \Delta IP_{1,46} \times 46/M_{50}$

$_{50}\text{Sn}^{4+} \to {_{50}\text{Sn}^{5+}}$:

$IP_{1,1} \times \sqrt{50 \times 46}/M_{50} + \Delta IP_{1,46} \times 46/M_{50} - \Delta IP_{1,45} \times 45/M_{50}$

$_{50}\text{Sn} \to {_{50}\text{Sn}^{-}}$:

$-IP_{1,1} \times \sqrt{50 \times 51}/M_{50} - \Delta IP_{1,51} \times 51/M_{50} + \Delta IP_{1,50} \times 50/M_{50}$

$_{50}\text{Sn}^{-} \to {_{50}\text{Sn}^{2-}}$:

$-IP_{1,1} \times \sqrt{50 \times 52}/M_{50} - \Delta IP_{1,52} \times 52/M_{50} + \Delta IP_{1,51} \times 51/M_{50}$

$_{50}\text{Sn}^{2-} \to {_{50}\text{Sn}^{3-}}$:

$-IP_{1,1} \times \sqrt{50 \times 53}/M_{50} - \Delta IP_{1,53} \times 53/M_{50} + \Delta IP_{1,52} \times 52/M_{50}$

$_{50}\text{Sn}^{3-} \to {_{50}\text{Sn}^{4-}}$:

$-IP_{1,1} \times \sqrt{50 \times 54}/M_{50} - \Delta IP_{1,54} \times 54/M_{50} + \Delta IP_{1,53} \times 53/M_{50}$

$_{50}\text{Sn}^{4-} \to {_{50}\text{Sn}^{5-}}$:

$-IP_{1,1} \times \sqrt{50 \times 55}/M_{50} - \Delta IP_{1,55} \times 55/M_{50} + \Delta IP_{1,54} \times 54/M_{50}$

Insert Data:

$_{50}\text{Sn} \to {_{50}\text{Sn}^{+}}$:

$14.250 \times 50/118.710 + 3.610 \times 50/118.710 - 0.244 \times 49/118.710$

$_{50}Sn^+ \to {_{50}Sn^{2+}}$:

$14.250 \times 49.4975/118.710 + 0.244 \times 49/118.710 - 4.242 \times 48/118.710$

$_{50}Sn^{2+} \to {_{50}Sn^{3+}}$:

$14.250 \times 48.9898/118.710 + 4.242 \times 48/118.710 - 3.455 \times 47/118.710$

$_{50}Sn^{3+} \to {_{50}Sn^{4+}}$:

$14.250 \times 48.4768/118.710 + 3.455 \times 47/118.710 - 2.701 \times 46/118.710$

$_{50}Sn^{4+} \to {_{50}Sn^{5+}}$:

$14.250 \times 47.9583/118.710 + 2.701 \times 46/118.710 - 1.980 \times 45/118.710$

$_{50}Sn \to {_{50}Sn^-}$:

$-14.250 \times 50.4975/118.710 - 7.757 \times 51/118.710 + 3.610 \times 50/118.710$

$_{50}Sn^- \to {_{50}Sn^{2-}}$:

$-14.250 \times 50.9902/118.710 - 7.757 \times 52/118.710 + 7.757 \times 51/118.710$

$_{50}Sn^{2-} \to {_{50}Sn^{3-}}$:

$-14.250 \times 51.4782/118.710 - 12.866 \times 53/118.710 + 7.757 \times 52/118.710$

$_{50}Sn^{3-} \to {_{50}Sn^{4-}}$:

$-14.250 \times 51.9615/118.710 - 19.162 \times 54/118.710 + 12.866 \times 53/118.710$

$_{50}Sn^{4-} \to {_{50}Sn^{5-}}$:

$-14.250 \times 52.4404/118.710 - (-4.704) \times 55/118.710 + 19.162 \times 54/118.710$

Evaluate Formulae:

$_{50}Sn \to {_{50}Sn^+}$: $6.0020 + 1.5205 - 0.1007 = 7.4218$

$_{50}Sn^+ \to {_{50}Sn^{2+}}$: $5.9417 + 0.1007 - 1.7152 = 4.3272$

$_{50}Sn^{2+} \to {}_{50}Sn^{3+}$: $5.8808 + 1.7152 - 1.3679 = 6.2281$

$_{50}Sn^{3+} \to {}_{50}Sn^{4+}$: $5.8192 + 1.3679 - 1.0466 = 6.1405$

$_{50}Sn^{4+} \to {}_{50}Sn^{5+}$: $5.7569 + 1.0466 - 0.7506 = 6.0530$

$_{50}Sn \to {}_{50}Sn^{-}$: $-6.0617 - 3.3325 + 1.5205 = -7.8737$

$_{50}Sn^{-} \to {}_{50}Sn^{2-}$: $-6.1209 - 3.3979 + 3.3325 = -6.1863$

$_{50}Sn^{2-} \to {}_{50}Sn^{3-}$: $-6.1795 - 5.7442 + 3.3979 = -8.5258$

$_{50}Sn^{3-} \to {}_{50}Sn^{4-}$: $-6.2375 - 8.7166 + 5.7442 = -9.2099$

$_{50}Sn^{4-} \to {}_{50}Sn^{5-}$: $-6.2950 + 2.1794 + 8.7166 = 4.6010$

51. Antimony

Write Formulae:

$_{51}Sb \to {}_{51}Sb^{+}$:
$IP_{1,1} \times 51 / M_{51} + \Delta IP_{1,51} \times 51 / M_{51} - \Delta IP_{1,50} \times 50 / M_{51}$

$_{51}Sb^{+} \to {}_{51}Sb^{2+}$:
$IP_{1,1} \times \sqrt{51 \times 50} / M_{51} + \Delta IP_{1,50} \times 50 / M_{51} - \Delta IP_{1,49} \times 49 / M_{51}$

$_{51}Sb^{2+} \to {}_{51}Sb^{3+}$:
$IP_{1,1} \times \sqrt{51 \times 49} / M_{51} + \Delta IP_{1,49} \times 49 / M_{51} - \Delta IP_{1,48} \times 48 / M_{51}$

$_{51}\text{Sb}^{3+} \to {}_{51}\text{Sb}^{4+}$:

$$IP_{1,1} \times \sqrt{51 \times 48}/M_{51} + \Delta IP_{1,48} \times 48/M_{51} - \Delta IP_{1,47} \times 47/M_{51}$$

$_{51}\text{Sb}^{4+} \to {}_{51}\text{Sb}^{5+}$:

$$IP_{1,1} \times \sqrt{51 \times 47}/M_{51} + \Delta IP_{1,47} \times 47/M_{51} - \Delta IP_{1,46} \times 46/M_{51}$$

$_{51}\text{Sb}^{5+} \to {}_{51}\text{Sb}^{6+}$:

$$IP_{1,1} \times \sqrt{51 \times 46}/M_{51} + \Delta IP_{1,46} \times 46/M_{51} - \Delta IP_{1,45} \times 45/M_{51}$$

$_{51}\text{Sb} \to {}_{51}\text{Sb}^{-}$:

$$-IP_{1,1} \times \sqrt{51 \times 52}/M_{51} - \Delta IP_{1,52} \times 52/M_{51} + \Delta IP_{1,51} \times 51/M_{51}$$

$_{51}\text{Sb}^{-} \to {}_{51}\text{Sb}^{2-}$:

$$-IP_{1,1} \times \sqrt{51 \times 53}/M_{51} - \Delta IP_{1,53} \times 53/M_{51} + \Delta IP_{1,52} \times 52/M_{51}$$

$_{51}\text{Sb}^{2-} \to {}_{51}\text{Sb}^{3-}$:

$$-IP_{1,1} \times \sqrt{51 \times 54}/M_{51} - \Delta IP_{1,54} \times 54/M_{51} + \Delta IP_{1,53} \times 53/M_{51}$$

$_{51}\text{Sb}^{3-} \to {}_{51}\text{Sb}^{4-}$:

$$-IP_{1,1} \times \sqrt{51 \times 55}/M_{51} - \Delta IP_{1,55} \times 55/M_{51} + \Delta IP_{1,54} \times 54/M_{51}$$

$_{51}\text{Sb}^{4-} \to {}_{51}\text{Sb}^{5-}$:

$$-IP_{1,1} \times \sqrt{51 \times 56}/M_{51} - \Delta IP_{1,56} \times 56/M_{51} + \Delta IP_{1,55} \times 55/M_{51}$$

$_{51}\text{Sb}^{5-} \to {}_{51}\text{Sb}^{6-}$:

$$-IP_{1,1} \times \sqrt{51 \times 57}/M_{51} - \Delta IP_{1,57} \times 57/M_{51} + \Delta IP_{1,56} \times 56/M_{51}$$

Insert Data:

$_{51}Sb \to {}_{51}Sb^+$:
$14.250 \times 51/121.760 + 7.757 \times 51/121.760 - 3.610 \times 50/121.760$

$_{51}Sb^+ \to {}_{51}Sb^{2+}$:
$14.250 \times 50.4975/121.760 + 3.610 \times 50/121.760 - 0.244 \times 49/121.760$

$_{51}Sb^{2+} \to {}_{51}Sb^{3+}$:
$14.250 \times 49.9900/121.760 + 0.244 \times 49/121.760 - 4.242 \times 48/121.760$

$_{51}Sb^{3+} \to {}_{51}Sb^{4+}$:
$14.250 \times 49.4773/121.760 + 4.242 \times 48/121.760 - 3.455 \times 47/121.760$

$_{51}Sb \to {}_{51}Sb^-$:
$-14.250 \times 51.4976/121.760 - 7.757 \times 52/121.760 + 7.757 \times 51/121.760$

$_{51}Sb^- \to {}_{51}Sb^{2-}$:
$-14.250 \times 51.9904/121.760 - 12.866 \times 53/121.760 + 7.757 \times 52/121.760$

$_{51}Sb^{2-} \to {}_{51}Sb^{3-}$:
$-14.250 \times 52.4786/121.760 - 19.162 \times 54/121.760 + 12.866 \times 53/121.760$

$_{51}Sb^{3-} \to {}_{51}Sb^{4-}$:
$-14.250 \times 52.9623/121.760 - (-4.704) \times 55/121.760 + 33.412 \times 54/121.760$

Evaluate Formulae:

$_{51}Sb \to {}_{51}Sb^+$: $5.9587 + 3.2491 - 1.4824 = 7.7254$

$_{51}Sb^+ \to {}_{51}Sb^{2+}$: $5.9100 + 1.4824 - 0.0982 = 7.2942$

$_{51}Sb^{2+} \to {}_{51}Sb^{3+}$: $5.8505 + 0.0982 - 1.6723 = 4.2764$

$_{51}Sb^{3+} \to {}_{51}Sb^{4+}$: $5.7905 + 1.6723 - 1.3336 = 6.1292$

$_{51}Sb \to {}_{51}Sb^-$: $-6.0269 - 3.3128 + 3.2491 = -6.0906$

$_{51}Sb^- \to {}_{51}Sb^{2-}$: $-6.0846 - 5.6003 + 3.3128 = -8.3721$

$_{51}Sb^{2-} \to {}_{51}Sb^{3-}$: $-6.1418 - 8.4983 + 5.6003 = -9.0398$

$_{51}Sb^{3-} \to {}_{51}Sb^{4-}$: $-6.1984 + 2.1248 + 14.8181 = 10.7445$

54. Xenon

Write Formulae:

$_{54}Xe \to {}_{54}Xe^+$:
$$IP_{1,1} \times 54 / M_{54} + \Delta IP_{1,54} \times 54 / M_{54} - \Delta IP_{1,53} \times 53 / M_{54}$$

$_{54}Xe^+ \to {}_{54}Xe^{2+}$:
$$IP_{1,1} \times \sqrt{54 \times 53}/M_{54} + \Delta IP_{1,53} \times 53 / M_{54} - \Delta IP_{1,52} \times 52 / M_{54}$$

$_{54}Xe^{2+} \to {}_{54}Xe^{3+}$:
$$IP_{1,1} \times \sqrt{54 \times 52}/M_{54} + \Delta IP_{1,52} \times 52 / M_{54} - \Delta IP_{1,51} \times 51 / M_{54}$$

$_{54}Xe^{3+} \to {}_{54}Xe^{4+}$:
$$IP_{1,1} \times \sqrt{54 \times 51}/M_{54} + \Delta IP_{1,51} \times 51 / M_{54} - \Delta IP_{1,50} \times 50 / M_{54}$$

$_{54}Xe \to {}_{54}Xe^-$:
$$-IP_{1,1} \times \sqrt{54 \times 55}/M_{54} - \Delta IP_{1,55} \times 55 / M_{54} + \Delta IP_{1,54} \times 54 / M_{54}$$

$_{54}Xe^- \rightarrow {}_{54}Xe^{2-}$:

$-IP_{1,1} \times \sqrt{54 \times 56}/M_{54} - \Delta IP_{1,56} \times 56/M_{54} + \Delta IP_{1,55} \times 55/M_{54}$

$_{54}Xe^{2-} \rightarrow {}_{54}Xe^{3-}$:

$-IP_{1,1} \times \sqrt{54 \times 57}/M_{54} - \Delta IP_{1,57} \times 57/M_{54} + \Delta IP_{1,56} \times 56/M_{54}$

$_{54}Xe^{3-} \rightarrow {}_{54}Xe^{4-}$:

$-IP_{1,1} \times \sqrt{54 \times 58}/M_{54} - \Delta IP_{1,56} \times 56/M_{54} + \Delta IP_{1,57} \times 57/M_{54}$

Insert Data:

$_{54}Xe \rightarrow {}_{54}Xe^+$:
$14.250 \times 54/131.290 + 19.162 \times 54/131.290 - 12.866 \times 53/131.290$

$_{54}Xe^+ \rightarrow {}_{54}Xe^{2+}$:
$14.250 \times 53.4977/131.290 + 12.866 \times 53/131.290 - 7.757 \times 52/131.290$

$_{54}Xe^{2+} \rightarrow {}_{54}Xe^{3+}$:
$14.250 \times 52.9906/131.290 + 7.757 \times 52/131.290 - 7.757 \times 51/131.290$

$_{54}Xe^{3+} \rightarrow {}_{54}Xe^{4+}$:
$14.250 \times 52.4786/131.290 + 7.757 \times 51/131.290 - 3.610 \times 50/131.290$

$_{54}Xe \rightarrow {}_{54}Xe^-$:
$-14.250 \times 54.4977/131.290 - (-4.704) \times 55/131.290 + 19.162 \times 54/131.290$

$_{54}Xe^- \rightarrow {}_{54}Xe^{2-}$:
$-14.250 \times 54.9909/131.290 - (-1.192) \times 56/131.290 + (-4.704) \times 55/131.290$

$_{54}Xe^{2-} \rightarrow {}_{54}Xe^{3-}$:
$-14.250 \times 55.4797/131.290 - (-1.857) \times 57/131.290 + (-1.192) \times 56/131.290$

$_{54}Xe^{3-} \to {_{54}}Xe^{4-}$:

$-14.250 \times 55.9643/131.290 - (-1.192) \times 56/131.290 + (-1.857) \times 57/131.290$

Evaluate Formulae:

$_{54}Xe \to {_{54}}Xe^{+}$: $5.8611 + 7.8814 - 5.1938 = 8.5487$

$_{54}Xe^{+} \to {_{54}}Xe^{2+}$: $5.8066 + 5.1938 - 3.0723 = 7.9281$

$_{54}Xe^{2+} \to {_{54}}Xe^{3+}$: $5.7515 + 3.0723 - 3.0132 = 5.8106$

$_{54}Xe^{3+} \to {_{54}}Xe^{4+}$: $5.6959 + 3.0132 - 1.3748 = 7.3343$

$_{54}Xe \to {_{54}}Xe^{-}$: $-5.9151 + 1.9706 + 7.8814 = 3.9369$

$_{54}Xe^{-} \to {_{54}}Xe^{2-}$: $-5.9686 + 0.5084 - 1.9706 = -7.4308$

$_{54}Xe^{2-} \to {_{54}}Xe^{3-}$: $-6.0217 + 0.8062 - 0.5084 = -5.7239$

$_{54}Xe^{3-} \to {_{54}}Xe^{4-}$: $-6.0743 + 0.5084 - 0.8062 = -6.3721$

55. Cesium

Write Formulae:

$_{55}Cs \to {_{55}}Cs^{+}$:
$IP_{1,1} \times 55/M_{55} + \Delta IP_{1,55} \times 55/M_{55} - \Delta IP_{1,54} \times 54/M_{55}$

$_{55}Cs^{+} \to {_{55}}Cs^{2+}$:
$IP_{1,1} \times \sqrt{55 \times 54}/M_{55} + \Delta IP_{1,54} \times 54/M_{55} - \Delta IP_{1,53} \times 53/M_{55}$

$_{55}Cs^{2+} \to {}_{55}Cs^{3+}$:

$IP_{1,1} \times \sqrt{55 \times 53}\big/M_{55} + \Delta IP_{1,53} \times 53/M_{55} - \Delta IP_{1,52} \times 52/M_{55}$

$_{55}Cs^{3+} \to {}_{55}Cs^{4+}$:

$IP_{1,1} \times \sqrt{55 \times 54}\big/M_{55} + \Delta IP_{1,52} \times 52/M_{55} - \Delta IP_{1,51} \times 51/M_{55}$

$_{55}Cs \to {}_{55}Cs^{-}$:

$-IP_{1,1} \times \sqrt{55 \times 56}\big/M_{55} - \Delta IP_{1,56} \times 56/M_{55} + \Delta IP_{1,55} \times 55/M_{55}$

$_{55}Cs^{-} \to {}_{55}Cs^{2-}$:

$-IP_{1,1} \times \sqrt{55 \times 57}\big/M_{55} - \Delta IP_{1,57} \times 57/M_{55} + \Delta IP_{1,56} \times 56/M_{55}$

$_{55}Cs^{2-} \to {}_{55}Cs^{3-}$:

$-IP_{1,1} \times \sqrt{55 \times 58}\big/M_{55} - \Delta IP_{1,58} \times 58/M_{55} + \Delta IP_{1,57} \times 57/M_{55}$

$_{55}Cs^{3-} \to {}_{55}Cs^{4-}$:

$-IP_{1,1} \times \sqrt{55 \times 59}\big/M_{55} - \Delta IP_{1,59} \times 59/M_{55} + \Delta IP_{1,58} \times 58/M_{55}$

Insert Data:

$_{55}Cs \to {}_{55}Cs^{+}$:
$14.250 \times 55/132.905 + (-4.704) \times 55/132.905 - 19.162 \times 54/132.905$

$_{55}Cs^{+} \to {}_{55}Cs^{2+}$:
$14.250 \times 54.4977/132.905 + 19.162 \times 54/132.905 - 12.866 \times 53/132.905$

$_{55}Cs^{2+} \to {}_{55}Cs^{3+}$:
$14.250 \times 53.9907/132.905 + 12.866 \times 53/132.905 - 7.757 \times 52/132.905$

$_{55}Cs^{3+} \rightarrow {_{55}Cs^{4+}}$:

$14.250 \times 54.4977/132.905 + 7.757 \times 52 / 132.905 - 7.757 \times 51 / 132.905$

$_{55}Cs \rightarrow {_{55}Cs^{-}}$:

$-14.250 \times 55.4977/132.905 - (-1.192) \times 56 / 132.905 + (-4.704) \times 55 / 132.905$

$_{55}Cs^{-} \rightarrow {_{55}Cs^{2-}}$:

$-14.250 \times 55.9911/132.905 - (-1.857) \times 57 / 132.905 + (-1.192) \times 56 / 132.905$

$_{55}Cs^{2-} \rightarrow {_{55}Cs^{3-}}$:

$-14.250 \times 56.4801/132.905 - (-1.667) \times 58 / 132.905 + (-1.857) \times 57 / 132.905$

$_{55}Cs^{3-} \rightarrow {_{55}Cs^{4-}}$:

$-14.250 \times 56.9649/132.905 - (-1.474) \times 59 / 132.905 + (-1.667) \times 58 / 132.905$

Evaluate Formulae:

$_{55}Cs \rightarrow {_{55}Cs^{+}}$: $5.8971 - 1.9467 - 7.7856 = -3.8352$

$_{55}Cs^{+} \rightarrow {_{55}Cs^{2+}}$: $5.8432 + 7.7856 - 5.1307 = 8.4981$

$_{55}Cs^{2+} \rightarrow {_{55}Cs^{3+}}$: $5.7889 + 5.1307 - 3.0350 = 7.8846$

$_{55}Cs^{3+} \rightarrow {_{55}Cs^{4+}}$: $5.8432 + 3.0350 - 2.9766 = 5.9016$

$_{55}Cs \rightarrow {_{55}Cs^{-}}$: $-5.9504 + 0.5023 - 1.9467 = -7.3948$

$_{55}Cs^{-} \rightarrow {_{55}Cs^{2-}}$: $-6.0033 + 0.7964 - 0.5023 = -5.7092$

$_{55}Cs^{2-} \rightarrow {_{55}Cs^{3-}}$: $-6.0558 + 0.7275 - 0.7964 = -6.1247$

$_{55}Cs^{3-} \rightarrow {_{55}Cs^{4-}}$: $-6.1077 + 0.6543 - 0.7275 = -6.1809$

57. Lanthanum

Write Formulae:

$_{57}La \to {}_{57}La^+$:

$IP_{1,1} \times 57 / M_{57} + \Delta IP_{1,57} \times 57 / M_{57} - \Delta IP_{1,56} \times 56 / M_{57}$

$_{57}La^+ \to {}_{57}La^{2+}$:

$IP_{1,1} \times \sqrt{57 \times 56}/M_{57} + \Delta IP_{1,56} \times 56 / M_{57} - \Delta IP_{1,55} \times 55 / M_{57}$

$_{57}La \to {}_{57}La^-$:

$-IP_{1,1} \times \sqrt{57 \times 58}/M_{57} - \Delta IP_{1,58} \times 58 / M_{57} + \Delta IP_{1,57} \times 57 / M_{57}$

$_{57}La^- \to {}_{57}La^{2-}$:

$-IP_{1,1} \times \sqrt{57 \times 59}/M_{57} - \Delta IP_{1,59} \times 59 / M_{57} + \Delta IP_{1,58} \times 58 / M_{57}$

Insert Data:

$_{57}La \to {}_{57}La^+$:
$14.250 \times 57 / 138.906 + (-1.857) \times 57 / 138.906 - (-1.193) \times 56 / 138.906$

$_{57}La^+ \to {}_{57}La^{2+}$:
$14.250 \times 56.4978/138.906 + (-1.193) \times 56 / 138.906 - (-4.704) \times 55 / 138.906$

$_{57}La \to {}_{57}La^-$:
$-14.250 \times 57.498/138.906 - (-1.667) \times 58 / 138.906 + (-1.587) \times 57 / 138.906$

$_{57}La^- \to {}_{57}La^{2-}$:
$-14.250 \times 57.9914/138.906 - (-1.474) \times 59 / 138.906 + (-1.667) \times 58 / 138.906$

Evaluate Formulae:

$_{57}La \to {}_{57}La^+$: $5.8475 - 0.7620 + 0.4810 = 5.5665$

$_{57}\text{La}^+ \to {}_{57}\text{La}^{2+}$: $5.7960 - 0.4810 + 1.8626 = 7.1776$

$_{57}\text{La} \to {}_{57}\text{La}^-$: $-5.8986 + 0.6961 - 0.6512 = -5.8537$

$_{57}\text{La}^- \to {}_{57}\text{La}^{2-}$: $-5.9492 + 0.6261 - 0.6961 = -6.0192$

74. Tungsten

Write Formulae:

$_{74}\text{W} \to {}_{74}\text{W}^+$:
$IP_{1,1} \times 74 / M_{74} + \Delta IP_{1,74} \times 74 / M_{74} - \Delta IP_{1,73} \times 73 / M_{74}$

$_{74}\text{W}^+ \to {}_{74}\text{W}^{2+}$:
$IP_{1,1} \times \sqrt{74 \times 73}/M_{74} + \Delta IP_{1,73} \times 73 / M_{74} - \Delta IP_{1,72} \times 72 / M_{74}$

$_{74}\text{W}^{2+} \to {}_{74}\text{W}^{3+}$:
$IP_{1,1} \times \sqrt{74 \times 72}/M_{74} + \Delta IP_{1,72} \times 72 / M_{74} - \Delta IP_{1,71} \times 71 / M_{74}$

$_{74}\text{W}^{3+} \to {}_{74}\text{W}^{4+}$:
$IP_{1,1} \times \sqrt{74 \times 71}/M_{74} + \Delta IP_{1,71} \times 71 / M_{74} - \Delta IP_{1,70} \times 70 / M_{74}$

$_{74}\text{W}^{4+} \to {}_{74}\text{W}^{5+}$:
$IP_{1,1} \times \sqrt{74 \times 70}/M_{74} + \Delta IP_{1,70} \times 70 / M_{74} - \Delta IP_{1,69} \times 69 / M_{74}$

$_{74}\text{W}^{5+} \to {}_{74}\text{W}^{6+}$:
$IP_{1,1} \times \sqrt{74 \times 69}/M_{74} + \Delta IP_{1,69} \times 69 / M_{74} - \Delta IP_{1,68} \times 68 / M_{74}$

$_{74}\text{W} \to {}_{74}\text{W}^-$:
$-IP_{1,1} \times \sqrt{74 \times 75}/M_{74} + \Delta IP_{1,75} \times 75 / M_{74} - \Delta IP_{1,74} \times 74 / M_{74}$

$_{74}W^- \to {}_{74}W^{2-}$:

$-IP_{1,1} \times \sqrt{74 \times 76} / M_{74} + \Delta IP_{1,76} \times 76 / M_{74} - \Delta IP_{1,75} \times 75 / M_{74}$

$_{74}W^{2-} \to {}_{74}W^{3-}$:

$-IP_{1,1} \times \sqrt{74 \times 77} / M_{74} + \Delta IP_{1,77} \times 77 / M_{74} - \Delta IP_{1,76} \times 76 / M_{74}$

$_{74}W^{3-} \to {}_{74}W^{4-}$:

$-IP_{1,1} \times \sqrt{74 \times 78} / M_{74} + \Delta IP_{1,78} \times 78 / M_{74} - \Delta IP_{1,77} \times 77 / M_{74}$

$_{74}W^{4-} \to {}_{74}W^{5-}$:

$-IP_{1,1} \times \sqrt{74 \times 79} / M_{74} + \Delta IP_{1,79} \times 79 / M_{74} - \Delta IP_{1,78} \times 78 / M_{74}$

$_{74}W^{5-} \to {}_{74}W^{6-}$:

$-IP_{1,1} \times \sqrt{74 \times 80} / M_{74} + \Delta IP_{1,80} \times 80 / M_{74} - \Delta IP_{1,79} \times 79 / M_{74}$

Insert Data:

$_{74}W \to {}_{74}W^+$:
$14.250 \times 74 / 183.840 + 6.434 \times 74 / 183.840 - 5.446 \times 73 / 183.840$

$_{74}W^+ \to {}_{74}W^{2+}$:
$14.250 \times 73.4983 / 183.840 + 5.446 \times 73 / 183.840 - 4.505 \times 72 / 183.840$

$_{74}W^{2+} \to {}_{74}W^{3+}$:
$14.250 \times 72.9932 / 183.840 + 4.505 \times 72 / 183.840 - 3.610 \times 71 / 183.840$

$_{74}W^{3+} \to {}_{74}W^{4+}$:
$14.250 \times 72.4845 / 183.840 + 3.610 \times 71 / 183.840 - 0.627 \times 70 / 183.840$

$_{74}W^{4+} \to {}_{74}W^{5+}$:
$14.250 \times 71.9722 / 183.840 + 0.627 \times 70 / 183.840 - 0.403 \times 69 / 183.840$

$_{74}W^{5+} \to {}_{74}W^{6+}$:

$14.250 \times 71.4563/183.840 + 0.403 \times 69 / 183.840 - 0.181 \times 68 / 183.840$

$_{74}W \to {}_{74}W^{-}$:

$-14.250 \times 74.4983/183.840 + 7.471 \times 75 / 183.840 - 6.434 \times 74 / 183.840$

$_{74}W^{-} \to {}_{74}W^{2-}$:

$-14.250 \times 74.9933/183.840 + 7.471 \times 76 / 183.840 - 7.471 \times 75 / 183.840$

$_{74}W^{2-} \to {}_{74}W^{3-}$:

$-14.250 \times 75.4851/183.840 + 8.560 \times 77 / 183.840 - 7.471 \times 76 / 183.840$

$_{74}W^{3-} \to {}_{74}W^{4-}$:

$-14.250 \times 75.9737/183.840 + 9.705 \times 78 / 183.840 - 8.560 \times 77 / 183.840$

$_{74}W^{4-} \to {}_{74}W^{5-}$:

$-14.250 \times 76.4591/183.840 + 10.906 \times 79 / 183.840 - 9.705 \times 78 / 183.840$

$_{74}W^{5-} \to {}_{74}W^{6-}$:

$-14.250 \times 76.9415/183.840 + 12.168 \times 80 / 183.840 - 10.906 \times 79 / 183.840$

Evaluate Formulae:

$_{74}W \to {}_{74}W^{+}$: $5.7360 + 2.5898 - 2.1625 = 6.1633$

$_{74}W^{+} \to {}_{74}W^{2+}$: $5.6971 + 2.1625 - 1.7644 = 6.0952$

$_{74}W^{2+} \to {}_{74}W^{3+}$: $5.6579 + 1.7644 - 1.3942 = 6.0281$

$_{74}W^{3+} \to {}_{74}W^{4+}$: $5.6185 + 1.3942 - 0.2387 = 6.7740$

$_{74}W^{4+} \to {}_{74}W^{5+}$: $5.5788 + 0.2387 - 0.1513 = 5.6662$

$_{74}W^{5+} \rightarrow {}_{74}W^{6+}$: $5.5388 + 0.1513 - 0.06695 = 5.6232$

$_{74}W \rightarrow {}_{74}W^-$: $-5.7746 - 3.0479 + 2.5898 = -6.2327$

$_{74}W^- \rightarrow {}_{74}W^{2-}$: $-5.8130 - 3.0885 + 3.0478 = -5.8536$

$_{74}W^{2-} \rightarrow {}_{74}W^{3-}$: $-5.8511 - 3.5853 + 3.0885 = -6.3479$

$_{74}W^{3-} \rightarrow {}_{74}W^{4-}$: $-5.8890 - 4.1177 + 3.5853 = -6.4214$

$_{74}W^{4-} \rightarrow {}_{74}W^{5-}$: $-5.9266 + 4.6865 - 4.1177 = -5.3578$

$_{74}W^{5-} \rightarrow {}_{74}W^{6-}$: $-5.9640 + 5.2950 - 4.6865 = -5.3555$

78. Platinum

Write Formulae:

$_{78}Pt \rightarrow {}_{78}Pt^+$:
$IP_{1,1} \times 78 / M_{78} + \Delta IP_{1,78} \times 78 / M_{78} - \Delta IP_{1,77} \times 77 / M_{78}$

$_{78}Pt^+ \rightarrow {}_{78}Pt^{2+}$:
$IP_{1,1} \times \sqrt{78 \times 77} / M_{78} + \Delta IP_{1,77} \times 77 / M_{78} - \Delta IP_{1,76} \times 76 / M_{78}$

$_{78}Pt^{2+} \rightarrow {}_{78}Pt^{3+}$:
$IP_{1,1} \times \sqrt{78 \times 76} / M_{78} + \Delta IP_{1,76} \times 76 / M_{78} - \Delta IP_{1,75} \times 75 / M_{78}$

$_{78}Pt^{3+} \rightarrow {}_{78}Pt^{4+}$:
$IP_{1,1} \times \sqrt{78 \times 75} / M_{78} + \Delta IP_{1,75} \times 75 / M_{78} - \Delta IP_{1,74} \times 74 / M_{78}$

$_{78}\text{Pt} \to {}_{78}\text{Pt}^-$:

$-IP_{1,1} \times \sqrt{78 \times 79}/M_{78} - \Delta IP_{1,79} \times 79/M_{78} + \Delta IP_{1,78} \times 78/M_{78}$

$_{78}\text{Pt}^- \to {}_{78}\text{Pt}^{2-}$:

$-IP_{1,1} \times \sqrt{78 \times 80}/M_{78} - \Delta IP_{1,80} \times 80/M_{78} + \Delta IP_{1,79} \times 79/M_{78}$

$_{78}\text{Pt}^{2-} \to {}_{78}\text{Pt}^{3-}$:

$-IP_{1,1} \times \sqrt{78 \times 81}/M_{78} - \Delta IP_{1,81} \times 81/M_{78} + \Delta IP_{1,80} \times 80/M_{78}$

$_{78}\text{Pt}^{3-} \to {}_{78}\text{Pt}^{4-}$:

$-IP_{1,1} \times \sqrt{78 \times 82}/M_{78} - \Delta IP_{1,82} \times 82/M_{78} + \Delta IP_{1,81} \times 81/M_{78}$

Insert Data:

$_{78}\text{Pt} \to {}_{78}\text{Pt}^+$:
$14.250 \times 78/195.076 + 9.705 \times 78/195.076 - 8.560 \times 77/195.076$

$_{78}\text{Pt}^+ \to {}_{78}\text{Pt}^{2+}$:
$14.250 \times 77.4984/195.076 + 8.560 \times 77/195.076 - 7.471 \times 76/195.076$

$_{78}\text{Pt}^{2+} \to {}_{78}\text{Pt}^{3+}$:
$14.250 \times 76.9935/195.076 + 7.471 \times 76/195.076 - 7.471 \times 75/195.076$

$_{78}\text{Pt}^{3+} \to {}_{78}\text{Pt}^{4+}$:
$14.250 \times 76.4853/195.076 + 7.471 \times 75/195.076 - 6.434 \times 74/195.076$

$_{78}\text{Pt} \to {}_{78}\text{Pt}^-$:
$-14.250 \times 78.4984/195.076 - 10.906 \times 79/195.076 + 9.705 \times 78/195.076$

$_{78}\text{Pt}^- \to {}_{78}\text{Pt}^{2-}$:
$-14.250 \times 78.9937/195.076 - 12.168 \times 80/195.076 + 10.906 \times 79/195.076$

$_{78}Pt^{2-} \to {}_{78}Pt^{3-}$:

$-14.250 \times 79.4858/195.076 - 2.265 \times 81/195.076 + 12.168 \times 80/195.076$

$_{78}Pt^{3-} \to {}_{78}Pt^{4-}$:

$-14.250 \times 79.9750/195.076 - 5.446 \times 82/195.076 + 2.265 \times 81/195.076$

Evaluate Formulae:

$_{78}Pt \to {}_{78}Pt^{+}$: $5.6978 + 3.8805 - 3.3788 = 6.1995$

$_{78}Pt^{+} \to {}_{78}Pt^{2+}$: $5.6611 + 3.3788 - 2.9106 = 6.1293$

$_{78}Pt^{2+} \to {}_{78}Pt^{3+}$: $5.6243 + 2.9106 - 2.8723 = 5.6626$

$_{78}Pt^{3+} \to {}_{78}Pt^{4+}$: $5.5871 + 2.8723 - 2.4407 = 6.0187$

$_{78}Pt \to {}_{78}Pt^{-}$: $-5.7342 - 4.4166 + 3.8805 = -6.2703$

$_{78}Pt^{-} \to {}_{78}Pt^{2-}$: $-5.7704 - 4.9901 + 4.4166 = -6.3439$

$_{78}Pt^{2-} \to {}_{78}Pt^{3-}$: $-5.8063 - 0.9405 + 4.9901 = -1.7567$

$_{78}Pt^{3-} \to {}_{78}Pt^{4-}$: $-5.8421 - 2.2892 + 0.9405 = -7.1908$

79. Gold

Write Formulae:

$_{79}Au \to {}_{79}Au^{+}$:

$IP_{1,1} \times 79 / M_{79} + \Delta IP_{1,79} \times 79 / M_{79} - \Delta IP_{1,78} \times 78 / M_{79}$

$_{79}\text{Au}^+ \to {}_{79}\text{Au}^{2+}$:

$IP_{1,1} \times \sqrt{79 \times 78}/M_{79} + \Delta IP_{1,78} \times 78/M_{79} - \Delta IP_{1,77} \times 77/M_{79}$

$_{79}\text{Au}^{2+} \to {}_{79}\text{Au}^{3+}$:

$IP_{1,1} \times \sqrt{79 \times 77}/M_{79} + \Delta IP_{1,77} \times 77/M_{79} - \Delta IP_{1,76} \times 76/M_{79}$

$_{79}\text{Au}^{3+} \to {}_{79}\text{Au}^{4+}$:

$IP_{1,1} \times \sqrt{79 \times 76}/M_{79} + \Delta IP_{1,76} \times 76/M_{79} - \Delta IP_{1,75} \times 75/M_{79}$

$_{79}\text{Au}^{4+} \to {}_{79}\text{Au}^{5+}$:

$IP_{1,1} \times \sqrt{79 \times 75}/M_{79} + \Delta IP_{1,75} \times 75/M_{79} - \Delta IP_{1,74} \times 74/M_{79}$

$_{79}\text{Au}^{5+} \to {}_{79}\text{Au}^{6+}$:

$IP_{1,1} \times \sqrt{79 \times 74}/M_{79} + \Delta IP_{1,74} \times 74/M_{79} - \Delta IP_{1,73} \times 73/M_{79}$

$_{79}\text{Au} \to {}_{79}\text{Au}^{-}$:

$-IP_{1,1} \times \sqrt{79 \times 80}/M_{79} - \Delta IP_{1,80} \times 80/M_{79} + \Delta IP_{1,79} \times 79/M_{79}$

$_{79}\text{Au}^{-} \to {}_{79}\text{Au}^{2-}$:

$-IP_{1,1} \times \sqrt{79 \times 81}/M_{79} - \Delta IP_{1,81} \times 81/M_{79} + \Delta IP_{1,80} \times 80/M_{79}$

$_{79}\text{Au}^{2-} \to {}_{79}\text{Au}^{3-}$:

$-IP_{1,1} \times \sqrt{79 \times 82}/M_{79} - \Delta IP_{1,82} \times 82/M_{79} + \Delta IP_{1,81} \times 81/M_{79}$

$_{79}\text{Au}^{3-} \to {}_{79}\text{Au}^{4-}$:

$-IP_{1,1} \times \sqrt{79 \times 83}/M_{79} - \Delta IP_{1,83} \times 83/M_{79} + \Delta IP_{1,82} \times 82/M_{79}$

$_{79}\text{Au}^{4-} \to {}_{79}\text{Au}^{5-}$:

$-IP_{1,1} \times \sqrt{79 \times 84}/M_{79} - \Delta IP_{1,84} \times 84/M_{79} + \Delta IP_{1,83} \times 83/M_{79}$

$_{79}\text{Au}^{5-} \to {}_{79}\text{Au}^{6-}$:

$-IP_{1,1} \times \sqrt{79 \times 85}/M_{79} - \Delta IP_{1,85} \times 85/M_{79} + \Delta IP_{1,84} \times 84/M_{79}$

Insert Data:

$_{79}\text{Au} \to {}_{79}\text{Au}^{+}$:
$14.250 \times 79/196.967 + 10.906 \times 79/196.967 - 9.705 \times 78/196.967$

$_{79}\text{Au}^{+} \to {}_{79}\text{Au}^{2+}$:
$14.250 \times 78.4984/196.967 + 9.705 \times 78/196.967 - 8.560 \times 77/196.967$

$_{79}\text{Au}^{2+} \to {}_{79}\text{Au}^{3+}$:
$14.250 \times 77.9936/196.967 + 8.560 \times 77/196.967 - 7.471 \times 76/196.967$

$_{79}\text{Au}^{3+} \to {}_{79}\text{Au}^{4+}$:
$14.250 \times 77.4855/196.967 + 7.471 \times 76/196.967 - 7.471 \times 75/196.967$

$_{79}\text{Au}^{4+} \to {}_{79}\text{Au}^{5+}$:
$14.250 \times 76.9740/196.967 + 7.471 \times 75/196.967 - 6.434 \times 74/196.967$

$_{79}\text{Au}^{5+} \to {}_{79}\text{Au}^{6+}$:
$14.250 \times 76.4591/196.967 + 6.434 \times 74/196.967 - 5.446 \times 73/196.967$

$_{79}\text{Au} \to {}_{79}\text{Au}^{-}$:
$-14.250 \times 79.4984/196.967 - 12.168 \times 80/196.967 + 10.906 \times 79/196.967$

$_{79}\text{Au}^{-} \to {}_{79}\text{Au}^{2-}$:
$-14.250 \times 79.9937/196.967 - 16.515 \times 81/196.967 + 12.168 \times 80/196.967$

$_{79}Au^{2-} \rightarrow {}_{79}Au^{3-}$:

$-14.250 \times 80.4860/196.967 - 5.446 \times 82/196.967 + 2.265 \times 81/196.967$

$_{79}Au^{3-} \rightarrow {}_{79}Au^{4-}$:

$-14.250 \times 80.9753/196.967 - 9.240 \times 83/196.967 + 5.446 \times 82/196.967$

$_{79}Au^{4-} \rightarrow {}_{79}Au^{5-}$:

$-14.250 \times 81.4616/196.967 - 9.240 \times 84/196.967 + 9.240 \times 83/196.967$

$_{79}Au^{5-} \rightarrow {}_{79}Au^{6-}$:

$-14.250 \times 81.9451/196.967 - 13.765 \times 85/196.967 + 9.240 \times 84/196.967$

Evaluate Formulae:

$_{79}Au \rightarrow {}_{79}Au^{+}$: $5.7154 + 4.3742 - 3.8432 = 6.2464$

$_{79}Au^{+} \rightarrow {}_{79}Au^{2+}$: $5.6791 + 3.8432 - 3.3463 = 6.1759$

$_{79}Au^{2+} \rightarrow {}_{79}Au^{3+}$: $5.6426 + 3.3463 - 2.8827 = 6.1062$

$_{79}Au^{3+} \rightarrow {}_{79}Au^{4+}$: $5.6059 + 2.8827 - 2.8448 = 5.6438$

$_{79}Au^{4+} \rightarrow {}_{79}Au^{5+}$: $5.5316 + 2.4172 - 2.0184 = 5.9304$

$_{79}Au^{5+} \rightarrow {}_{79}Au^{6+}$: $5.5688 + 2.8448 - 2.4172 = 5.9963$

$_{79}Au \rightarrow {}_{79}Au^{-}$: $-5.7515 - 4.9421 + 4.3742 = -6.3194$

$_{79}Au^{-} \rightarrow {}_{79}Au^{2-}$: $-5.7873 - 6.7916 + 4.9421 = -7.6368$

$_{79}Au^{2-} \rightarrow {}_{79}Au^{3-}$: $-5.8229 - 2.2672 + 0.9315 = -7.1586$

$_{79}\text{Au}^{3-} \to {}_{79}\text{Au}^{4-}$: $-5.8583 - 3.8936 + 2.2672 = -7.4847$

$_{79}\text{Au}^{4-} \to {}_{79}\text{Au}^{5-}$: $-5.8935 - 3.9406 + 3.8936 = -5.9405$

$_{79}\text{Au}^{5-} \to {}_{79}\text{Au}^{6-}$: $-5.9285 - 5.9402 + 3.9406 = -7.9281$

80. Mercury

Write Formulae:

$_{80}\text{Hg} \to {}_{80}\text{Hg}^{+}$:
$$IP_{1,1} \times 80 / M_{80} + \Delta IP_{1,80} \times 80 / M_{80} - \Delta IP_{1,79} \times 79 / M_{80}$$

$_{80}\text{Hg}^{+} \to {}_{80}\text{Hg}^{2+}$:
$$IP_{1,1} \times \sqrt{80 \times 79}\big/ M_{80} + \Delta IP_{1,79} \times 79 / M_{80} - \Delta IP_{1,78} \times 78 / M_{80}$$

$_{80}\text{Hg}^{2+} \to {}_{80}\text{Hg}^{3+}$:
$$IP_{1,1} \times \sqrt{80 \times 78}\big/ M_{80} + \Delta IP_{1,78} \times 78 / M_{80} - \Delta IP_{1,77} \times 77 / M_{80}$$

$_{80}\text{Hg}^{3+} \to {}_{80}\text{Hg}^{4+}$:
$$IP_{1,1} \times \sqrt{80 \times 77}\big/ M_{80} + \Delta IP_{1,77} \times 77 / M_{80} - \Delta IP_{1,78} \times 78 / M_{80}$$

$_{80}\text{Hg}^{4+} \to {}_{80}\text{Hg}^{5+}$:
$$IP_{1,1} \times \sqrt{80 \times 76}\big/ M_{80} + \Delta IP_{1,76} \times 76 / M_{80} - \Delta IP_{1,77} \times 77 / M_{80}$$

$_{80}\text{Hg}^{5+} \to {}_{80}\text{Hg}^{6+}$:
$$IP_{1,1} \times \sqrt{80 \times 75}\big/ M_{80} + \Delta IP_{1,75} \times 75 / M_{80} - \Delta IP_{1,76} \times 76 / M_{80}$$

$_{80}\text{Hg} \to {_{80}\text{Hg}^-}$:

$-IP_{1,1} \times \sqrt{80 \times 81}/M_{80} - \Delta IP_{1,81} \times 81/M_{80} + \Delta IP_{1,80} \times 80/M_{80}$

$_{80}\text{Hg}^- \to {_{80}\text{Hg}^{2-}}$:

$-IP_{1,1} \times \sqrt{80 \times 82}/M_{80} - \Delta IP_{1,82} \times 82/M_{80} + \Delta IP_{1,81} \times 81/M_{80}$

$_{80}\text{Hg}^{2-} \to {_{80}\text{Hg}^{3-}}$:

$-IP_{1,1} \times \sqrt{80 \times 83}/M_{80} - \Delta IP_{1,83} \times 83/M_{80} + \Delta IP_{1,82} \times 82/M_{80}$

$_{80}\text{Hg}^{3-} \to {_{80}\text{Hg}^{4-}}$:

$-IP_{1,1} \times \sqrt{80 \times 84}/M_{80} - \Delta IP_{1,84} \times 84/M_{80} + \Delta IP_{1,83} \times 83/M_{80}$

$_{80}\text{Hg}^{4-} \to {_{80}\text{Hg}^{5-}}$:

$-IP_{1,1} \times \sqrt{80 \times 85}/M_{80} - \Delta IP_{1,85} \times 85/M_{80} + \Delta IP_{1,84} \times 84/M_{80}$

$_{80}\text{Hg}^{5-} \to {_{80}\text{Hg}^{6-}}$:

$-IP_{1,1} \times \sqrt{80 \times 86}/M_{80} - \Delta IP_{1,86} \times 86/M_{80} + \Delta IP_{1,85} \times 85/M_{80}$

Insert Data:

$_{80}\text{Hg} \to {_{80}\text{Hg}^+}$:

$14.250 \times 80/200.530 + 12.168 \times 80/200.530 - 10.906 \times 79/200.530$

$_{80}\text{Hg}^+ \to {_{80}\text{Hg}^{2+}}$:

$14.250 \times 79.4984/200.530 + 10.906 \times 79/200.530 - 9.705 \times 78/200.530$

$_{80}\text{Hg}^{2+} \to {_{80}\text{Hg}^{3+}}$:

$14.250 \times 78.9937/200.530 + 9.705 \times 78/200.530 - 8.560 \times 77/200.530$

$_{80}Hg^{3+} \to {}_{80}Hg^{4+}$:

$14.250 \times 78.4857/200.530 + 8.560 \times 77/200.530 - 9.705 \times 78/200.530$

$_{80}Hg^{4+} \to {}_{80}Hg^{5+}$:

$14.250 \times 77.9744/200.530 + 7.471 \times 76/200.530 - 8.560 \times 77/200.530$

$_{80}Hg^{5+} \to {}_{80}Hg^{6+}$:

$14.250 \times 77.4597/200.530 + 7.471 \times 75/200.530 - 7.471 \times 76/200.530$

$_{80}Hg \to {}_{80}Hg^{-}$:

$-14.250 \times 80.4984/200.530 - 2.265 \times 81/200.530 + 12.168 \times 80/200.530$

$_{80}Hg^{-} \to {}_{80}Hg^{2-}$:

$-14.250 \times 80.9938/200.530 - 5.446 \times 82/200.530 + 2.265 \times 81/200.530$

$_{80}Hg^{2-} \to {}_{80}Hg^{3-}$:

$-14.250 \times 81.4862/200.530 - 9.240 \times 83/200.530 + 5.446 \times 82/200.530$

$_{80}Hg^{3-} \to {}_{80}Hg^{4-}$:

$-14.250 \times 81.9756/200.530 - 9.240 \times 84/200.530 + 9.240 \times 83/200.530$

$_{80}Hg^{4-} \to {}_{80}Hg^{5-}$:

$-14.250 \times 82.4621/200.530 - 13.765 \times 85/200.530 + 9.240 \times 84/200.530$

$_{80}Hg^{5-} \to {}_{80}Hg^{6-}$:

$-14.250 \times 82.9458/200.530 - 19.164 \times 86/200.530 + 13.765 \times 85/200.530$

Evaluate Formulae:

$_{80}Hg \to {}_{80}Hg^{+}$: $5.6849 + 4.8543 - 4.2965 = 6.2427$

$_{80}Hg^{+} \to {}_{80}Hg^{2+}$: $5.7682 + 4.2965 - 3.7749 = 6.2898$

$_{80}Hg^{2+} \to {}_{80}Hg^{3+}$: $5.6134 + 3.7749 - 3.2689 = 6.1014$

$_{80}Hg^{3+} \to {}_{80}Hg^{4+}$: $5.5773 + 3.2869 - 3.7749 = 5.0893$

$_{80}Hg^{4+} \to {}_{80}Hg^{5+}$: $5.5410 + 2.8315 - 3.2869 = 5.0856$

$_{80}Hg^{5+} \to {}_{80}Hg^{6+}$: $5.5044 + 2.7942 - 2.8315 = 5.4671$

$_{80}Hg \to {}_{80}Hg^{-}$: $-5.7204 - 0.9149 + 4.8543 = -1.7810$

$_{80}Hg^{-} \to {}_{80}Hg^{2-}$: $-5.7556 - 2.2270 + 0.9149 = -7.0677$

$_{80}Hg^{2-} \to {}_{80}Hg^{3-}$: $-5.7905 - 3.8245 + 2.2270 = -7.3880$

$_{80}Hg^{3-} \to {}_{80}Hg^{4-}$: $-5.8253 - 3.8705 + 3.8245 = -5.8713$

$_{80}Hg^{4-} \to {}_{80}Hg^{5-}$: $-5.8599 - 5.8347 + 3.8705 = -7.8241$

$_{80}Hg^{5-} \to {}_{80}Hg^{6-}$: $-5.8943 - 8.2187 + 5.8347 = -8.2783$

81. Thallium

Write Formulae:

$_{81}Tl \to {}_{81}Tl^{+}$:
$IP_{1,1} \times 81 / M_{81} + \Delta IP_{1,81} \times 81 / M_{81} - \Delta IP_{1,80} \times 80 / M_{81}$

$_{81}Tl^{+} \to {}_{81}Tl^{2+}$:
$IP_{1,1} \times \sqrt{81 \times 80} / M_{81} + \Delta IP_{1,80} \times 80 / M_{81} - \Delta IP_{1,79} \times 79 / M_{81}$

$_{81}Tl^{2+} \to {}_{81}Tl^{3+}$:

$IP_{1,1} \times \sqrt{81 \times 79}/M_{81} + \Delta IP_{1,79} \times 79/M_{81} - \Delta IP_{1,78} \times 78/M_{81}$

$_{81}Tl^{3+} \to {}_{81}Tl^{4+}$:

$IP_{1,1} \times \sqrt{81 \times 78}/M_{81} + \Delta IP_{1,78} \times 78/M_{81} - \Delta IP_{1,77} \times 77/M_{81}$

$_{81}Tl^{4+} \to {}_{81}Tl^{5+}$:

$IP_{1,1} \times \sqrt{81 \times 77}/M_{81} + \Delta IP_{1,77} \times 77/M_{81} - \Delta IP_{1,76} \times 76/M_{81}$

$_{81}Tl^{5+} \to {}_{81}Tl^{6+}$:

$IP_{1,1} \times \sqrt{81 \times 76}/M_{81} + \Delta IP_{1,76} \times 76/M_{81} - \Delta IP_{1,75} \times 75/M_{81}$

$_{81}Tl \to {}_{81}Tl^{-}$:

$-IP_{1,1} \times \sqrt{81 \times 82}/M_{81} - \Delta IP_{1,82} \times 82/M_{81} + \Delta IP_{1,81} \times 81/M_{81}$

$_{81}Tl^{-} \to {}_{81}Tl^{2-}$:

$-IP_{1,1} \times \sqrt{81 \times 83}/M_{81} - \Delta IP_{1,83} \times 83/M_{81} + \Delta IP_{1,82} \times 82/M_{81}$

$_{81}Tl^{2-} \to {}_{81}Tl^{3-}$:

$-IP_{1,1} \times \sqrt{81 \times 84}/M_{81} - \Delta IP_{1,84} \times 84/M_{81} + \Delta IP_{1,83} \times 83/M_{81}$

$_{81}Tl^{3-} \to {}_{81}Tl^{4-}$:

$-IP_{1,1} \times \sqrt{81 \times 85}/M_{81} - \Delta IP_{1,85} \times 85/M_{81} + \Delta IP_{1,84} \times 84/M_{81}$

$_{81}Tl^{4-} \to {}_{81}Tl^{5-}$:

$-IP_{1,1} \times \sqrt{81 \times 86}/M_{81} - \Delta IP_{1,86} \times 86/M_{81} + \Delta IP_{1,85} \times 85/M_{81}$

$_{81}Tl^{5-} \to {}_{81}Tl^{6-}$:
$$-IP_{1,1} \times \sqrt{81 \times 87}/M_{81} - \Delta IP_{1,87} \times 87/M_{81} + \Delta IP_{1,86} \times 86/M_{81}$$

Insert Data:

$_{81}Tl \to {}_{81}Tl^+$:
$14.250 \times 81/204.383 + 2.265 \times 81/204.383 - 26.418 \times 80/204.383$

$_{81}Tl^+ \to {}_{81}Tl^{2+}$:
$14.250 \times 80.4984/204.383 + 12.168 \times 80/204.383 - 10.906 \times 79/204.383$

$_{81}Tl^{2+} \to {}_{81}Tl^{3+}$:
$14.250 \times 79.9937/204.383 + 10.906 \times 79/204.383 - 9.705 \times 78/204.383$

$_{81}Tl \to {}_{81}Tl^-$:
$-14.250 \times 81.4985/204.383 - 5.446 \times 82/204.383 + 2.265 \times 81/204.383$

$_{81}Tl^- \to {}_{81}Tl^{2-}$:
$-14.250 \times 81.9939/204.383 - 9.240 \times 83/204.383 + 5.446 \times 82/204.383$

$_{81}Tl^{2-} \to {}_{81}Tl^{3-}$:
$-14.250 \times 82.4864/204.383 - 9.240 \times 84/204.383 + 9.240 \times 83/204.383$

Evaluate Formulae:

$_{81}Tl \to {}_{81}Tl^+$: $5.6475 + 0.8977 - 10.3406 = -3.7954$

$_{81}Tl^+ \to {}_{81}Tl^{2+}$: $5.5778 + 4.7628 - 4.2155 = 6.1251$

$_{81}Tl^{2+} \to {}_{81}Tl^{3+}$: $5.5773 + 4.2155 - 3.7038 = 6.0890$

$_{81}Tl \to {}_{81}Tl^-$: $-5.6475 - 2.1850 + 0.8977 = -6.9348$

$_{81}Tl^- \to {}_{81}Tl^{2-}$: $-5.7168 - 3.7524 + 2.1850 = -7.2842$

$_{81}Tl^{2-} \to {}_{81}Tl^{3-}$: $-5.7511 - 3.7976 + 3.7524 = -5.7963$

82. Lead

Write Formulae:

$_{82}Pb \to {}_{82}Pb^+$:
$IP_{1,1} \times 82 / M_{82} + \Delta IP_{1,82} \times 82 / M_{82} - \Delta IP_{1,81} \times 81 / M_{82}$

$_{82}Pb^+ \to {}_{82}Pb^{2+}$:
$IP_{1,1} \times \sqrt{82 \times 81}/M_{82} + \Delta IP_{1,81} \times 81 / M_{82} - \Delta IP_{1,80} \times 80 / M_{82}$

$_{82}Pb^{2+} \to {}_{82}Pb^{3+}$:
$IP_{1,1} \times \sqrt{82 \times 80}/M_{82} + \Delta IP_{1,80} \times 80 / M_{82} - \Delta IP_{1,79} \times 79 / M_{82}$

$_{82}Pb^{3+} \to {}_{82}Pb^{4+}$:
$IP_{1,1} \times \sqrt{82 \times 79}/M_{82} + \Delta IP_{1,79} \times 79 / M_{82} - \Delta IP_{1,78} \times 78 / M_{82}$

$_{82}Pb^{4+} \to {}_{82}Pb^{5+}$:
$IP_{1,1} \times \sqrt{82 \times 78}/M_{82} + \Delta IP_{1,78} \times 78 / M_{82} - \Delta IP_{1,77} \times 77 / M_{82}$

$_{82}Pb \to {}_{82}Pb^-$:
$-IP_{1,1} \times \sqrt{82 \times 83}/M_{82} - \Delta IP_{1,83} \times 83 / M_{82} + \Delta IP_{1,82} \times 82 / M_{82}$

$_{82}Pb^- \to {}_{82}Pb^{2-}$:
$-IP_{1,1} \times \sqrt{82 \times 84}/M_{82} - \Delta IP_{1,84} \times 84 / M_{82} + \Delta IP_{1,83} \times 83 / M_{82}$

$_{82}\text{Pb}^{2-} \to {}_{82}\text{Pb}^{3-}$:

$-IP_{1,1} \times \sqrt{82 \times 85}/M_{82} - \Delta IP_{1,85} \times 85/M_{82} + \Delta IP_{1,84} \times 84/M_{82}$

$_{82}\text{Pb}^{3-} \to {}_{82}\text{Pb}^{4-}$:

$-IP_{1,1} \times \sqrt{82 \times 86}/M_{82} - \Delta IP_{1,86} \times 86/M_{82} + \Delta IP_{1,85} \times 85/M_{82}$

$_{82}\text{Pb}^{4-} \to {}_{82}\text{Pb}^{5-}$:

$-IP_{1,1} \times \sqrt{82 \times 87}/M_{82} - \Delta IP_{1,87} \times 87/M_{82} + \Delta IP_{1,86} \times 86/M_{82}$

Insert Data:

$_{82}\text{Pb} \to {}_{82}\text{Pb}^{+}$:
$14.250 \times 82/207.200 + 5.446 \times 82/207.200 - 2.265 \times 81/207.200$

$_{82}\text{Pb}^{+} \to {}_{82}\text{Pb}^{2+}$:
$14.250 \times 81.4985/207.200 + 2.265 \times 81/207.200 - 12.168 \times 80/207.200$

$_{82}\text{Pb}^{2+} \to {}_{82}\text{Pb}^{3+}$:
$14.250 \times 80.9938/207.200 + 12.168 \times 80/207.200 - 10.906 \times 79/207.200$

$_{82}\text{Pb}^{3+} \to {}_{82}\text{Pb}^{4+}$:
$14.250 \times 80.4860/207.200 + 10.906 \times 79/207.200 - 9.705 \times 78/207.200$

$_{82}\text{Pb}^{4+} \to {}_{82}\text{Pb}^{5+}$:
$14.250 \times 79.9750/207.200 + 23.955 \times 78/207.200 - 8.560 \times 77/207.200$

$_{82}\text{Pb} \to {}_{82}\text{Pb}^{-}$:
$-14.250 \times 82.4985/207.200 - 9.240 \times 83/207.200 + 5.446 \times 82/207.200$

$_{82}\text{Pb}^{-} \to {}_{82}\text{Pb}^{2-}$:
$-14.250 \times 82.9940/207.200 - 9.240 \times 84/207.200 + 9.240 \times 83/207.200$

$_{82}Pb^{2-} \to {}_{82}Pb^{3-}$:

$-14.250 \times 83.4865/207.200 - 13.765 \times 85/207.200 + 9.240 \times 84/207.200$

$_{82}Pb^{3-} \to {}_{82}Pb^{4-}$:

$-14.250 \times 83.9761/207.200 - 19.164 \times 86/207.200 + 13.765 \times 85/207.200$

$_{82}Pb^{4-} \to {}_{82}Pb^{5-}$:

$-14.250 \times 84.4630/207.200 - (-4.704) \times 87/207.200 + 33.412 \times 86/207.200$

Evaluate Formulae:

$_{82}Pb \to {}_{82}Pb^{+}$: $5.6395 + 2.1553 - 0.8854 = 6.9094$

$_{82}Pb^{+} \to {}_{82}Pb^{2+}$: $5.6050 + 0.8854 - 4.6981 = 1.7923$

$_{82}Pb^{2+} \to {}_{82}Pb^{3+}$: $5.5703 + 4.6981 - 4.1582 = 6.1102$

$_{82}Pb^{3+} \to {}_{82}Pb^{4+}$: $5.5354 + 4.1582 - 3.6534 = 6.0402$

$_{82}Pb^{4+} \to {}_{82}Pb^{5+}$: $5.5002 + 9.0178 - 3.1811 = 11.3369$

$_{82}Pb \to {}_{82}Pb^{-}$: $-5.6738 - 3.7014 + 2.1553 = -7.2199$

$_{82}Pb^{-} \to {}_{82}Pb^{2-}$: $-5.7078 - 3.7459 + 3.7014 = -5.7523$

$_{82}Pb^{2-} \to {}_{82}Pb^{3-}$: $-5.7417 - 5.6468 + 3.7459 = -7.6426$

$_{82}Pb^{3-} \to {}_{82}Pb^{4-}$: $-5.7754 - 7.9542 + 5.6468 = -8.0828$

$_{82}Pb^{4-} \to {}_{82}Pb^{5-}$: $-5.8089 + 1.9751 + 13.8679 = 10.0341$

83. Bismuth

Write Formulae:

$_{83}Bi \rightarrow {_{83}Bi^+}$:

$IP_{1,1} \times 83 / M_{83} + \Delta IP_{1,83} \times 83 / M_{83} - \Delta IP_{1,82} \times 82 / M_{83}$

$_{83}Bi^+ \rightarrow {_{83}Bi^{2+}}$:

$IP_{1,1} \times \sqrt{83 \times 82} / M_{83} + \Delta IP_{1,82} \times 82 / M_{83} - \Delta IP_{1,81} \times 81 / M_{83}$

$_{83}Bi^{2+} \rightarrow {_{83}Bi^{3+}}$:

$IP_{1,1} \times \sqrt{83 \times 81} / M_{83} + \Delta IP_{1,81} \times 81 / M_{83} - \Delta IP_{1,80} \times 80 / M_{83}$

$_{83}Bi^{3+} \rightarrow {_{83}Bi^{4+}}$:

$IP_{1,1} \times \sqrt{83 \times 80} / M_{83} + \Delta IP_{1,80} \times 80 / M_{83} - \Delta IP_{1,79} \times 79 / M_{83}$

$_{83}Bi \rightarrow {_{83}Bi^-}$:

$-IP_{1,1} \times \sqrt{83 \times 84} / M_{83} - \Delta IP_{1,84} \times 84 / M_{83} + \Delta IP_{1,83} \times 83 / M_{83}$

$_{83}Bi^- \rightarrow {_{83}Bi^{2-}}$:

$-IP_{1,1} \times \sqrt{83 \times 85} / M_{83} - \Delta IP_{1,85} \times 85 / M_{83} + \Delta IP_{1,84} \times 84 / M_{83}$

$_{83}Bi^{2-} \rightarrow {_{83}Bi^{3-}}$:

$-IP_{1,1} \times \sqrt{83 \times 86} / M_{83} - \Delta IP_{1,86} \times 86 / M_{83} + \Delta IP_{1,85} \times 85 / M_{83}$

$_{83}Bi^{3-} \rightarrow {_{83}Bi^{4-}}$:

$-IP_{1,1} \times \sqrt{83 \times 87} / M_{83} - \Delta IP_{1,87} \times 87 / M_{83} + \Delta IP_{1,86} \times 86 / M_{83}$

Insert Data:

$_{83}Bi \to {_{83}Bi^+}$:

$14.250 \times 83/208.980 + 9.240 \times 83/208.980 - 5.446 \times 82/208.980$

$_{83}Bi^+ \to {_{83}Bi^{2+}}$:

$14.250 \times 82.4985/208.980 + 5.446 \times 82/208.980 - 2.265 \times 81/208.980$

$_{83}Bi^{2+} \to {_{83}Bi^{3+}}$:

$14.250 \times 81.9939/208.980 + 2.265 \times 81/208.980 - 12.168 \times 80/208.980$

$_{83}Bi^{3+} \to {_{83}Bi^{4+}}$:

$14.250 \times 81.4862/208.980 + 12.168 \times 80/208.980 - 10.906 \times 79/208.980$

$_{83}Bi \to {_{83}Bi^-}$:

$-14.250 \times 83.4985/208.980 - 9.240 \times 84/208.980 + 9.240 \times 83/208.980$

$_{83}Bi^- \to {_{83}Bi^{2-}}$:

$-14.250 \times 83.9940/208.980 - 13.765 \times 85/208.980 + 9.240 \times 84/208.980$

$_{83}Bi^{2-} \to {_{83}Bi^{3-}}$:

$-14.250 \times 84.4867/208.980 - 19.164 \times 86/208.980 + 13.765 \times 85/208.980$

$_{83}Bi^{3-} \to {_{83}Bi^{4-}}$:

$-14.250 \times 84.9765/208.980 + 4.704 \times 87/208.980 + 19.164 \times 86/208.980$

Evaluate Formulae:

$_{83}Bi \to {_{83}Bi^+}$: $5.6596 + 3.6698 - 2.1369 = 7.1925$

$_{83}Bi^+ \to {_{83}Bi^{2+}}$: $5.6254 + 2.1369 - 0.8779 = 6.8844$

$_{83}Bi^{2+} \to {_{83}Bi^{3+}}$: $5.5910 + 0.8779 - 4.6581 = 1.8108$

$_{83}\text{Bi}^{3+} \to {}_{83}\text{Bi}^{4+}$: $5.5564 + 4.6581 - 4.1228 = 6.0917$

$_{83}\text{Bi} \to {}_{83}\text{Bi}^{-}$: $-5.6936 - 3.7140 + 3.6698 = -5.7378$

$_{83}\text{Bi}^{-} \to {}_{83}\text{Bi}^{2-}$: $-5.7274 - 5.5987 + 3.7140 = -7.6121$

$_{83}\text{Bi}^{2-} \to {}_{83}\text{Bi}^{3-}$: $-5.7610 - 7.8864 + 5.5987 = -8.0487$

$_{83}\text{Bi}^{3-} \to {}_{83}\text{Bi}^{4-}$: $-5.7944 + 1.9583 + 7.8864 = 4.0503$

84. Polonium

Write Formulae:

$_{84}\text{Po} \to {}_{84}\text{Po}^{+}$:
$$IP_{1,1} \times 84/M_{84} + \Delta IP_{1,84} \times 84/M_{84} - \Delta IP_{1,83} \times 83/M_{84}$$

$_{84}\text{Po}^{+} \to {}_{84}\text{Po}^{2+}$:
$$IP_{1,1} \times \sqrt{84 \times 83}/M_{84} + \Delta IP_{1,83} \times 83/M_{84} - \Delta IP_{1,82} \times 82/M_{84}$$

$_{84}\text{Po}^{2+} \to {}_{84}\text{Po}^{3+}$:
$$IP_{1,1} \times \sqrt{84 \times 82}/M_{84} + \Delta IP_{1,82} \times 82/M_{84} - \Delta IP_{1,81} \times 81/M_{84}$$

$_{84}\text{Po}^{3+} \to {}_{84}\text{Po}^{4+}$:
$$IP_{1,1} \times \sqrt{84 \times 81}/M_{84} + \Delta IP_{1,81} \times 81/M_{84} - \Delta IP_{1,80} \times 80/M_{84}$$

$_{84}\text{Po}^{4+} \to {}_{84}\text{Po}^{5+}$:
$$IP_{1,1} \times \sqrt{84 \times 80}/M_{84} + \Delta IP_{1,80} \times 80/M_{84} - \Delta IP_{1,79} \times 79/M_{84}$$

$_{84}\text{Po}^{5+} \to {}_{84}\text{Po}^{6+}$:

$$IP_{1,1} \times \sqrt{84 \times 79}\big/M_{84} + \Delta IP_{1,79} \times 79/M_{84} - \Delta IP_{1,78} \times 78/M_{84}$$

$_{84}\text{Po}^{6+} \to {}_{84}\text{Po}^{7+}$:

$$IP_{1,1} \times \sqrt{84 \times 78}\big/M_{84} + \Delta IP_{1,78} \times 78/M_{84} - \Delta IP_{1,77} \times 77/M_{84}$$

$_{84}\text{Po} \to {}_{84}\text{Po}^{-}$:

$$-IP_{1,1} \times \sqrt{84 \times 85}\big/M_{84} - \Delta IP_{1,85} \times 85/M_{84} + \Delta IP_{1,84} \times 84/M_{84}$$

$_{84}\text{Po}^{-} \to {}_{84}\text{Po}^{2-}$:

$$-IP_{1,1} \times \sqrt{84 \times 86}\big/M_{84} - \Delta IP_{1,86} \times 86/M_{84} + \Delta IP_{1,85} \times 85/M_{84}$$

$_{84}\text{Po}^{2-} \to {}_{84}\text{Po}^{3-}$:

$$-IP_{1,1} \times \sqrt{84 \times 87}\big/M_{84} - \Delta IP_{1,87} \times 87/M_{84} + \Delta IP_{1,86} \times 86/M_{84}$$

$_{84}\text{Po}^{3-} \to {}_{84}\text{Po}^{4-}$:

$$-IP_{1,1} \times \sqrt{84 \times 88}\big/M_{84} - \Delta IP_{1,88} \times 88/M_{84} + \Delta IP_{1,87} \times 87/M_{84}$$

$_{84}\text{Po}^{4-} \to {}_{84}\text{Po}^{5-}$:

$$-IP_{1,1} \times \sqrt{84 \times 89}\big/M_{84} - \Delta IP_{1,89} \times 89/M_{84} + \Delta IP_{1,88} \times 88/M_{84}$$

$_{84}\text{Po}^{5-} \to {}_{84}\text{Po}^{6-}$:

$$-IP_{1,1} \times \sqrt{84 \times 90}\big/M_{84} - \Delta IP_{1,90} \times 90/M_{84} + \Delta IP_{1,89} \times 89/M_{84}$$

$_{84}\text{Po}^{6-} \to {}_{84}\text{Po}^{7-}$:

$$-IP_{1,1} \times \sqrt{84 \times 91}\big/M_{84} - \Delta IP_{1,91} \times 91/M_{84} + \Delta IP_{1,90} \times 90/M_{84}$$

Insert Data:

$_{84}Po \rightarrow {}_{84}Po^+$:
$14.250 \times 84 / 209.000 + 9.240 \times 84 / 209.000 - 9.240 \times 83 / 209.000$

$_{84}Po^+ \rightarrow {}_{84}Po^{2+}$:
$14.250 \times 83.4985/209.000 + 9.240 \times 83 / 209.000 - 5.446 \times 82 / 209.000$

$_{84}Po^{2+} \rightarrow {}_{84}Po^{3+}$:
$14.250 \times 82.9940/209.000 + 5.446 \times 82 / 209.000 - 2.265 \times 81 / 209.000$

$_{84}Po^{3+} \rightarrow {}_{84}Po^{4+}$:
$14.250 \times 82.4864/209.000 + 2.265 \times 81 / 209.000 - 12.168 \times 80 / 209.000$

$_{84}Po^{4+} \rightarrow {}_{84}Po^{5+}$:
$14.250 \times 81.9756/209.000 + 12.168 \times 80 / 209.000 - 10.906 \times 79 / 209.000$

$_{84}Po^{5+} \rightarrow {}_{84}Po^{6+}$:
$14.250 \times 81.4616/209.000 + 10.906 \times 79 / 209.000 - 9.705 \times 78 / 209.000$

$_{84}Po^{6+} \rightarrow {}_{84}Po^{7+}$:
$14.250 \times 80.9444/209.000 + 9.705 \times 78 / 209.000 - 8.560 \times 77 / 209.000$

$_{84}Po \rightarrow {}_{84}Po^-$:
$-14.250 \times 84.4985/209.000 - 13.765 \times 85 / 209.000 + 9.240 \times 84 / 209.000$

$_{84}Po^- \rightarrow {}_{84}Po^{2-}$:
$-14.250 \times 84.9941/209.000 - 19.164 \times 86 / 209.000 + 13.765 \times 85 / 209.000$

$_{84}Po^{2-} \rightarrow {}_{84}Po^{3-}$:
$-14.250 \times 85.4868/209.000 - (-4.704) \times 87 / 209.000 + 19.164 \times 86 / 209.000$

$_{84}Po^{3-} \to {}_{84}Po^{4-}$:

$-14.250 \times 85.9767/209.000 - (-1.193) \times 88/209.000 + (-4.704) \times 87/209.000$

$_{84}Po^{4-} \to {}_{84}Po^{5-}$:

$-14.250 \times 86.4639/209.000 - (-1.857) \times 89/209.000 + (-1.193) \times 88/209.000$

$_{84}Po^{5-} \to {}_{84}Po^{6-}$:

$-14.25 \times 86.9483/209.000 - (-1.667) \times 90/209.000 + (-.857) \times 89/209.000$

Evaluate Formulae:

$_{84}Po \to {}_{84}Po^{+}$: $5.7273 + 3.7137 - 3.6695 = 5.7715$

$_{84}Po^{+} \to {}_{84}Po^{2+}$: $5.6931 + 3.669 - 2.1367 = 7.2254$

$_{84}Po^{2+} \to {}_{84}Po^{3+}$: $5.6587 + 2.1367 - 0.8778 = 6.9176$

$_{84}Po^{3+} \to {}_{84}Po^{4+}$: $5.6241 + 0.8778 - 4.6576 = 1.8443$

$_{84}Po^{4+} \to {}_{84}Po^{5+}$: $5.5892 + 4.6576 - 4.1224 = 6.1244$

$_{84}Po^{5+} \to {}_{84}Po^{6+}$: $5.5542 + 4.1224 - 3.6220 = 6.0546$

$_{84}Po^{6+} \to {}_{84}Po^{7+}$: $5.5189 + 3.6220 - 3.1537 = 5.9872$

$_{84}Po \to {}_{84}Po^{-}$: $-5.7613 - 5.5982 + 3.7137 = -7.6458$

$_{84}Po^{-} \to {}_{84}Po^{2-}$: $-5.7951 - 7.8857 + 5.5982 = -8.0826$

$_{84}Po^{2-} \to {}_{84}Po^{3-}$: $-5.8286 + 1.9581 + 7.8857 = 4.0152$

$_{84}Po^{3-} \to {}_{84}Po^{4-}$: $-5.8620 + 0.5023 - 1.9581 = -7.3178$

$_{84}Po^{4-} \rightarrow {}_{84}Po^{5-}$: $-5.8952 + 0.7908 - 0.5023 = -5.6067$

86. Radon

Write Formulae:

$_{86}Rn \rightarrow {}_{86}Rn^+$:
$IP_{1,1} \times 86 / M_{86} + \Delta IP_{1,86} \times 86 / M_{86} - \Delta IP_{1,85} \times 85 / M_{86}$

$_{86}Rn^+ \rightarrow {}_{86}Rn^{2+}$:
$IP_{1,1} \times \sqrt{86 \times 85} / M_{86} + \Delta IP_{1,85} \times 85 / M_{86} - \Delta IP_{1,84} \times 84 / M_{86}$

$_{86}Rn^{2+} \rightarrow {}_{86}Rn^{3+}$:
$IP_{1,1} \times \sqrt{86 \times 84} / M_{86} + \Delta IP_{1,84} \times 84 / M_{86} - \Delta IP_{1,83} \times 83 / M_{86}$

$_{86}Rn^{3+} \rightarrow {}_{86}Rn^{4+}$:
$IP_{1,1} \times \sqrt{86 \times 83} / M_{86} + \Delta IP_{1,83} \times 83 / M_{86} - \Delta IP_{1,82} \times 82 / M_{86}$

$_{86}Rn \rightarrow {}_{86}Rn^-$:
$-IP_{1,1} \times \sqrt{86 \times 87} / M_{18} - \Delta IP_{1,87} \times 87 / M_{86} + \Delta IP_{1,86} \times 86 / M_{86}$

$_{86}Rn^- \rightarrow {}_{86}Rn^{--}$:
$-IP_{1,1} \times \sqrt{86 \times 88} / M_{86} - \Delta IP_{1,88} \times 88 / M_{86} + \Delta IP_{1,87} \times 87 / M_{86}$

$_{86}Rn^{--} \rightarrow {}_{86}Rn^{3-}$:
$-IP_{1,1} \times \sqrt{86 \times 89} / M_{86} - \Delta IP_{1,89} \times 89 / M_{86} + \Delta IP_{1,88} \times 88 / M_{86}$

$_{86}Rn^{3-} \rightarrow {}_{86}Rn^{4-}$:
$-IP_{1,1} \times \sqrt{86 \times 90} / M_{86} - \Delta IP_{1,90} \times 90 / M_{86} + \Delta IP_{1,89} \times 89 / M_{86}$

Insert Data:

$_{86}Rn \rightarrow {}_{86}Rn^+$:
$14.250 \times 86 / 222.000 + 19.164 \times 86 / 222.000 - 13.765 \times 85 / 222.000$

$_{86}Rn^+ \rightarrow {}_{86}Rn^{2+}$:
$14.250 \times 85.4985/222.000 + 13.765 \times 85 / 222.000 - 9.240 \times 84 / 222.000$

$_{86}Rn^{2+} \rightarrow {}_{86}Rn^{3+}$:
$14.250 \times 84.9941/222.000 + 9.240 \times 84 / 222.000 - 9.240 \times 83 / 222.000$

$_{86}Rn^{3+} \rightarrow {}_{86}Rn^{4+}$:
$14.250 \times 84.4867/222.000 + 9.240 \times 83 / 222.000 - 5.446 \times 82 / 222.000$

$_{86}Rn \rightarrow {}_{86}Rn^-$:
$-14.250 \times 86.4986/222.000 - (-4.704) \times 87 / 222.000 + 19.164 \times 86 / 222.000$

$_{86}Rn^- \rightarrow {}_{86}Rn^{2-}$:
$-14.250 \times 86.9943/222.000 - (-1.193) \times 88 / 222.000 + (-4.704) \times 87 / 222.000$

$_{86}Rn^{2-} \rightarrow {}_{86}Rn^{3-}$:
$-14.250 \times 87.4871/222.000 - (-1.857) \times 89 / 222.000 + (-1.193) \times 88 / 222.000$

$_{86}Rn^{3-} \rightarrow {}_{86}Rn^{4-}$:
$-14.250 \times 87.9773/222.000 - (-1.667) \times 90 / 222.000 + (-1.857) \times 89 / 222.000$

Evaluate Formulae:

$_{86}Rn \rightarrow {}_{86}Rn^+$: $5.5203 + 7.4239 - 5.2704 = 7.6738$

$_{86}Rn^+ \rightarrow {}_{86}Rn^{2+}$: $5.4881 + 5.2704 - 3.4962 = 7.2623$

$_{86}Rn^{2+} \rightarrow {}_{86}Rn^{3+}$: $5.4557 + 3.4962 - 3.4546 = 5.4973$

$_{86}Rn^{3+} \rightarrow {}_{86}Rn^{4+}$: $5.4231 + 3.4546 - 2.0116 = 6.8661$

$_{86}Rn \rightarrow {}_{86}Rn^-$: $-5.5523 + 1.8435 + 7.4239 = 3.7151$

$_{86}Rn^- \rightarrow {}_{86}Rn^{2-}$: $-5.5841 + 0.4729 - 1.8435 = -6.9547$

$_{86}Rn^{2-} \rightarrow {}_{86}Rn^{3-}$: $-5.6157 + 0.7445 - 0.4729 = -5.3441$

$_{86}Rn^{3-} \rightarrow {}_{86}Rn^{4-}$: $-5.6472 + 0.6758 - 0.7445 = -5.7159$

87. Francium

Write Formulae:

$_{87}Fr \rightarrow {}_{87}Fr^+$:
$$IP_{1,1} \times 87 / M_{87} + \Delta IP_{1,87} \times 87 / M_{87} - \Delta IP_{1,86} \times 86 / M_{87}$$

$_{87}Fr^+ \rightarrow {}_{87}Fr^{2+}$:
$$IP_{1,1} \times \sqrt{87 \times 86}/M_{87} + \Delta IP_{1,86} \times 86 / M_{87} - \Delta IP_{1,85} \times 85 / M_{87}$$

$_{87}Fr^{2+} \rightarrow {}_{87}Fr^{3+}$:
$$IP_{1,1} \times \sqrt{87 \times 85}/M_{87} + \Delta IP_{1,85} \times 85 / M_{87} - \Delta IP_{1,84} \times 84 / M_{87}$$

$_{87}Fr^{3+} \rightarrow {}_{87}Fr^{4+}$:
$$IP_{1,1} \times \sqrt{87 \times 84}/M_{87} + \Delta IP_{1,84} \times 84 / M_{87} - \Delta IP_{1,83} \times 83 / M_{87}$$

$_{87}Fr \rightarrow {}_{87}Fr^-$:
$$-IP_{1,1} \times \sqrt{87 \times 88}/M_{87} - \Delta IP_{1,88} \times 88 / M_{87} + \Delta IP_{1,87} \times 87 / M_{87}$$

$_{87}\text{Fr}^- \to {}_{87}\text{Fr}^{2-}$:

$-IP_{1,1} \times \sqrt{87 \times 89}/M_{87} - \Delta IP_{1,89} \times 89 / M_{87} + \Delta IP_{1,88} \times 88 / M_{87}$

$_{87}\text{Fr}^{2-} \to {}_{87}\text{Fr}^{3-}$:

$-IP_{1,1} \times \sqrt{87 \times 90}/M_{87} - \Delta IP_{1,90} \times 90 / M_{87} + \Delta IP_{1,89} \times 89 / M_{87}$

$_{87}\text{Fr}^{3-} \to {}_{87}\text{Fr}^{4-}$:

$-IP_{1,1} \times \sqrt{87 \times 91}/M_{87} - \Delta IP_{1,91} \times 91 / M_{87} + \Delta IP_{1,90} \times 90 / M_{87}$

Insert Data:

$_{87}\text{Fr} \to {}_{87}\text{Fr}^+$:
$14.250 \times 87 / 223.000 + (-4.704) \times 87 / 223.000 - 19.164 \times 86 / 223.000$

$_{87}\text{Fr}^+ \to {}_{87}\text{Fr}^{2+}$:
$14.250 \times 86.4986/223.000 + 19.164 \times 86 / 223.000 - 13.765 \times 85 / 223.000$

$_{87}\text{Fr}^{2+} \to {}_{87}\text{Fr}^{3+}$:
$14.250 \times 85.9942/223.000 + 13.765 \times 85 / 223.000 - 9.240 \times 84 / 223.00$

$_{87}\text{Fr}^{3+} \to {}_{87}\text{Fr}^{4+}$:
$14.259 \times 85.4868/223.000 + 9.240 \times 84 / 223.000 - 9.240 \times 83 / 223.000$

$_{87}\text{Fr} \to {}_{87}\text{Fr}^-$:
$-14.250 \times 87.4986/223.000 - (-1.193) \times 88 / 223.000 + (-4.704) \times 87 / 223.000$

$_{87}\text{Fr}^- \to {}_{87}\text{Fr}^{2-}$:
$-14.250 \times 87.9943/223.000 - (-1.857) \times 89 / 223.000 + (-1.193) \times 88 / 223.000$

$_{87}\text{Fr}^{2-} \to {}_{87}\text{Fr}^{3-}$:
$-14.250 \times 88.4873/223.000 - (-1.667) \times 90 / 223.000 + (-1.857) \times 89 / 223.000$

$_{87}Fr^{3-} \rightarrow {}_{87}Fr^{4-}$:

$-14.250 \times 88.9775/223.000 - (-1.474) \times 91/223.000 + (-1.667) \times 90/223.000$

Evaluate Formulae:

$_{87}Fr \rightarrow {}_{87}Fr^{+}$: $5.5594 - 1.8352 - 7.3906 = -3.666$

$_{87}Fr^{+} \rightarrow {}_{87}Fr^{2+}$: $5.5274 + 7.3906 - 5.2467 = 7.6713$

$_{87}Fr^{2+} \rightarrow {}_{87}Fr^{3+}$: $5.4951 + 5.2467 - 3.4805 = 7.2613$

$_{87}Fr^{3+} \rightarrow {}_{87}Fr^{4+}$: $5.4798 + 3.4805 - 3.4391 = 5.5212$

$_{87}Fr \rightarrow {}_{87}Fr^{-}$: $-5.5913 + 0.4708 - 1.8352 = -6.9557$

$_{87}Fr^{-} \rightarrow {}_{87}Fr^{2-}$: $-5.6230 + 0.7411 - 0.4708 = -5.3527$

$_{87}Fr^{2-} \rightarrow {}_{87}Fr^{3-}$: $-5.6545 + 0.6728 - 0.7411 = -5.7228$

$_{87}Fr^{3-} \rightarrow {}_{87}Fr^{4-}$: $-5.6858 + 0.6015 - 0.6728 = -5.7571$

89. Actinium

Write Formulae:

$_{89}Ac \rightarrow {}_{89}Ac^{+}$:
$IP_{1,1} \times 89/M_{89} + \Delta IP_{1,89} \times 89/M_{89} - \Delta IP_{1,88} \times 88/M_{89}$

$_{89}Ac^{+} \rightarrow {}_{89}Ac^{2+}$:
$IP_{1,1} \times \sqrt{89 \times 88}/M_{89} + \Delta IP_{1,88} \times 88/M_{89} - \Delta IP_{1,87} \times 87/M_{89}$

$_{89}Ac \to {}_{89}Ac^-$:

$-IP_{1,1} \times \sqrt{89 \times 90}/M_{89} - \Delta IP_{1,90} \times 90/M_{89} + \Delta IP_{1,89} \times 89/M_{89}$

$_{89}Ac^- \to {}_{89}Ac^{2-}$:

$-IP_{1,1} \times \sqrt{89 \times 91}/M_{89} - \Delta IP_{1,91} \times 91/M_{89} + \Delta IP_{1,90} \times 90/M_{89}$

Insert Data:

$_{89}Ac \to {}_{89}Ac^+$:
$14.250 \times 89/227.000 + (-1.857) \times 89/227.000 - (-1.193) \times 88/227.000$

$_{89}Ac^+ \to {}_{89}Ac^{2+}$:
$14.250 \times 88.4986/227.000 + (-1.193) \times 88/227.000 - (-4.704) \times 87/227.000$

$_{89}Ac \to {}_{89}Ac^-$:
$-14.250 \times 89.4986/227.000 - (-1.667) \times 90/227.000 + (-1.857) \times 89/227.000$

$_{89}Ac^- \to {}_{89}Ac^{2-}$:
$-14.250 \times 89.9944/227.000 - (-1.474) \times 91/227.000 + (-1.667) \times 90/227.000$

Evaluate Formulae:

$_{89}Ac \to {}_{89}Ac^+$: $5.5870 - 0.7281 + 0.4625 = 5.3214$

$_{89}Ac^+ \to {}_{89}Ac^{2+}$: $5.5555 - 0.4625 + 1.8029 = 6.8959$

$_{89}Ac \to {}_{89}Ac^-$: $-5.6183 + 0.6609 - 0.7281 = -5.6855$

$_{89}Ac^- \to {}_{89}Ac^{2-}$: $-5.6494 + 0.5909 - 0.6609 = -5.7194$

Chapter 3

IONS AND STATES OF MATTER

ABSTRACT

This Chapter investigates the relationships between the macro states of matter – solid, liquid, gas, and plasma – and the micro states of ionization – neutral, singly ionized, doubly ionized, and so on. Readily available data show that boiling points and melting points follow a pattern, related to the pattern that first-order ionization potentials follow. Such patterning suggests that observed boiling and melting are related to hidden changes in ion populations. That would mean that the macroscopic states of matter are related to the microscopic states of atoms. This Chapter poses a hypothesis about the relationship, and investigates the hypothesis in a qualitative way. The story involves the Planck energy distribution for black body radiation as a background for transitions between ionization states. The background is a source for photons of energy appropriate to provoke transitions of ionization state, and it is a dumping ground for waste heat from spontaneous transitions of ionization state. The state changes themselves often have a cascade character: the background supports the first change, and then some subsequent changes occur spontaneously. That scenario can make macroscopic state changes look as abrupt as they do.

INTRODUCTION

This story goes far back to antiquity. Aristotle identified his "elements" as Earth, Air, Fire, and Water. Seen in retrospect, that is an amazingly good categorization, not of "elements" as we use the word today, but of macroscopic states of matter. The modern view has those same four macro states, but reordered as Earth, Water, Air, and Fire, and renamed as Solid, Liquid, Gas, and Plasma. But there are now also lots of sub-categories that are specifically acknowledged. A Solid can be Conductor, a Super Conductor, a Semi Conductor, an Insulator… A Liquid can be a Fluid, a Super Fluid, a solvent, an Oil, a Plastic, a Pyroclastic, a Glass, a Conductor or an Insulator, a Solution or a Suspension,… A Gas can be Inert, Explosive,… A Plasma can be Hot, Cold,…

This Chapter is mainly about the transitions between the four macroscopic states of matter. Starting from the solid state, we have melting to the liquid state; in reverse, we have freezing from the liquid state to the solid state. Starting from the liquid state, we have boiling to the gas state; in reverse we have condensation from the gas state to the liquid state. Starting from the gas state, we have excitation to the plasma state; in reverse, we have relaxation from the plasma state to the gas state.

1. State Change Temperatures and First Order Ionization Potentials

It is reasonable to consider the possibility that macroscopic states of matter have something to do with ionization states, since one of them, namely the plasma state, has absolutely everything to do with ionization states.

What is the key similarity between macroscopic states of matter and ionization states? It is energy. In the case of macroscopic states of matter, the difference between one state and another is thermal energy, and, possibly, some pressure × volume energy. In the case of ionization states, the difference between one state and another is electromagnetic energy, and, possibly, some thermal energy. To accomplish a change of ionization state, electromagnetic work may be invested, and heat may be dumped. The final energy of an ionization state is the cumulative sum of work increments invested and heat increments dumped in getting from the neutral state to the ionized state.

Data about macroscopic state changes are usually provided as point values of temperatures: melting points and boiling points. (As for excitation to the plasma state, that seems more difficult to document with temperature data.) Figure 1 shows reported melting points and boiling points, in comparison to first-order ionization potentials. The horizontal axis is the nuclear charge Z of the elements. The vertical axis is electron volts (eV) for ionization potentials, and degrees Kelvin (K) for melting points and freezing points.

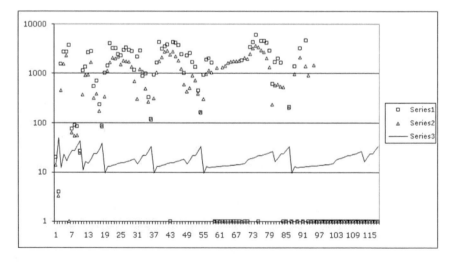

Figure 1. Correlation between state-change temperatures and ionization potentials.

As in Chapter 1, the vertical scale is logarithmic to accommodate a large dynamic range. Starting at the top, the squares are the boiling points, and then the triangles are the melting points. The line at the bottom reproduces the first order IP's, or M/Z-scaled ionization potentials, from the Appendix to Chapter 1.

The dynamic range of boiling temperatures and melting temperatures is huge compared to the dynamic range of IP's: we need four decades for temperatures, *vs.* one decade for IP's. The temperature range for the liquid state varies from nearly a factor of 5.5 (*i.e.* very substantial liquid range) down to nearly a factor of 1 (*i.e.*, no liquid range at all).

Despite the huge difference in dynamic range, the temperature data are clearly correlated with the IP data. The correlations are rough, but very compelling. What is actually going on here? To investigate this question, we require a hypothesis to test. My candidate hypothesis is that:

- The solid state involves ions and/or radicals existing in pairs that, together, are in negative energy states overall;
- The liquid state involves neutral atoms and/or molecules, along with some ions existing in pairs that are mostly in negative energy states;
- The gas state involves neutral atoms and/or molecules, along with some ions existing in pairs that are mostly in positive energy states;
- The plasma state is composed significantly of ions existing in pairs that are in positive energy states.

2. RELATIONSHIPS BETWEEN SOME ELEMENT PAIRS

To begin to explore the hypothesis, this Section explores some specific relationships between some pairs of elements using the data from Chapter 2 to It offers a few interesting observations about the temperature ranges over which pairs of neighboring elements are in the liquid state. The range of examples is large, and it is sometimes appropriate to rename the melting points and boiling points using other words. For example, substances that are solids at temperatures comfortable to humans melt, and then boil, whereas substances that are liquids at temperatures comfortable to humans freeze or boil, while substances that are gasses at temperatures comfortable to humans condense, and then freeze. As for the plasma state, it does not usually exist at temperatures comfortable to humans.

2.1. Hydrogen and Helium

Hydrogen and Helium are gasses at temperatures comfortable to humans. Compare the following transitions:

$$2\,_1\text{H} \rightarrow \,_1\text{H}^+ + \,_1\text{H}^- : 14.1369 - 90.6769 = -76.540\,\text{eV},$$

very easy.

$$2\,_2\text{He} \rightarrow \,_2\text{He}^+ + \,_2\text{He}^- : 24.9189 + 10.4141 = 35.333\,\text{eV},$$

very difficult.

So Hydrogen gas should condense to liquid at a temperature much higher than the temperature at which Helium gas condenses to liquid. And what do we find? Hydrogen boils (or condenses to liquid) at 20.23 K, whereas Helium boils (or condenses to liquid) at 4.056 K.

What about freezing? The $_1\text{H}^+$ ion cannot become more ionized than it already is. By contrast, $_1\text{H}^-$ can become more ionized, if it can just find a donor of another electron. So we can look at the transition

$$3\,_1\text{H} \rightarrow 2\,_1\text{H}^+ + \,_1\text{H}^{2-} : 2 \times 14.1369 + 51.4986 = 79.7724\,\text{eV},$$

very difficult.

Helium can readily become more ionized. We can look at the energy increment to get from

$$_2\text{He}^+ + \,_2\text{He}^- \text{ to } \,_2\text{He}^{2+} + \,_2\text{He}^{2-} :$$

$$5.0343 - 20.4735 = -15.4392\,\text{eV},$$

very easy.

So one would expect it to be difficult to freeze Hydrogen, but easy to freeze Helium. What actually happens? Hydrogen freezes at 13.8 K, whereas Helium freezes at 3.3 K. The temperature reduction factor over which Hydrogen remains liquid is $13.8 / 20.23 = 0.68$, whereas the temperature reduction factor over which Helium remains liquid is $3.3 / 4.056 = 0.81$. The temperature reduction factor to get to freezing is indeed more significant for Hydrogen.

2.2. Lithium and Beryllium

Lithium and Beryllium are solids at temperatures comfortable to humans. Compare the following transitions:

$2\,_3\text{Li} \to \,_3\text{Li}^+ + \,_3\text{Li}^-$: $-4.8758 - 13.1124 = -17.9882$ eV,

easy;

$_3\text{Li}^+ + \,_3\text{Li}^- \to \,_3\text{Li}^{2+} + \,_3\text{Li}^{2-}$: $15.2929 - 4.7408 = 10.5521$ eV,

difficult.

$2\,_4\text{Be} \to \,_4\text{Be}^+ + \,_4\text{Be}^-$: $10.9466 - 4.5983 = 6.3483$ eV,

difficult;

$_4\text{Be}^+ + \,_4\text{Be}^- \to \,_4\text{Be}^{2+} + \,_4\text{Be}^{2-}$:

$$-6.7203 - 10.9844 = -17.7047 \text{ eV},$$

easy.

So one would expect Lithium to remain liquid over a relatively larger temperature range than Beryllium does. What actually happens? Lithium melts at 453 K, and boils at 1599 K, whereas Beryllium melts at 1556 K, and boils at 2773 K. The temperature rise factor over which Lithium remains liquid is $1599 / 453 = 3.5298$, whereas the temperature rise factor over which Beryllium remains liquid is $2773 / 1556 = 1.7821$. The temperature rise factor needed to get from melting to boiling is indeed much larger for Lithium.

2.3. Nitrogen and Oxygen

Nitrogen and Oxygen are gasses at temperatures comfortable to humans.

$2\,_7\text{N} \to \,_7\text{N}^+ + \,_7\text{N}^-$: $10.498 - 8.5432 = 1.9548$ eV,

slightly inconvenient;

$_7\text{N}^+ + \,_7\text{N}^- \to \,_7\text{N}^{2+} + \,_7\text{N}^{2-}$: $8.7277 - 13.646 = -4.9183$ eV,

easy.

$2\,_8\text{O} \to \,_8\text{O}^+ + \,_8\text{O}^-$: $7.9399 - 12.4354 = -4.4955$ eV, easy,

$$_8O^+ + {_8O^-} \rightarrow {_8O^{2+}} + {_8O^{2-}} : 9.6214 - 14.9434 = -5.322 \text{ eV},$$

easy.

One might expect that Nitrogen would be slightly difficult to condense, but not so difficult to freeze, whereas nothing about Oxygen would be difficult. So the temperature rise factor over which Nitrogen is liquid should be less than the temperature rise factor over which Oxygen is liquid.

What actually happens? Nitrogen melts at 62.99 K, and boils at 77.18 K. Oxygen melts at 54.24 K and boils at 90.03 K. So the temperature rise factor for Nitrogen is $77.18 / 62.99 = 1.2253$, while the temperature rise factor for Oxygen is $90.03 / 54.24 = 1.6598$. The temperature rise factor needed to get to the boiling point is indeed less for Nitrogen. It is starting to look as if the correlation between melting and boiling temperatures and the information about IP's adds up to some coherent story. So it is time to become more quantitative about that story.

3. STATES OF MATTER AND STATES OF IONIZATION

Figure 2 offers a conceptual structure for the problem at hand. The horizontal axis represents temperature, ranging from absolute zero to some very high temperature, represented by the '1' at the right end. The vertical axis represents the population fraction of atom pairs in different ionization states. The energy of an ionized state is the cumulative result of work increments required and heat increments dumped in getting from the neutral state to the ionized state. It can be negative, positive, or zero. The three curves represent these three regimes of ionization state energy. The left curve, consistently descending with temperature, represents the fraction of ion pairs that are to be found in negative energy states. The right curve, consistently increasing with temperature, represents the fraction of ion pairs that are to be found in positive energy states.

The middle curve, first increasing with temperature and then decreasing with temperature, represents the fraction of atom pairs to be found in the neutral, un-ionized state.

Figure 2 is quite generic. Its particular realization for a particular substance may be shifted right or left, or have the middle crossing point higher or lower, with the middle bump correspondingly lower or higher. *Much* higher is typical; you know this is true if you think about water: the proportion of H^+ is only about 1×10^{-7}. The curves may also have some non-smooth detail, such as precipitous drops or rises. You could think of Figure 2 as an average over all elements, or even all compounds, which suppresses details associated with particular elements. You can expect that the fundamental character of the picture will be the same for all elements and compounds: a declining curve associated with negative-energy ionization states, a rising curve associated with positive-energy ionization states, and a middle bump curve associated with zero-energy, un-ionized states.

Ions and States of Matter

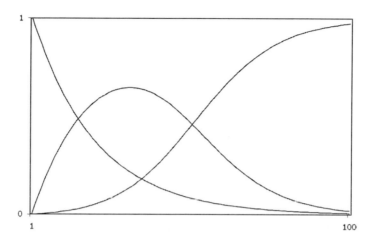

Figure 2. Population fractions of three regimes of ionization state energy. Negative dominates on the right, positive dominates on the left, and neutral dominates in the middle. (First published in [1]).

The three curves exhibit three crossing points, and they thereby define four temperature ranges. This situation invites consideration of the possibility that these four ranges of temperature correlate with the four top-level states of matter: solid, liquid, gas, and plasma.

Why is that a reasonable association? Well, consider the following four questions:

- At one extreme, what does it take to be in the solid state, which consists mostly of ionization states with negative energies? The temperature has to be so low that thermal agitation cannot promote a negative-energy ion pair to the neutral, unionized state, much less to a positive-energy state. That puts the solid state on the left side of the picture.
- At the other extreme, what does it take to be in the plasma state, which consists mostly of ionization states with positive energies? The temperature has to be so high that thermal agitation will prevent a positive-energy ion pair from falling back to the neutral, unionized state, much less to a negative-energy state. That puts the plasma state on the right side of the picture.
- In between, what does it take to be in the liquid state, which consists of neutral atoms, zero-energy ionization states, mixed with mostly negative-energy ionization states? The temperature has to be middling low: low enough to avoid promoting a neutral-energy atom pair into a positive-energy ion state, but high enough to keep a negative-energy ion pair from falling into a more negative-energy state. That puts the liquid state in the center left part of the picture.
- Also in between, what does it take to be in the gas state, which consists of neutral atoms, zero-energy ionization states, mixed with mostly positive-energy ionization states. The temperature has to be middling high: high enough to keep a neutral-energy atom pair falling into a negative-energy state, but low enough to avoid promoting a positive-energy ion pair into a more positive-energy state. That puts the gas state in the center right part of the picture.

Here is a puzzle to think about: can there exist a substance for which, or a circumstance under which, not all four states of matter clearly occur? That is, can the bump curve in the middle of the picture fall upon or below the intersection of the two monotonic curves? I will come back to this question later.

For now, let us try to develop more quantitative relationships between the temperature regimes defined by the curves on Figure 2 and the numerical data about the possible ionization states for the element. For starters, what about the macroscopic state transitions? What can we find out from available data?

3.1. Melting Points

Melting points (and boiling points too) are typically reported in degrees centigrade. Why? It is so because the centigrade scale of temperature is keyed to the human comfort zone of temperatures. The zero point corresponds to the freezing point of water, and the 100-degree point corresponds to the boiling point of water, and we humans are made mainly of water. But that fact is quite irrelevant for the physics of general macroscopic state changes in relation to changes of ionization states. So in this book, all temperatures are quoted in degrees Kelvin. The zero of this temperature scale is absolute zero, the lowest temperature allowed in Nature. Kelvin is a physically meaningful temperature scale. It comes from Thermodynamics, not from human circumstances.

Let us focus first on the solid/liquid transitions, and on the far left column of the Periodic Table: Hydrogen and the alkali metals. The melting points in degrees Kelvin are:

- Hydrogen $_1H$, 13.8 K;
- Lithium $_3Li$, 453 K;
- Sodium $_{11}Na$, 371 K;
- Potassium $_{19}K$, 336.4 K;
- Rubidium $_{37}Rb$, 311.8 K;
- Cesium $_{55}Cs$, K, 301.7 K;
- Francium $_{87}Fr$, ~297 K*.

Hydrogen $_1H$ is clearly not really meant to be in a group with the alkali metals. We will learn a little about why shortly. For now, just set Hydrogen aside. Francium $_{87}Fr$ is also a little different from the other alkali metals, in that its melting point was not actually available in my handbook for me to quote. That melting point given above for $_{87}Fr$ is estimated by extrapolation from the smooth pattern displayed by the other elements. See Figure 3. The horizontal axis refers to the elements: #1 is $_3Li$, #2 is $_{11}Na$, #3 is $_{19}K$, #4 is $_{37}Rb$, #5 is $_{55}Cs$, and #6 is $_{87}Fr$. The vertical axis refers to the melting temperatures, reported for elements #1 – #5, and estimated for element #6, expressed in degrees Kelvin. The scale is logarithmic. The estimate ~297 K for $_{87}$Fr is based on the data for $_{55}$Cs and $_{37}$Rb. It scales down from $_{55}$Cs by a factor that is half as strong as the factor by which $_{55}$Cs scales down from $_{37}$Rb: $301.7 \times \left[(301.7 + 311.8) / (2 \times 311.8) \right]$.

Clearly, there is a regular pattern to the melting points for the alkali metals. The curve is consistently declining and leveling. Maybe there is a corresponding pattern to the energies associated with ionization states. That possibility is investigated next. Thinking about ionizations, we can understand right away why Hydrogen was not like the other elements originally listed, and so had to be set aside; all the other elements allow several positively charged ionization states, whereas Hydrogen allows only one. That condition is severely limiting for Hydrogen.

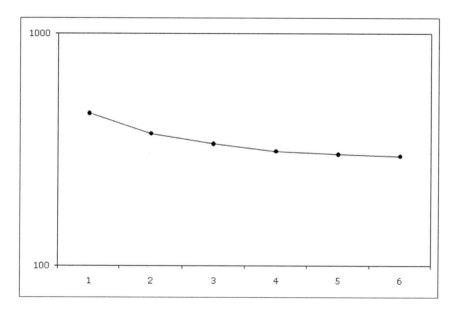

Figure 3. Melting temperatures for alkali metals.

Now what about solid/liquid transitions of the other elements? The information we will use all comes from the Appendix to Chapter 2, and consists of the following:

Lithium $_3$Li :

$2\,_3\text{Li} \rightarrow\, _3\text{Li}^+ +\, _3\text{Li}^-$: $-4.8758 - 13.1124 = -17.9882$ eV

$_3\text{Li}^+ +\, _3\text{Li}^- \rightarrow\, _3\text{Li}^{2+} +\, _3\text{Li}^{2-}$: $15.2929 - 4.7408 = 10.5521$ eV

$_3\text{Li}^{2+} +\, _3\text{Li}^{2-} \rightarrow\, _3\text{Li}^{3+} +\, _3\text{Li}^{3-}$: $3.5560 - 13.0171 = -9.4611$ eV

Sodium $_{11}$Na :

$2\,_{11}\text{Na} \to {}_{11}\text{Na}^+ + {}_{11}\text{Na}^-$: $-7.5643 - 9.9278 = -17.4921$ eV

$_{11}\text{Na}^+ + {}_{11}\text{Na}^- \to {}_{11}\text{Na}^{2+} + {}_{11}\text{Na}^{2-}$: $11.3563 - 6.5844 = 4.7719$ eV

$_{11}\text{Na}^{2+} + {}_{11}\text{Na}^{2-} \to {}_{11}\text{Na}^{3+} + {}_{11}\text{Na}^{3-}$: $9.5617 - 2.0694 = 7.4923$ eV

$_{11}\text{Na}^{3+} + {}_{11}\text{Na}^{3-} \to {}_{11}\text{Na}^{4+} + {}_{11}\text{Na}^{4-}$:
$$6.3813 - 11.4234 = -5.0421 \text{ eV}$$

Potassium $_{19}$K :

$2\,_{19}\text{K} \to {}_{19}\text{K}^+ + {}_{19}\text{K}^-$: $-6.3807 - 8.7804 = -15.1611$ eV

$_{19}\text{K}^+ + {}_{19}\text{K}^- \to {}_{19}\text{K}^{2+} + {}_{19}\text{K}^{2-}$: $10.8280 - 7.2498 = 3.5782$ eV

$_{19}\text{K}^{2+} + {}_{19}\text{K}^{2-} \to {}_{19}\text{K}^{3+} + {}_{19}\text{K}^{3-}$: $9.5448 - 10.0649 = -0.5201$ eV

$_{19}\text{K}^{3+} + {}_{19}\text{K}^{3-} \to {}_{19}\text{K}^{4+} + {}_{19}\text{K}^{4-}$: $6.6008 - 7.9599 = -1.3591$ eV

Rubidium $_{37}$Rb :

$2\,_{37}\text{Rb} \to {}_{37}\text{Rb}^+ + {}_{37}\text{Rb}^-$: $-3.9386 - 7.7578 = -11.6964$ eV

$_{37}\text{Rb}^+ + {}_{37}\text{Rb}^- \to {}_{37}\text{Rb}^{2+} + {}_{37}\text{Rb}^{2-}$:
$$8.8874 - 6.3195 = 2.5679 \text{ eV}$$

$_{37}\text{Rb}^{2+} + {}_{37}\text{Rb}^{2-} \to {}_{37}\text{Rb}^{3+} + {}_{37}\text{Rb}^{3-}$:
$$8.1829 - 6.6722 = 1.5107 \text{ eV}$$

$_{37}\text{Rb}^{3+} + {}_{37}\text{Rb}^{3-} \to {}_{37}\text{Rb}^{4+} + {}_{37}\text{Rb}^{4-}$:
$$3.7306 - 6.7774 = -3.0468 \text{ eV}$$

Cesium $_{55}$Cs:

$$2\,_{55}\text{Cs} \rightarrow \,_{55}\text{Cs}^{+} + \,_{55}\text{Cs}^{-}: \;-3.8352 - 7.3948 = -11.230 \text{ eV}$$

$$_{55}\text{Cs}^{+} + \,_{55}\text{Cs}^{-} \rightarrow \,_{55}\text{Cs}^{2+} + \,_{55}\text{Cs}^{2-}: \; 8.4981 - 5.7092 = 2.7889 \text{ eV}$$

$$_{55}\text{Cs}^{2+} + \,_{55}\text{Cs}^{2-} \rightarrow \,_{55}\text{Cs}^{3+} + \,_{55}\text{Cs}^{3-}:$$
$$7.8846 - 6.1247 = 1.7599 \text{ eV}$$

$$_{55}\text{Cs}^{3+} + \,_{55}\text{Cs}^{3-} \rightarrow \,_{55}\text{Cs}^{4+} + \,_{55}\text{Cs}^{4-}:$$
$$5.9016 - 6.1809 = -0.2793 \text{ eV}$$

Francium $_{87}$Fr:

$$2\,_{87}\text{Fr} \rightarrow \,_{87}\text{Fr}^{+} + \,_{87}\text{Fr}^{-}: \;-3.666 - 6.9557 = -10.6217 \text{ eV}$$

$$_{87}\text{Fr}^{+} + \,_{87}\text{Fr}^{-} \rightarrow \,_{87}\text{Fr}^{2+} + \,_{87}\text{Fr}^{2-}: \; 7.6713 - 5.3527 = 2.3186 \text{ eV}$$

$$_{87}\text{Fr}^{2+} + \,_{87}\text{Fr}^{2-} \rightarrow \,_{87}\text{Fr}^{3+} + \,_{87}\text{Fr}^{3-}: \; 7.2613 - 5.7228 = 1.5385 \text{ eV}$$

$$_{87}\text{Fr}^{3+} + \,_{87}\text{Fr}^{3-} \rightarrow \,_{87}\text{Fr}^{4+} + \,_{87}\text{Fr}^{4-}:$$
$$5.5212 - 5.7571 = -0.2359 \text{ eV}$$

The transitions listed above all move toward greater ionization. The ones that have negative energies associated with them occur spontaneously. All of the first ionization steps occur spontaneously. All of the second ionization steps and some of the third ionization steps have positive energies associated. They require an energy input in order to occur. The fact that such positive requirements do exist means that, surprisingly, some steps in a really deep freezing process can require some energy input. But such energy is in the form of work, not heat.

The transitions required for producing melting of these metals move in the reverse direction, toward lesser ionization. They are the reverse of the transitions listed, and the energies associated have reversed signs. The energies that are negative in freezing become the positive energies that need to be supplied to cause the melting. And the energies that are positive in freezing become negative, and represent energies actually released during a melting process.

It is generally assumed that a 'melting point' is the same thing as a 'freezing point', so that both melting and freezing processes are occurring simultaneously, and they have to be exactly in balance. So energy requirements for both freezing and melting have to be met. So we should look at sums of the absolute values of energy requirements.

Figure 4 shows the cumulative absolute energy requirements for ionization and deionization of alkali metals. The horizontal axis refers to the elements: #1 is $_3$Li, #2 is $_{11}$Na, #3 is $_{19}$K, #4 is $_{37}$Rb, #5 is $_{55}$Cs, and #6 is $_{87}$Fr. The vertical axis is eV. The lowest curve includes just the single ionizations, neutral to singly ionized and singly ionized to neutral. The successively higher curves include successively more ionizations.

Figure 4 shows that the alkali metals exhibit a rough convergence to a consistent decrease in the cumulative energy requirements for melting/freezing, similar to the consistent decrease in melting temperature shown in Figure 3. Why is the correlation rough? In part, because no account has yet been taken of different population numbers in the different ionization states, and no justification has been offered for stopping at just four ionizations. These issues are addressed later.

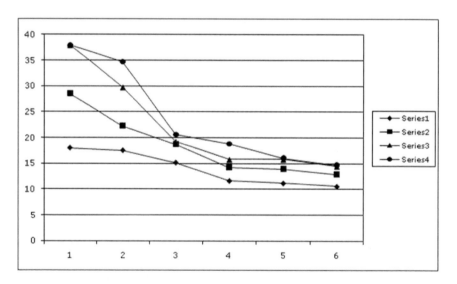

Figure 4. Energy requirements for balanced melting/freezing of alkali metals.

For now, let us investigate whether some such rough correlation exists in another case. Let us investigate another column of elements on the Periodic Table: the noble gasses, on the far right. The melting temperatures are:

- Helium $_2$He, 3.3 K;
- Neon $_{10}$Ne, 24.41 K;
- Argon $_{18}$Ar, 83.8 K;
- Krypton $_{36}$Kr, 115.79 K;
- Xenon $_{54}$Xe, 161,1 K;

- Radon $_{86}Rn$, K; 202 K.

Figure 5 displays these data points. The horizontal axis refers to the elements: # 1 is $_2He$, #2 is $_{10}Ne$, #3 is $_{18}Ar$, #4 is $_{36}Kr$, #5 is $_{54}Xe$, #6 is $_{86}Rn$. The vertical axis is degrees K. As was the case for Figure 3, the scale is logarithmic.

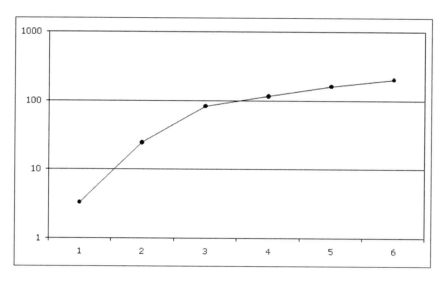

Figure 5. Melting temperatures for the noble gasses.

In stark opposition to Figure 3, the curve on Figure 5 is consistently rising and leveling. Do the ionization data for these elements behave in any correlated way? We can find out by conducting for them the same kind of exercise that was conducted previously for the alkali metals. The relevant information from the Appendix to Chapter 2 tells us:

Helium $_2He$:

$$2\,_2He \rightarrow \,_2He^+ + \,_2He^- : 24.9189 + 10.4141 = 35.3330 \text{ eV}$$

$$_2He^+ + \,_2He^- \rightarrow \,_2He^{2+} + \,_2He^{2-} : 5.0343 - 20.4735 = -15.4392 \text{ eV}$$

Neon $_{10}Ne$:

$$2\,_{10}Ne \rightarrow \,_{10}Ne^+ + \,_{10}Ne^- : 12.5928 + 4.5191 = 17.1119 \text{ eV}$$

$$_{10}Ne^+ + \,_{10}Ne^- \rightarrow \,_{10}Ne^{2+} + \,_{10}Ne^{2-} :$$
$$10.5662 - 10.9327 = -0.3665 \text{ eV}$$

$_{10}Ne^{2+} + {}_{10}Ne^{2-} \rightarrow {}_{10}Ne^{3+} + {}_{10}Ne^{3-}$:

$6.9617 - 7.1082 = -0.1465$ eV

$_{10}Ne^{3+} + {}_{10}Ne^{3-} \rightarrow {}_{10}Ne^{4+} + {}_{10}Ne^{4-}$:

$8.2518 - 11.1525 = -2.9007$ eV

Argon $_{18}Ar$:

$2{}_{18}Ar \rightarrow {}_{18}Ar^+ + {}_{18}Ar^-$: $10.4218 + 6.4257 = 16.8475$ eV

$_{18}Ar^+ + {}_{18}Ar^- \rightarrow {}_{18}Ar^{2+} + {}_{18}Ar^{2-}$: $9.1708 - 8.4082 = 0.7626$ eV

$_{18}Ar^{2+} + {}_{18}Ar^{2-} \rightarrow {}_{18}Ar^{3+} + {}_{18}Ar^{3-}$:

$6.2944 - 6.9353 = -0.6409$ eV

$_{18}Ar^{3+} + {}_{18}Ar^{3-} \rightarrow {}_{18}Ar^{4+} + {}_{18}Ar^{4-}$:

$7.8534 - 7.3886 = 0.4648$ eV

Krypton $_{36}Kr$:

$2{}_{36}Kr \rightarrow {}_{36}Kr^+ + {}_{36}Kr^-$: $13.1566 - 4.1026 = 9.0540$ eV

$_{36}Kr^+ + {}_{36}Kr^- \rightarrow {}_{36}Kr^{2+} + {}_{36}Kr^{2-}$: $8.2625 - 7.8254 = 0.4371$ eV

$_{10}Kr^{2+} + {}_{10}Kr^{2-} \rightarrow {}_{10}Kr^{3+} + {}_{10}Kr^{3-}$:

$6.0417 - 6.3575 = -0.3158$ eV

$_{10}Kr^{3+} + {}_{10}Kr^{3-} \rightarrow {}_{10}Kr^{4+} + {}_{10}Kr^{4-}$: $7.5373 - 6.7160 = 0.8213$ eV

Xenon $_{54}Xe$:

$2{}_{54}Xe \rightarrow {}_{54}Xe^+ + {}_{54}Xe^-$: $8.5487 + 3.9369 = 12.4856$ eV

$$_{54}Xe^+ + {_{54}Xe^-} \rightarrow {_{54}Xe^{2+}} + {_{54}Xe^{2-}} : 7.9281 - 7.4308 = 0.4973 \, eV$$

$$_{54}Xe^{2+} + {_{54}Xe^{2-}} \rightarrow {_{54}Xe^{3+}} + {_{54}Xe^{3-}} :$$

$$5.8106 - 5.7239 = 0.0867 \, eV$$

$$_{54}Xe^{3+} + {_{54}Xe^{3-}} \rightarrow {_{54}Xe^{4+}} + {_{54}Xe^{4-}} :$$

$$7.3343 - 6.3721 = 0.9622 \, eV$$

Radon $_{86}Rn$:

$$2 \, _{86}Rn \rightarrow {_{86}Rn^+} + {_{86}Rn^-} : 7.6738 + 3.7151 = 11.3889 \, eV$$

$$_{86}Rn^+ + {_{86}Rn^-} \rightarrow {_{86}Rn^{2+}} + {_{86}Rn^{2-}} :$$

$$7.2623 - 6.9547 = 0.3076 \, eV$$

$$_{86}Rn^{2+} + {_{86}Rn^{2-}} \rightarrow {_{86}Rn^{3+}} + {_{86}Rn^{3-}} :$$

$$5.4973 - 5.3441 = 0.1532 \, eV$$

$$_{86}Rn^{3+} + {_{86}Rn^{3-}} \rightarrow {_{86}Rn^{4+}} + {_{86}Rn^{4-}} :$$

$$6.8661 - 5.7159 = 1.1502 \, eV$$

Figure 6 shows the cumulative energy requirements for ionization and deionization of noble gasses. The horizontal axis refers to the elements: # 1 is $_2He$, #2 is $_{10}Ne$, #3 is $_{18}Ar$, #4 is $_{36}Kr$, #5 is $_{54}Xe$, #6 is $_{86}Rn$.

The vertical axis is electron volts. The lowest curve includes just the single ionizations, neutral to singly ionized and singly ionized to neutral. The successively higher curves include successively more ionizations.

This time, the rough correlation between melting temperatures and energy requirements for ionizations/de-ionizations is clearly negative. What could possibly explain this fact? Well, note that the alkali metals are all solid within the human comfort zone, so the data reported for them are probably from experiments that start in the solid state, which has more ions, and progress to the liquid state, which has more neutral atoms. By contrast, the data reported for the noble gasses probably comes from experiments that start with the gas state, which has fewer ions, progress through the liquid state, and then into the solid state, which has more ions: a process of increasing ionization.

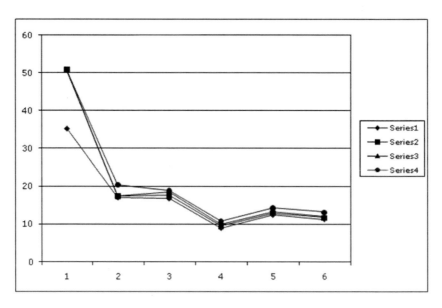

Figure 6. Energy requirements for balanced melting/freezing of noble gasses.

The fact that the melting point is climbing with atomic number may really mean that, as atomic number increases, it is becoming easier to *freeze* these substances. The declining energy requirements for balanced melting/freezing reflect the declining effort required to produce the freezing.

Note that it is even possible that the melting point and the freezing point could be different from each other; *i.e.*, the temperature data obtained could depend on direction of approach. Situations like that occur a lot in Nature. In engineering terminology, they are referred by the term 'hysteresis'. Observed hysteresis is often a clue that something about the system observed is not linear. But at this point, we cannot be certain about this issue in regard to melting points and freezing points.

These observations suggest another possible reason for Hydrogen to be so different from the alkali metals in regard to reported melting temperature. The starting point for Hydrogen is not the solid state, not even the liquid state, but rather the gas state. So the data point recorded for Hydrogen would be about freezing, not melting.

It would be nice to carry out an exercise like the ones above for some elements that are in the liquid state within the human comfort zone. But there are apparently only two such elements: Bromine $_{17}Br$ and Mercury $_{80}Hg$. They aren't even very close to each other: Bromine occurs in the column of halogens, and Mercury occurs among metals. The circumstances that make them liquids are not the same.

For now, let us take a more global view. Is there a pattern to reported melting points over the entire Periodic Table? To approach this problem, let us put all readily available melting-point data right onto the Mendeleyev-style Periodic Table. Table 1 does this.

Table 1. The Mendeleyev-Style Periodic Table With Melting Points Added

$_1H_{13.8}$ $_2He_{3.3}$
$_3Li_{453}$ $_4Be_{1556}$ 5 $_{10}Ne_{24.41}$
$_{11}Na_{371}$ $_{12}Mg_{923}$ col- $_{18}Ar_{83.8}$
$_{19}K_{336.4}$ $_{20}Ca_{1123}$ 10 col-   umns   of 6 $_{36}Kr_{115.79}$
$_{37}Rb_{311.8}$ $_{38}Sr_{1043}$ umns of   of 6   rows $_{54}Xe_{161.1}$
$_{55}Cs_{301.7}$ $_{56}Ba_{977}$ 14 columns of   4 rows   rows $_{86}Rn_{202}$
$_{87}Fr_?$ $_{88}Ra_{973}$ 2 rows each   each   each $_{118}??_?$

The fourteen columns of two rows each are:

$_{57}La_{1153}$ $_{58}Ce_{1048}$ $_{59}Pr_{1297}$ $_{60}Nd_?$ $_{61}Pm_{1325}$ $_{62}Sm_{1423}$ $_{63}Eu_{1623}$
$_{89}Ac_?$ $_{90}Th_{2113}$ $_{91}Pa_{1406}$ $_{92}U_{913}$ $_{93}Np_?$ $_{94}Pu_{1473}$ $_{95}Am_?$

and

$_{64}Gd_{1623}$ $_{65}Tb_{1723}$ $_{66}Dy_{1773}$ $_{67}Ho_{1773}$ $_{68}Er_{1798}$ $_{69}Tm_?$ $_{70}Yb_{2073}$
$_{96}Cm_?$ $_{97}Bk_?$ $_{98}Cf_?$ $_{99}Es_?$ $_{100}Fm_?$ $_{101}Md_?$ $_{102}No_?$

The ten columns of four rows each are are:

$_{21}Sc_{1673}$ $_{22}Ti_{2085}$ $_{23}V_{2003}$ $_{24}Cr_{2176}$ $_{25}Mn_{1517}$
$_{39}Y_{1773}$ $_{40}Zr_{2125}$ $_{41}Nb_{2760}$ $_{42}Mo_{2883}$ $_{43}Tc_{2413}$
$_{71}Lu_{1973}$ $_{72}Hf_{2500}$ $_{73}Ta_{3250}$ $_{74}W_{3653}$ $_{75}Re_{3420}$
$_{103}Lr_?$ $_{104}Rf_?$ $_{105}Db_?$ $_{106}Sg_?$ $_{107}Bh_?$

and

$_{26}Fe_{1808}$ $_{27}Co_{1765}$ $_{28}Ni_{1726}$ $_{29}Cu_{1356}$ $_{30}Zn_{692.5}$
$_{44}Ru_{2773}$ $_{45}Rh_{2233}$ $_{46}Pd_{1825}$ $_{47}Ag_{1233.8}$ $_{48}Cd_{593.9}$
$_{76}Os_{2973}$ $_{77}Ir_{2716}$ $_{78}Pt_{2042.3}$ $_{79}Au_{1336}$ $_{80}Hg_{134.13}$
$_{108}Hs_?$ $_{109}Mt_?$ $_{110}Ds_?$ $_{111}Rg_?$ $_{112}??_?$

The five columns of six rows each are:

176   Cynthia Kolb Whitney

$_5B_{2313}$	$_6C_{3773}$	$_7N_{62.99}$	$_8O_{54.24}$	$_9F_{55.04}$
$_{13}Al_{933.1}$	$_{14}Si_{1683}$	$_{15}P_{317.2}$	$_{16}S_{385.8}$	$_{17}Cl_{172}$
$_{31}Ga_{302.78}$	$_{32}Ge_{1210.6}$	$_{33}As_{1090}$	$_{34}Se_{490.4}$	$_{35}Br_{265.7}$
$_{49}In_{429.4}$	$_{50}Sn_{504.9}$	$_{51}Sb_{903.5}$	$_{52}Te_{723}$	$_{53}I_{386.6}$
$_{81}Tl_{576.6}$	$_{82}Pb_{600.3}$	$_{83}Bi_{544}$	$_{84}Po_{527}$	$_{85}At_{?}$
$_{113}??_{?}$	$_{114}??_{?}$	$_{115}??_{?}$	$_{116}??_{?}$	$_{117}??_{?}$

Note this detail about Table 1: some elements actually have more than one solid form, and hence more than one melting point. These include Carbon $_6C$, Phosphorus $_{15}P$, Sulfur $_{16}S$, Arsenic $_{33}As$, Selenium $_{34}Se$, and Tin $_{50}Sn$. Only the lower melting points are displayed here. My problem here is the same as an important problem dealt with in the Science of Merchandising: optimum allocation of shelf space!

There may be a pattern to Table 1, but it is hard to see. The pattern visibility problem is caused by the rectangular layout of the Periodic Table: it automatically leads to too many columns (thirty two of them), and columns that are too short:(ten of them with only four elements, and fourteen of them with just two elements - and most of those elements with incomplete information).

We need a different way of looking at the available information. Figure 7 offers a candidate for the different information layout needed for this problem. Instead of a Periodic Table, it is a Periodic Arch. Instead of straight rectangular blocks, it has curving arches. Instead of thirty-two separated columns, it has interconnection in every direction.

Figure 7. The Periodic Arch of the Elelments. The labels from 7 down to 1 on the lower left refer to the rows in the traditional Periodic Table of the Elements.

Why an Arch? Recall that the Prolog to this Part I of the Book said that there exist many novel displays of the Periodic Table, and speculated that each one probably reflects the soul of its author. This one has nested arches. That is, it has nested parabolas. Think conic sections. Think rocket science. This is what you get from a researcher who worked nearly two decades for a NASA contractor.

Observe that Hydrogen is in the middle of the Periodic Arch. Hydrogen isn't really grouped with any elements more than any others. Instead, its connection to all the other elements is radial. This is a real advantage. In [2], Eric Scerri mentioned the existence of at least *four* different candidate places for Hydrogen: Group 1 (alkali metals - Lithium, *etc.*), Group 17 (halogens - Fluorine, *etc.*), Group 14 (Carbon, *etc.*), or off the Periodic Table entirely, because it is so odd! With the 'Periodic Arch' display, Hydrogen is pretty much dead center. It is within touching range of the three element connections mentioned, as well as the 'ground' underneath the arch, corresponding to the fourth option mentioned: removal of the troublesome element from the main body of the Periodic Table. Scerri also noted similar issues about Helium, inviting at least two possible placements: with noble gases, or with alkaline earths. On the Periodic Arch, Helium falls between noble gasses on one side, and alkaline earths close by on the other side, and it is also near the 'ground'; *i.e.* the 'off the Table' limbo to which troublesome elements might be banished. And Scerri noted even more issues about elements in the Lanthanide and Actinide series, two of which, all of which, or none of which, could be grouped with Scandium Yttrium, with or without Lutetium and Lawrencium, in 'Group 3'. On the Periodic Arch, the Lanthanides and the Actinides are running up the left side of the Periodic Arch, causing no problem to anybody. It does not matter if they belong to Group 3 or not. So overall, the Periodic Arch type of display of the elements, along with the Algebraic Chemistry type of calculation approach, completely avoids the issues that otherwise arise over where to place some elements in the Periodic Table. Hydrogen and Helium are both in the middle of the PA display, near to everything mentioned as deserving their proximity. And the Lanthanides and Actinides fall neatly into place on the left side of the PA. Another big advantage that the Periodic Arch has in comparison to the Periodic Table is its visibility all at once, as an integrated whole. This quality becomes important as we add the information on melting temperatures. Figure 8 displays the result. It is the job of Science to offer some predictions for future testing. From the display that Figure 8 provides, one can begin to detect some patterns, and make some reasonable predictions about data points that are presently missing. The elements with question marks are all in Periods 6 and 7. Let us now develop some reasonable estimates for those melting points.

- In Period 7, the elements $_{89}Ac$ and $_{93}Np$ are in situations similar to that of $_{87}Fr$, with two neighbor elements close inboard. We can estimate melting temperature for $_{89}Ac$ and $_{93}Np$ similarly, using their inboard neighbors. So we have:
- Francium $_{87}Fr$, $301.7 \times \left[(301.7 + 311.8)/(2 \times 311.8)\right]$: $297 K$,
- Actinium $_{89}Ac$, $1153 \times \left[(1153 + 1773)/(2 \times 1773)\right]$: $951 K$,
- Neptunium $_{93}Np$, $1325 \times \left[(1325 + 2883)/(2 \times 2883)\right]$: $967 K$.
- For $_{69}Tm$ and $_{60}Nd$ in Period 6, the only close neighbors are not inboard, but adjacent, so we have to rely on just the adjacent neighbors:

- Thulium $_{69}$Tm, $[(1798+2073)/2]$: 1935 K,
- Neodymium $_{60}$Nd, $[(1297+1325)/2]$: 1311 K.
- In Period 6, $_{85}$At is well endowed with neighbors on three sides. We can use the inboard neighbors for:
- Astatine $_{85}$At, $386.6 \times [(386.6+265.7)/(2 \times 265.7)]$: 474 K.

Or we can use the adjacent neighbors for:

- Astatine $_{85}$At, $[(527+202)/2]$: 365 K.

Or we can use all of the available information for:

- Astatine $_{85}$At, $[(474+365)/2]$: 419 K.

					$_{101}$Md	$_{102}$No	$_{103}$Lr			
					?	?	?			
			$_{100}$Fm						$_{104}$Rf	
			?	$_{68}$Er	$_{69}$Tm	$_{70}$Yb	$_{71}$Lu	$_{72}$Hf	?	
				1798	?	2073	1973	2500		
		$_{99}$Es	$_{67}$Ho					$_{73}$Ta	$_{105}$Db	
		?	1773					3250	?	
	$_{98}$Cf	$_{66}$Dy					$_{74}$W	$_{106}$Sg		
	?	1773					3653	?		
$_{97}$Bk	$_{65}$Tb					$_{75}$Re	$_{107}$Bh			
?	1723			$_{45}$Rh		3420	?			
$_{96}$Cm	$_{64}$Gd	$_{44}$Ru		2233		$_{46}$Pd	$_{76}$Os	$_{108}$Hs		
?	1623	2773				1825	2973	?		
$_{95}$Am	$_{63}$Eu	$_{43}$Tc				$_{47}$Ag	$_{77}$Ir	$_{109}$Mt		
?	1623	2413				1233.8	2716	?		
$_{94}$Pu	$_{62}$Sm	$_{25}$Mn		$_{27}$Co		$_{29}$Cu	$_{78}$Pt	$_{110}$Ds		
1473	1423	1517		1765		1356	2042.3	?		
$_{93}$Np	$_{61}$Pm	$_{42}$Mo	$_{26}$Fe		$_{28}$Ni	$_{48}$Cd	$_{79}$Au	$_{111}$Rg		
?	1325	2883	1808		1726	593.9	1336	?		
$_{92}$U	$_{60}$Nd	$_{24}$Cr				$_{30}$Zn	$_{80}$Hg	$_{112}$??		
913	?	2176				692.5	134.13	?		
$_{91}$Pa	$_{59}$Pr	$_{41}$Nb	$_{23}$V	$_{14}$Si		$_{31}$Ga	$_{49}$In	$_{81}$Tl	$_{113}$??	
1406	1297	2760	2003	1683		302.78	429.4	576.6	?	
$_{90}$Th	$_{58}$Ce	$_{40}$Zr	$_{22}$Ti	$_{12}$Mg $_{13}$Al	$_{6}$C $_{15}$P	$_{16}$S	$_{32}$Ge	$_{50}$Sn	$_{82}$Pb	$_{114}$??
2113	1048	2125	2985	923 933.1	3773 317.2	385.8	1210.6	504.9	600.3	?
$_{89}$Ac	$_{57}$La	$_{39}$Y	$_{21}$Sc	$_{5}$B	$_{7}$N		$_{33}$As	$_{51}$Sb	$_{83}$Bi	$_{115}$??
?	1153	1773	1673	2313	62.99		1090	903.5	544	?
$_{88}$Ra	$_{56}$Ba	$_{38}$Sr	$_{20}$Ca	$_{4}$Be	$_{8}$O		$_{34}$Se	$_{52}$Te	$_{84}$Po	$_{116}$??
973	977	1043	1123	1556	54.24		490.4	723	527	?
$_{87}$Fr	$_{55}$Cs	$_{37}$Rb	$_{19}$K	$_{11}$Na $_{3}$Li	$_{1}$H	$_{9}$F	$_{17}$Cl	$_{35}$Br	$_{53}$I	$_{85}$At $_{117}$??
?	301.7	311.8	336.4	371 453	13.8	55.04	172	265.7	386.8	? ?
period:						$_{2}$He	$_{10}$Ne	$_{18}$Ar	$_{36}$Kr	$_{54}$Xe $_{86}$Rn $_{118}$??
7	6	5	4	3 2	1	3.3	24.41	83.8	115.79	161.1 202 ?

Figure 8. The Periodic Arch with Data on Melting Points Added.

This author would bet on the 419 K being the best estimate.

- Finally there is the long string of elements in Period 7 going from $_{95}$Am on. Those elements are without any melting point data at all, and they have just one close neighbor, the one inboard in Period 6. We have to rely on that one neighbor. What we know in the case of IP's is that there exists no significant difference between the elements of Period 7 and those of Period 6. So for the case of melting points, we guess that all of them just follow the one inboard neighbor, probably with a slight reduction. So for example, we estimate
- Americium $_{95}$Am, < 1623 K.

3.2. Boiling Points

This Sub-Section develops some more exercises like the ones in the last Sub-Section, but focusing on boiling points instead of melting points. As data, boiling points are a bit less reliable than melting points. Most of them are at very high temperatures that are difficult to achieve, and difficult to measure. One can see in the raw data, which are reported in degrees centigrade rather than degrees Kelvin, that many data points are reported in fifty degree increments, suggesting that they are that rough.

Nevertheless, we can tell a little bit of a story about boiling points, and that is the mission of this Subsection.

For the leftmost column of the Periodic Table, the alkali metals, we have boiling points temperatures:

- Hydrogen $_1$H, 20.23 K;
- Lithium $_3$Li, 1599 K;
- Sodium $_{11}$Na, 1162 K;
- Potassium $_{19}$K, 1030 K;
- Rubidium $_{37}$Rb, 952 K;
- Cesium $_{55}$Cs, K; 963 K;
- Francium $_{87}$Fr, ~952 K*.

Again, Hydrogen $_1$H does not belong. And again, we have to use an estimate for $_{87}$Fr.

Figure 9 shows that, except for the actual magnitudes involved, the boiling points for alkali metals follow a pattern much like that followed by their melting points shown on Figure 3. Again, the horizontal axis refers to the elements. #1 is $_3$Li, #2 is $_{11}$Na, #3 is $_{19}$K, #4 is $_{37}$Rb, #5 is $_{55}$Cs, and #6 is $_{87}$Fr. The vertical axis refers to the boiling temperatures, reported for elements #1 – #5, and estimated for element #6, expressed in degrees Kelvin. The scale is logarithmic, this time with two decades.

For the noble gasses, the boiling points are:

- Helium $_2$He, 4.056 K;
- Neon $_{10}$Ne, 27.1 K;
- Argon $_{18}$Ar, 87.13 K;
- Krypton $_{36}$Kr, 119.77 K;
- Xenon $_{54}$Xe, 164.9 K;
- Radon $_{86}$Rn, K; 211.2 K.

These are displayed on Figure 10. This Figure is visually indistinguishable from Figure, 5 for the melting points of noble gasses. Again, the horizontal axis refers to the elements: # 1 is $_2$He, #2 is $_{10}$Ne, #3 is $_{18}$Ar, #4 is $_{36}$Kr, #5 is $_{54}$Xe, #6 is $_{86}$Rn. The vertical axis is degrees K, and the scale is logarithmic. Why do we have so much similarity of shape between curves for melting points and curves for boiling points (or for freezing points and for condensation points)? I believe it is because all such macroscopic state changes are related to microscopic changes of ionization states, and the same changes occur, in balance, in both of the macro state changes: melting/freezing, and condensing/boiling.

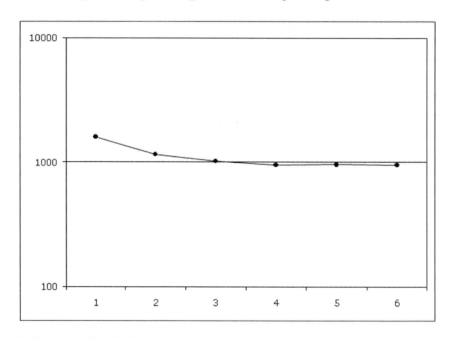

Figure 9. Boiling points for alkali metals.

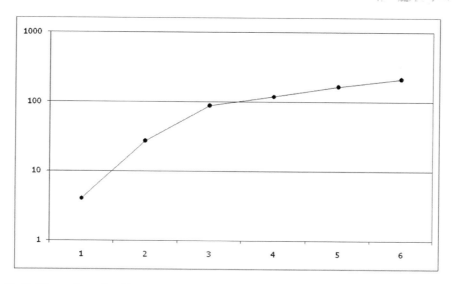

Figure 10. Boiling points of noble gasses.

But boiling points are dodgy to measure for a reason even beyond those already cited. The truth is: temperature T is not the only independent variable. There is always at least one other. Pressure P can be taken as the other one. So macroscopic state changes are not adequately characterized as a point over a one-dimensional temperature axis; they need a line on a two-dimensional P, T phase diagram.

3.3. Phase Diagrams

Figure 10 shows what a generic phase diagram looks like. The horizontal axis represents temperature T, and the vertical axis represents pressure P. The 1 on the horizontal axis corresponds to absolute zero temperature. The 101 means the plot is constructed from 101 temperature data points. The 0 on the vertical axis means zero pressure, and the 1 means the maximum pressure plotted, whatever that might actually be.

The left side of the picture corresponds to low temperature, and hence the solid state. The middle part of the picture represents higher temperature and hence the liquid state. The bottom of the picture represents low pressure, and hence the gas state.

The short curved line segment in the lower left marks transitions directly between solid and gas states, called 'sublimation' in the direction of solid-to-gas. Sublimation is an example of something mentioned as a thought problem in Sect. 4: a situation wherein not all four states of matter occur. At low pressure, the liquid state does not occur; the solid state goes directly to the gas state. The nearly vertical line marks the more usual transitions between solid and liquid. Observe that over most of the pressure range, these transitions occur at nearly the same temperature. So that is why the idea of 'melting point', or 'freezing point', is a pretty reliable one to put a single number to. However, pressure does affect melting temperature a little bit. You know that this is true, if you ever go ice-skating. The second curved line segment marks transitions between liquid and gas. It depends noticeably on pressure. You know that this is true, if you ever go camping at various mountain altitudes. That is why the idea of 'boiling point' or 'condensation point' is tough to put a single number to.

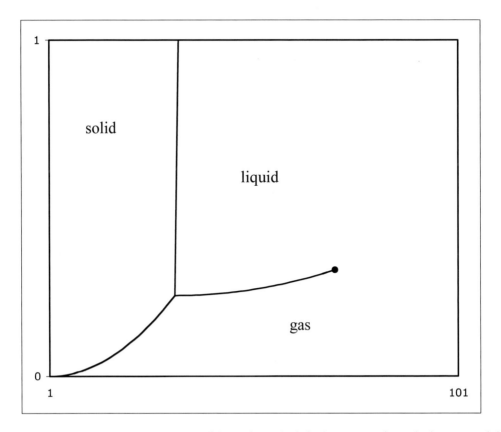

Figure 11. A generic phase diagram: the solid state is on the left, the gas state is on the bottom, and the liquid state is in the upper right, but clearly distinguishable from the gas state only up to the critical point, indicated by the black circle. (First published in [1]).

The point on Figure 11 where the three lines meet is called the 'triple point'. Three states of matter, solid, liquid, and gas, can co-exist there.

The point on Figure 11 marked by a big black dot is called the 'critical point'. The distinction between liquid and gas simply disappears there. Matter in this condition simply goes 'opalescent'. This is a big mystery, worthy of much research. It begs the questions: what about points not beyond, but beside the critical point? Where exactly does the distinction between liquid and gas reappear?

Figure 11 overall begs an even bigger question: what about the plasma state? Unfortunately, it hasn't been customary in the past to include the plasma state on phase diagrams. So, Dear Reader, there is a big research topic potentially waiting for you.

Again in the spirit of offering predictions that can be tested later, let me say this. I believe the stories of the triple point and the critical point are just fragments from a larger, more integrated, story. If the plasma state were included on phase diagrams, where might the transitions to plasma state go? I suspect a nexus to the critical point. I don't really believe Nature does a 'termination' at the critical point; it more likely does another triple point, this one involving liquid, gas, and plasma.

If so, where would the transition lines for liquid/plasma and gas/plasma likely go? It seems clear that for temperatures high enough, the plasma state trumps the gas state, and for pressures high enough, the solid state trumps the plasma state. So I expect the lines missing from Figure 11 to go generally diagonal, downward to the right, as illustrated by the new straight line shown on Figure 12. I do not mean to imply that the new lines should actually form this straight line; they could curve, they could change direction sharply at the critical point, or do whatever one can imagine; at present, we have no knowledge about such details. We only know that we should now go looking for them.

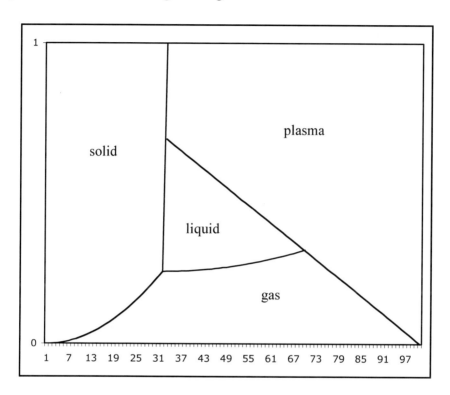

Figure 12. A phase diagram with the plasma state included. (First published in [1]).

Observe that Figure 12 shows the liquid state limited to the totally surrounded, approximately triangular, area in the center. This means that, over the cosmological range of temperatures and pressures out there, the liquid state is something very rare. You know that this is true, if you are interested in space exploration, and follow NASA's search for water.

Observe too that Figure 12 suggests yet another triple point, somewhere near the left top of the Figure. This third triple point involves the solid, liquid, and plasma states. It is my prediction that such a triple point exists, and will some day be observed.

Observe too that there are two more transitions of the sublimation type: a state of matter being skipped. Besides the transition from solid to gas, there is a transition from solid to plasma and a transition from liquid to plasma.

How can all this be accounted for in terms of variations on populations of ionization states? Can we draw Figures like Figure 2 to account for triple points? Truly, it is easier to draw a Figure that accounts for one 'quadruple point' that allows co-existence of all four states of matter! Figure 13 shows how that would work.

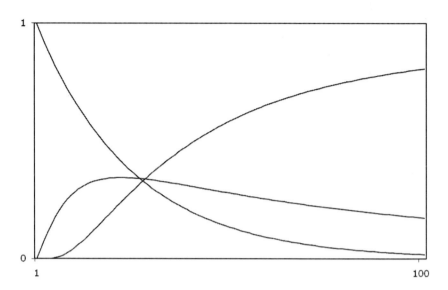

Figure 13. The special case of Figure 2 that would cause a 'quadruple point'.

As on Figure 2, the horizontal axis represents temperature, ranging from absolute zero to some very high temperature, represented by the '1' at the right end. The vertical axis represents the population fraction of atom pairs in the different ionization states: negative energy on the left, positive energy on the right, and neutral in the middle. All of the lines cross at the same temperature point. All of the macroscopic states of matter are accessible at that point.

Separating the fictional quadruple point into three possibly real triple points requires more detail. First of all, we need to acknowledge that the lines seen as functions of temperature are actually surfaces that are functions of both temperature and pressure. Secondly, the surfaces are not necessarily smooth. They can have abrupt discontinuities. Why? It is so because one change of ionization state can be the beginning of a cascade of ionization state changes.

Consider, for example, the case of Phosphorus, $_{15}P$. From the data provided in the Appendix to Chapter 2, we can see that the ionization state transition requirements are:

$$2\,_{15}P \rightarrow \,_{15}P^+ + \,_{15}P^- : 9.4701 - 7.4379 = 2.0322 \text{ eV}$$

$$_{15}P^+ + \,_{15}P^- \rightarrow \,_{15}P^{2+} + \,_{15}P^{2-} : 8.4745 - 11.1265 = -2.6520 \text{ eV}$$

$$_{15}P^{++} + \,_{15}P^{--} \rightarrow \,_{15}P^{3+} + \,_{15}P^{3-} :$$
$$5.8100 - 13.1799 = -7.3699 \text{ eV}$$

$$_{15}P^{3+} + \,_{15}P^{3-} \rightarrow \,_{15}P^{4+} + \,_{15}P^{4-} : 8.2555 + 9.0220 = 17.2775 \text{ eV}$$

Observe that, starting from neutral, the first transition is blocked: it requires an external investment of 2.0322 eV to get started on ionization. But then the next two transitions go spontaneously. That is a 'cascade'. (This spontaneous cascade apparently stops at triple ionization: the transition to quadruple ionization then takes a big investment, 17.2775 eV.)

Check another element. Check any element. They are all like this: the path from one ionization state to another is always a sequence of barriers and cascades.

The first triple point to account for is the familiar solid-liquid-gas one. Here is how its occurrence can be accounted for in terms of population fractions of ionization states and cascades that change them. At the appropriate pressure and temperature, the fraction of negative energy ionization states drops, and the other population fractions rise correspondingly, and the subsequent negative energy ionization fraction falls below the already low positive energy ionization fraction. That means that just below that temperature, we have both solid state and liquid state, whereas just above that temperature, we have gas state. The plasma state is the looser.

The second triple point to account for is the one previously called 'the critical point': liquid-gas-plasma. Here is how its occurrence can be accounted for in terms of population fractions and cascades. At the appropriate pressure and temperature, the fraction of positive energy ionization states soars, and the other population fractions drop correspondingly, and the subsequent positive energy ionization fraction stands above the declining negative-energy ionization fraction. That means that just above that temperature, we have both gas state and plasma state, whereas just below that temperature, we have liquid state. The solid state is the looser. The situation is approximately mirror image to the situation for the first triple point.

The third triple point to account for is the newly proposed solid-liquid-plasma one. Here is how its occurrence can be accounted for. At the temperature and pressure where the fractions of positive and negative energy ionization states are equal, the fraction of neutral energy unionized state drops abruptly, and the fractions of ionized states change correspondingly and equally, with positive energy states increasing and negative energy states decreasing. That means that just below this temperature, we have liquid state and solid state, and just above this temperature we have plasma state. The gas state is the looser.

Note: there is no triple point where the liquid state is the absent one. And so we move on.

4. How Temperature Drives Populations of Ionization States

The transitions between ionization states are what they are, and levy the same energy requirements regardless of temperature, but the steady-state populations of the different ionization states are totally determined by temperature. This Section discusses what the relationship has to be.

The difference between changes of ionization state and populations of ionization states is analogous to something that exists in the subject of Economics: the difference between deficit and debt. The deficit represents a current rate of over-spending, whereas the debt represents the cumulative result of over-spending over a long time.

The magnitudes of some kinds of money expenditures share a property with the energy expenditures for particular ionization state transitions: they are non-negotiable. So only the number of them that are accomplished per unit time can vary. And the variation of transition numbers per unit time controls what the steady state populations of different ionization states turn out to be.

A way of modeling the populations of different ionization states is provided by the discipline of Statistical Mechanics. There are two main ideas involved there. The first idea is that different states of any kind system are populated in proportion to an exponential factor of the form $\exp(-E/kT)$, where E is the state energy, T is the temperature, and k is the famous Boltzmann constant, which has units of energy per degree of temperature.

The other main idea from Statistical Mechanics is that all matter radiates, and absorbs, energy according to the famous Planck black-body spectrum, wherein energy $E = h\nu$, with ν being frequency and h being the famous Planck's constant. The development of the Planck black-body spectrum can be said to have launched twentieth-century quantum mechanics, the whole science of photons, atoms, molecules, and all things small.

A third main idea is introduced here. It is this: The way in which temperature drives transitions between the macroscopic states of matter is through the blackbody spectrum that the temperature creates: at the correct temperatures, the blackbody spectrum provides enough photons of the correct energies to drive particular changes of ionization states, which in turn allow cascades of more changes of ionization states, which ultimately result in the transitions of macroscopic state.

The following Subsections detail all these main ideas.

4.1. Boltzmann Factors

The forces of Nature will, on average, always push toward ever-increasing disorder. You know that this is true, if your household has any children in it!

One technical name for disorder is 'entropy'. This name is taken from the discipline of Thermodynamics. Thermodynamic entropy is one of a long list of thermodynamic variables that are 'extensive', meaning the numerical value grows with the number of particles in the system under discussion. Every such extensive variable has a partner variable that is 'intensive', meaning it is more about the environment than about the system. For entropy, the partner variable is temperature. Another such pair of partner variables is volume and pressure. Another is magnetization and magnetic field. Another is length and tension force. Another is concentration and chemical potential. And so it goes, on and on.

All such pairs of partner variables and 'conjugate', meaning their product is something important: an energy increment. The energy internal to a system is the sum of all such energy increments.

For the present discussion, we are particularly interested in the entropy × temperature increment, and in the chemical-potential × number increment, for a system wherein those increments arise from the population of ionization states. In such a system, the word 'particle' really means a *pair* of atoms. The atom pair can be in various states of balanced positive and negative ionization.

The different ionization states that an atom pair can be in are distinguished by an index i. The state i for a particular pair/particle is a matter of chance. The population fractions in the different states define a set of probabilities $\{p_i\}$. The use of probabilities allows us to look at internal energy and its contributing increments on a per-particle (per-atom-pair) basis. So we are looking at averages:

$$<E> = <S>T + <\mu>$$

where, for example,

$$<E> = \sum_i p_i E_i$$

The problem at hand is to determine what the p_i have to be, in order that prescribed physical conditions can be met.

For the entropy, a good formula comes from Statistical Communication Theory. There the 'information' content of a message received is defined as

$$I = -\sum_i p_i \log p_i$$

with the p_i here representing the statistics on the relative frequencies of different standard messages. In the $\log p_i$, the log is usually taken using base 2, because base 2 is convenient in disciplines that use computers a lot. With base 2, the information is measured in bits. The formula for I says that the information content in a particular message sent varies between zero, which occurs when only one message is even possible, to a lot of bits, which occurs when a lot of messages are possible, but only one message was actually chosen and sent.

The thermodynamic entropy per particle in a physical system can be defined in a similar way:

$$S = -k \sum_i p_i \log p_i$$

where k is Boltzmann's constant. Here the $\log p_i$ is usually taken using base $e = 2.71828...$ because that is the natural base for use in physics.

A candidate solution to the problem posed is that the p_i be proportional to $\exp(-E_i/kT)$. This kind of function is called a 'Boltzmann factor'. It is occurs in Statistical Mechanics, both classical and quantum.

The first condition to be met is obviously that the p_i sum to unity: $\sum_i p_i = 1$. That condition necessarily makes them

$$p_i = \exp(-E_i/kT) \Big/ \sum_i \exp(-E_i/kT).$$

In Statistical Mechanics, the sum in the denominator above is called the 'partition function' ζ. Various statistical averages of interest can be extracted from ζ via various derivatives of it. But here we are only interested in statistical averages as written out more explicitly.

The entropy term in the sum forming internal energy is:

$$ST = -kT \sum_i p_i \log p_i$$

$$= -kT \sum_i \frac{\exp(-E_i/kT)}{\sum_j \exp(-E_j/kT)} \left\{ -(E_i/kT) - \log\left[\sum_j \exp(-E_j/kT)\right] \right\}$$

$$= \sum_i \frac{\exp(-E_i/kT)}{\sum_j \exp(-E_j/kT)} \left\{ E_i + kT \log\left[\sum_j \exp(-E_j/kT)\right] \right\}$$

The first term in ST, just involving E_i is recognizable as equal to the average energy per particle due to ionization:

$$\sum_i \left[\frac{\exp(-E_i/kT)}{\sum_j \exp(-E_j/kT)} E_i \right] = \sum_i p_i E_i = <E_i>$$

The second term in ST, just involving $kT \log\left[\sum_j \exp(-E_j/kT)\right]$, resembles a typical chemical potential term in Thermodynamics. It has the logarithmic form, and it depends on an extensive variable (here the sum over i, there a volume V). It differs just in its sign, and in its per-particle definition. So it can be considered as the negative of the average chemical potential energy per particle $<\mu>$; that is:

$$\sum_i \frac{\exp(-E_i/kT)}{\sum_j \exp(-E_j/kT)} \left\{ kT \log\left[\sum_j \exp(-E_j/kT)\right] \right\}$$

$$= \left\{ kT \log\left[\sum_j \exp(-E_j/kT)\right] \right\} \sum_i \frac{\exp(-E_i/kT)}{\sum_j \exp(-E_j/kT)}$$

$$= kT \log\left[\sum_j \exp(-E_j/kT)\right] = -<\mu>$$

So then the net internal energy per particle, consisting of ST and $<\mu>$, becomes exactly equal to $<E_i>$. This neatly wraps up the story about why p_i is proportional to $\exp(-E_i/kT)$.

The factors $\exp(-E_i/kT)$ are Boltzmann factors. Historically, we saw such factors first in Classical Statistical Mechanics, in the classical kinetic theory of gasses. In that early problem, the energy involved was the kinetic energy $mv^2/2$, where m was particle mass and v was speed. The Boltzmann factor was $\exp(-mv^2/2kT)$. Since v was a continuous variable, not discrete like the index i, the math involved integration, not summation. The normalizing denominator came out to be $\sqrt{2\pi kT}$. The Boltzmann factor $\exp(-E_i/kT)$, normalized by $\sqrt{2\pi kT}$, was named the Maxwell-Boltzmann distribution. The average energy per particle did not come out $<mv^2/2> = kT$; it came out $<mv^2/2> = 3kT/2$. The numerator 3 arose for 3 dimensions of space, and hence three components of velocity vector. The denominator 2 arose because, in any one dimension, the differential for integration was dv, not $d(v^2)$, i.e. $2vdv$.

The Boltzmann factors came up again in Quantum Statistics. In Quantum Statistics, we have states that are discrete, and rules about the number of individual, but indistinguishable, particles that can be in each discrete state. There are two cases: Fermi-Dirac Statistics, with only 0, 1, or 2 particles in a state, and Bose-Einstein Statistics, with arbitrarily many particles in a state. The Fermi-Dirac case ties to the spin quantum number featured in Part II of this book. The Bose-Einstein case leads to the Planck black-body spectrum, discussed next.

4.2. The Planck Black-Body Spectrum

The Planck black-body spectrum is expressed in terms of the generic, non-dimensional energy ratio $E = h\nu/kT$. Here h is Planck's constant, ν is radiation frequency, k is Boltzmann's constant, and T is temperature. The Planck distribution gives the density of E for black-body radiation at frequency ν as the function

$$\frac{dE}{d\nu} = \frac{h}{c^3} \frac{8\pi\nu^3}{\left[\exp(h\nu/kT)-1\right]}.$$

This function tends to $dE/d\nu = 8\pi\nu^2 kT/c^3$ for $h\nu \ll kT$. This is the classical limit, which was first understood by thinking about heat radiation confined inside a spherical black cavity. But for $kT \ll h\nu$, the function tends to $dE/d\nu = (8\pi h\nu^3/c^3)\exp(-h\nu/kT)$. This is the quantum limit, fist understood by Planck in the early 20th century.

The $\left[\exp(h\nu/kT)-1\right]$ denominator that leads to these limits arises from the sum of an infinite series representing the infinite sum of Boltzmann factors corresponding to the infinite series of possible numbers of photons, and hence possible contributions to energy density $dE/d\nu$, for the state associated with energy $h\nu$:

$$\frac{1}{[\exp(h\nu/kT)-1]} = \frac{\exp(-h\nu/kT)}{[1-\exp(-h\nu/kT)]} =$$

$$\exp(-h\nu/kT) + \exp(-2h\nu/kT) + \exp(-3h\nu/kT)...$$

The formulation of the Planck energy distribution defined a milestone for physics. It resolved the famous problem known as 'the ultraviolet catastrophy': the classical function $8\pi\nu^2 kT/c^3$ increases indefinitely as ν increases. The Planck distribution is shown on Figure 14. Observe that shape of the Planck distribution is similar to the middle curve on Figure 2. The important difference is that, on Figure 2, the horizontal axis is proportional to temperature T, whereas on Figure 14, the horizontal axis is proportional to inverse temperature, $1/T$.

Figure 14 is constructed from 101 data points, representing a non-dimensional range of from 0 to 100. The frame top is at 0.02, so the area of the frame is $0.02 \times 100 = 2$. You, Dear Reader, can appreciate visually that the area under the displayed part of the Planck distribution is a little less than 1, and you can anticipate that, with the cut-off infinite tail included, the area would be exactly 1.

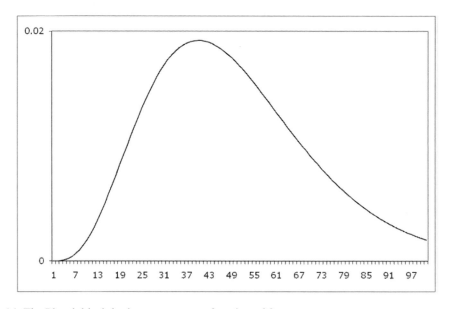

Figure 14. The Planck black-body spectrum as a function of frequency.

4.3. A Mechanism for Driving Macroscopic State Changes

It is customary to think about the Planck distribution as a function of frequency ν, with temperature T held constant. But it is additionally informative to think about it in the opposite way, as a function of inverse temperature, $1/T$, with frequency ν held constant.

Thought of in this way, the Planck distribution can suggest something more about the transitions between the four main states of matter.

We have posited that gross-matter state transitions reflect transitions between states of ionization of atom pairs. So the type of problem we need to solve is: given the energy requirement for a particular change of ionization state, what temperature do we need to have in order to generate black-body photons that meet the energy requirement, and are sufficient in number to make the change of ionization state occur and persist? The approach to this problem is to set v so that hv exactly meets the requirement, display the Planck distribution for dE vs. $1/T$, and strike a line through it at mid height, corresponding to its mean value. Figure 15 shows this construction.

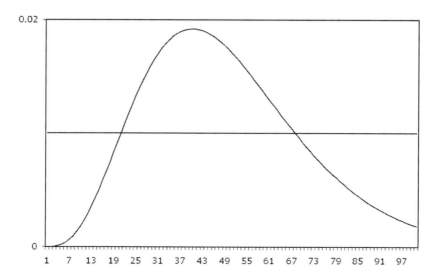

Figure 15. The Planck black-body spectrum as a function of inverse temperature.

On Figure 15, there are two intercepts, corresponding to two values of inverse temperature, $1/T$. Those values of $1/T$ correspond in turn to two values of temperature T. One can fairly say that between the two values of T, photons of the needed energy are plentiful, whereas beyond the two values of T, they are scarce.

Between the two solution values of $1/T$, the Planck distribution also identifies a third $1/T$ point potentially of interest: the point at which the peak occurs. These three $1/T$ points define three values of temperature T, and hence four intervals of T. The first interval is very low $1/T$, and hence very high T, whereas the last interval is very high $1/T$, and hence very low T. In the middle, we have two intervals of low to middling $1/T$ and middling to high $1/T$, and hence high to middling T and middling to low T. It is tempting to wonder if such a set of four temperature intervals can correspond to the four main states of matter: plasma, gas, liquid, and solid.

The idea developing here implies a physical relationship between the solid/liquid transition and the gas/plasma transition. What is that relationship? Both transitions involve changes in ionization states, and such changes impose the same energy requirements regardless of context. So whether a state transition is moving toward a population with more negative-energy ion pairs (solid), or toward a population with more positive-energy ion pairs (gas), it has to start the same way: with an ionic transition from neutral to singly ionized. That event can then lead to an ionic transition to a doubly ionized state, and so on. And although the final populations of ion pairs will turn out differently weighted, the transitions that have to occur along the way each pose the same energy requirements.

Figure 15 thus suggests the following possible similarities among different state transitions:

- Transitions from liquid to gas and transitions from gas to liquid. (no surprise);
- Transitions from liquid to solid and transitions from gas to plasma. (a surprise, if so);
- Transitions from solid to liquid and transitions from plasma to gas. (a surprise, if so).

The two potentially surprising cases arise because of similarity in the direction of change in ionization, one case being two transitions from more neutral to more ionized state, and the other case being two transitions from more ionized to more neutral state.

The interpretations in terms of direction of change in ionization state also indirectly warn about the following possible dissimilarities:

- Transitions from liquid to solid *vs.* transitions from solid to liquid. (a big surprise if so);
- Transitions from gas to plasma *vs.* transitions from plasma to gas. (a big surprise if so).

Each of these two cases contrasts a situation going from a more neutral state to a more ionized state, against a situation going from a more ionized state to a more neutral state. Remember, the energy requirements might not the same going in both directions.

Consider again the case of Phosphorus, $_{15}P$. From the state transition requirements, we can see that the actual ionization state energies are:

$$_{15}P^+ + {}_{15}P^- : 9.4701 - 7.4379 = 2.0322 \text{ eV}$$

$$_{15}P^{2+} + {}_{15}P^{2-} : 2.0322 - 2.6520 = -0.6198 \text{ eV}$$

$$_{15}P^{3+} + {}_{15}P^{3-} : -0.6198 - 7.3699 = -7.9897 \text{ eV}$$

$$_{15}P^{4+} + {}_{15}P^{4-} : -7.9897 + 17.2775 = 9.2878 \text{ eV}$$

The triply ionized state is definitely at the bottom of a potential well. Suppose the transition of interest is out of this well and back to neutral. An initial investment of $7.3699 + 2.6520 = 10.0219$ eV is needed, probably in two separate doses. Then the last step in de-ionization then happens spontaneously.

The two situations are completely different. If the first one is about freezing, and the second one is about melting, then the different energy requirements can mean different optimum temperatures for the transitions. The common assumption that the concepts of melting point and freezing point are the same thing may be yet another of those infernal conflations that bedevil and retard science.

Conclusion

This Chapter starts with ideas dating back to antiquity, then further develops several current ideas, and finally points into the future, with predictions about various data not yet in hand. Quite a few topics are undertaken. They include:

- A conceptual relationship between the energies of microscopic ionization states and the macroscopic states of matter;
- A warning about possible direction sensitivity in measurements of melting points and boiling points;
- A re-organization of data about melting points for easier pattern recognition: the Periodic Arch;
- An inference from data about boiling points to the need for phase diagrams;
- An extension of the idea of phase diagrams to include the plasma phase;
- Comments and predictions about triple points and critical points;
- An accounting for triple points in terms of populations fractions of ionization states;
- A discussion of Boltzmann factors in relation to the relative populations of ionization states that correlate to the macro states of matter;
- A discussion of Planck black body radiation in relation to transitions between ionization states, and hence macro states of matter.

Acknowledgments

Several Figures have been marked as "First published in [1]". Ref. [1] is my paper in the *International Journal of Molecular Sciences* Special Issue entitled *Atoms in Molecules and in Nanostructures* organized by Prof. Mihai V. Putz (2012, vol. 13, pp 1066-1094). I thank Prof. Putz for the opportunity to publish my paper in that Special Issue, and I thank IJMS for permission to reproduce material from it.

A Respite for Readers

In the Book Introduction, Section 3, I said that the story inherent in Chemistry was a drama, and in fact a comic one. The macroscopic states of matter provide a good example. Please enjoy thinking about them in the following totally irreverent way:

- A plasma is like a gang of teenagers: radiant with energy, and controlled only by unseen forces. There's no telling what a plasma will do. It might burn your house down.
- A gas is like a group of single young adults at work: the individuals just bounce off each other.
- A liquid is like a group of dating couples: liaisons are casual, quite fluid, and can switch around at will. The individuals pay a small cover charge at the temporary venue where they meet. It's a beach party with a tent. The couples give up some electrons to create that palpable surface tension that is the tent.
- A solid is like a group of married couples: they form a stable, settled society. They pay rent to the landlord of the apartment building in which they live: that is, they give up some electrons to create and defend a definite boundary all around – ceiling, walls, and floor. The apartment building is the crystal state; of course, an amorphous form of the solid state also exists. It is the sprawling suburbia.

REFERENCES

[1] Whitney, C. On the Several Molecules and Nanostructures of Water, *Int. J. Mol. Sci*, vol. 13, pp. 1066-1094, 2012.

[2] Scerri, E. Explaining the periodic table, and the role of chemical triads, *Found. Chem. Online* (28 January 2010).

Chapter 4

SINGULAR ELEMENTS

ABSTRACT

This Chapter takes the last Chapter one step further, employing the quantitative data that Algebraic Chemistry provides, and developing quantitative answers to interesting questions about a few especially unusual elements. We look at the macroscopic states of matter exhibited, and the reactions exhibited, and *not* exhibited, by these elements. The Chapter illustrates the techniques for analyzing these issues, and all other issues addressed in later Chapters: we look at possible ionic configurations, determine which ones are probable, and correlate them with the macroscopic states of matter and/or reactions observed.

INTRODUCTION

This Chapter refers to the Periodic Table from the Prolog to Part I and/or the Periodic Arch from the preceding Chapter. Most of the elements are solid metals. A handful of the elements are solid non-metals. Seven of the elements are keystone elements: elements that lie mid period, and tell a tale of chemical and/or nuclear technology in our civilization. Six of the elements are noble gasses: elements that avoid entanglements. Five of the elements are halogens: elements that will kill to get an electron. Two elements are liquid under the conditions in which humans live: one metal, and one halogen. This Chapter focuses on these unusual elements.

Section 1 recounts the stories of the keystone elements on the Periodic Arch.

Section 2 then focuses on two of the noble gasses, Helium and Neon. These are chosen because they attract a lot of current research. Two of the halogens figure in the story: Fluorine and Bromine.

Bromine is the more unusual halogen, because it is one of the two elements that exist in liquid state under the conditions in which humans live.

Two metals are discussed: Mercury and Gallium. Mercury is the other liquid element in our world, and Gallium is not far behind, being very low melting.

Bit parts in these stories are played by the elements Nitrogen, Oxygen, and Cesium.

1. KEYSTONE ELEMENTS

The Periodic Arch calls particular attention to the elements that run up the middle: Hydrogen, Carbon, Silicon, Cobalt, Rhodium, Ytterbium and Nobelium. I call them the 'keystone elements'. The terminology 'keystone' is suggestive of importance. Certainly all those elements are important facilitators of molecule formation, because they can give or take electrons in equal numbers. Note that the connection from Hydrogen to Carbon is not a connection to Group 14 in particular. The next element up is Silicon, also from Group 14, but above that come Cobalt and Rhodium, from Group 9, and then Ytterbium and Nobelium from Group 16. Group numbers mean nothing to me. What matters is function. The giving and taking of electrons is a physically meaningful function. More abstractly, Hydrogen is certainly the 'keystone' for all of present-day physical analysis of atoms. And Carbon is certainly the 'keystone' for all of organic chemistry and biological life. And Silicon is certainly the 'keystone' for present-day technological life. Cobalt is not so famous, but it lies between Iron and Nickel, and is functionally better than either of them: harder, more corrosion resistant, and more heat resistant. And, like Iron, it is much strengthened by the addition of a trace of Carbon, that other keystone element. We humans are currently more than three millennia into our 'Iron Age', which, with the help of Carbon and other trace additives, has morphed into our 'Steel Age'. Had Cobalt been more plentiful on this planet, this might have been our 'Cobalt/Steel Age'. But it is not plentiful, because it is the first of the elements not offered up by thermonuclear fusion in the last stage of the life of normal stars, like our Sun. Cobalt and all heavier natural elements come only from super-nova explosions. That fact alone merits the honorific 'keystone' for Cobalt. The next element up, Rhodium, is also not so famous, probably because it is not so plentiful, although it is known to be excellent for plating and alloying. The remaining keystone elements, Ytterbium, and Nobelium, probably offer some interesting properties that we don't completely know about yet.

2. NOBLE GASSES

All of the noble gasses are especially unusual, because they participate so unwillingly in Chemistry. They are largely inert. They keep largely to themselves. That accounts for their name.

Avoiding entanglements with other atoms, even of their own kind, these elements naturally appear as gasses under conditions that humans inhabit. But at low temperatures, they get very interesting.

2.1. Helium

The relevant information about Helium is:

$$_2He \rightarrow {_2He^+}: 24.9189 \text{ eV}$$

$_2\text{He} \to {}_2\text{He}^{2+}$: $24.9189 + 5.0343 = 29.9532$ eV

$_2\text{He} \to {}_2\text{He}^{-}$: 10.4141 eV

$_2\text{He} \to {}_2\text{He}^{2-}$: $10.4141 - 20.4735 = -10.0594$ eV

From this we infer:

$2\text{He} \to \text{He}^{+} + \text{He}^{-}$: $24.9189 + 10.4141 = 35.333$ eV

$2\text{He} \to \text{He}^{2+} + \text{He}^{2-}$: $29.9532 - 10.0594 = 19.8938$ eV

Both numbers are significantly positive. So Helium is simply unable to ionize with itself. It does not hold together at all. That accounts for it being a gas under most conditions.

When cooled enough, Helium liquefies. There are two named sub-states of liquid Helium: Helium I and Helium II. There is not much to distinguish Helium I from the gaseous state. Helium I has very low refraction, viscosity, and density, so a surface is hard to detect. Helium II is different: it exhibits what is called super-fluidity: it spreads out to form a film of a consistent 30 nm thickness. It climbs the walls of its container and escapes (whereupon it immediately heats up and vaporizes).

What is the nature of the difference that distinguishes liquid Helium I from liquid Helium II? Certainly temperature *makes* the difference, but it is difficult to say what *is* the difference. And any experimental investigation of the question is exceedingly difficult to conduct. So let us make a reasonable speculation based on Algebraic Chemistry: maybe ionic configurations in Helium I and Helium II are different. Suppose Helium I is a state analogous to that of a neutral molecule in liquid state, like water. Then suppose Helium II is a state analogous to that of a metallic superconductor, in which electrons are affiliated, not with individual atoms, but with the atom population as a whole. In the case of Helium, some of the liberated electrons can circulate around the perimeter of the liquid, with the stripped nuclei left in the area enclosed by the circuit. The electrons will repel each other, driving that perimeter to get longer, while the nuclei will repel each other, driving the enclosed area to get larger. Both effects together will then produce the super-fluid behavior.

There is a lot of current research about whether Helium can bond with anything. Certainly it cannot bond with itself on any but a very temporary basis. The best ionic configuration is $\text{He}^{2-} + \text{He}^{2+}$, and it takes $-10.0594 + 29.9532 = 19.8938$ eV. This fairly large positive energy has to come from some major intervention, like electric glow discharge, electron bombardment, or whatever else can put the Helium into its plasma state. The He_2 quickly falls apart again, losing energy to radiation.

There is a search going on for other Helium compounds too. For example, it is said that existing theory, presumably standard quantum theory, predicts a trimer molecule $HHeF$. What does Algebraic Chemistry say about $HHeF$?

The relevant information about Hydrogen is:

$_1H \rightarrow {_1H^+}$: 14.1369 eV

$_1H \rightarrow {_1H^-}$: −90.6769 eV

The relevant information about Fluorine is:

$_9F \rightarrow {_9F^+}$: 10.8584 eV

$_9F \rightarrow {_9F^{2+}}$: 10.8584 + 7.0505 = 17.9089 eV

$_9F \rightarrow {_9F^{3+}}$: 17.9089 + 8.4432 = 26.3521 eV

$_9F \rightarrow {_9F^-}$: −12.9914 eV

$_9F \rightarrow {_9F^{2-}}$: −12.9914 + 9.9413 = −3.0501 eV

$_9F \rightarrow {_9F^{3-}}$: −3.0501 − 11.1912 = −14.2413 eV

Suppose the ionic configuration of $HHeF$ is $H^- + He^{2+} + F^-$. That takes:

−90.6769 + 29.9532 − 12.9914 = −73.7151 eV,

which is indeed eminently feasible. The charge mirror image ionic configuration, $H^+ + He^{2-} + F^+$, takes 14.1369 − 20.4735 + 10.8584 = 4.5218 eV, which is positive, but not by very much, and so it might also be possible as a transient.

Another molecule that is predicted, and is being sought, is $CsFHeO$. What does Algebraic Chemistry say about $CsFHeO$? First of all, it isn't clear how the four atoms fit geometrically to make a molecule. It looks more plausible to have two molecules, CsF and HeO, that are both polarized, and so attract each other, but rather more weakly than an actual bond would do. The likely ionic configurations would be $Cs^+ + F^-$ and $He^{2+} + O^{2-}$.

The relevant information about Cesium is:

$$_{55}Cs \rightarrow {}_{55}Cs^+: -3.8352 \text{ eV}$$

$$_{55}Cs \rightarrow {}_{55}Cs^-: -7.3948 \text{ eV}$$

The relevant information about Fluorine is just:

$$_9F \rightarrow {}_9F^+: 10.8584 \text{ eV}$$

$$_9F \rightarrow {}_9F^-: -12.9914 \text{ eV}$$

The relevant information about Oxygen is:

$$_8O \rightarrow {}_8O^+: 7.9399 \text{ eV}$$

$$_8O \rightarrow {}_8O^{2+}: 7.9399 + 9.6214 = 17.5613 \text{ eV}$$

$$_8O \rightarrow {}_8O^-: -12.4354 \text{ eV}$$

$$_8O \rightarrow {}_8O^{2-}: -12.4354 - 14.9434 = -27.3788 \text{ eV}$$

Thus the ionic configuration $Cs^+ + F^-$ takes $-3.8352 - 12.9914 = -16.8266 \text{ eV}$, which is eminently feasible. The ionic configuration $He^{2+} + O^{2-}$ takes $29.9532 - 27.3788 = 2.5744 \text{ eV}$, which is slightly positive, but when paired with the -16.8266 eV leaves -14.2522 eV, making the set of all four ions together still eminently feasible. Even the other ionic configuration, $He^{2-} + O^{2+}$, takes only $-10.0594 + 17.5613 = 7.5019 \text{ eV}$, which, when paired with the -16.8266 eV, leaves -9.3247 eV, which is still eminently feasible. So this search indeed looks promising.

In fact, there may be a fairly robust generic formula here: Noble gas + Oxygen with Alkali Metal + Halogen. With five Halogens, six Alkali metals, and maybe Hydrogen as well, and six Noble gasses, there are hundreds of possibilities to investigate. It might take several hours to investigate them all.

Also predicted, and being sought, is the larger molecule $N(CH_3)FHeO$. Again, the molecule sought looks like two polarized molecules loosely attached. One of them is HeO, with ionic configuration $He^{2+} + O^{2-}$. The partner to HeO is $N(CH_3)$. It could look like:

$$\begin{array}{c} H \\ \bullet \\ H \cdot C \cdot N \cdot F \\ \bullet \\ H \end{array}$$

This molecule can have ionic configuration $N^{2+} + C^{4-} + 3H^+ + F^-$. The relevant information about Nitrogen is:

$_7N \rightarrow {_7N^+}$: 10.498 eV

$_7N \rightarrow {_7N^{2+}}$: 10.498 + 8.7277 = 19.2257 eV

$_7N \rightarrow {_7N^{3+}}$: 19.2257 + 4.4278 = 23.6535 eV

$_7N \rightarrow {_7N^-}$: −8.5432 eV

$_7N \rightarrow {_7N^{2-}}$: −8.5432 − 13.646 = −22.1892 eV

$_7N \rightarrow {_7N^{3-}}$: −22.1892 − 25.4814 = −47.6706 eV

The relevant information about Carbon is:

$_6C \rightarrow {_6C^+}$: 9.6074 eV

$_6C \rightarrow {_6C^{2+}}$: 9.6074 + 4.6431 = 14.2505 eV

$_6C \rightarrow {_6C^{3+}}$: 14.2505 + 9.2799 = 23.5304 eV

$_6C \to {}_6C^{4+}$: $23.5304 - 1.3434 = 22.187$ eV

$_6C \to {}_6C^-$: -11.6267 eV

$_6C \to {}_6C^{2-}$: $-11.6267 - 9.3046 = -20.9313$ eV

$_6C \to {}_6C^{3-}$: $-20.9313 - 15.2156 = -36.1469$ eV

$_6C \to {}_6C^{4-}$: $-36.1469 - 18.4835 = -54.6304$ eV

The proposed ionic configuration $N^{2+} + C^{4-} + 3H^+ + F^-$ takes $19.2257 - 54.6304 + 3 \times 14.1369 - 12.9914 = -5.9854$ eV, and more than compensates the 2.5744 eV taken by the $He^{2+} + O^{2-}$ (although it cannot quite compensate the 7.5019 eV taken by the $He^{2-} + O^{2+}$). This search too looks reasonably promising.

2.2. Neon

Neon is presently characterized as the only remaining element still thought to be actually inert. What does Algebraic Chemistry say about Neon? Since $He^{2+} + O^{2-}$ figured so prominently in the story about Helium, let us now look into $Ne^{2+} + O^{2-}$. The relevant information about Neon is probably no more than:

$_{10}Ne \to {}_{10}Ne^+$: 12.5928 eV

$_{10}Ne \to {}_{10}Ne^{2+}$: $12.5928 + 10.5662 = 23.159$ eV

$_{10}Ne \to {}_{10}Ne^-$: 4.5191 eV

$_{10}Ne \to {}_{10}Ne^{2-}$: $4.5191 - 10.9327 = -6.4136$ eV

The ionic configuration $Ne^{2+} + O^{2-}$ takes $23.159 - 27.3788 = -4.2198 \text{ eV}$. This is patently feasible to make. It looks to be only a matter of time until someone actually makes it.

3. HALOGENS

3.1. Bromine

Since Fluorine already figured prominently in the current research about Helium, let us turn now to another halogen, Bromine. It could add another candidate compound for Helium and/or Neon. Bromine is also one of only two elements that are liquids in the temperature/pressure range familiar to humans.

The relevant information about Bromine is:

$_{35}Br \rightarrow {_{35}Br^+}$: 8.5768 eV

$_{35}Br \rightarrow {_{35}Br^{2+}}$: $8.5768 + 6.2492 = 14.826$ eV

$_{35}Br \rightarrow {_{35}Br^{3+}}$: $14.826 + 7.8188 = 22.6448$ eV

$_{35}Br \rightarrow {_{35}Br^-}$: -9.3278 eV

$_{35}Br \rightarrow {_{35}Br^{2-}}$: $-9.3278 + 4.3938 = -4.934$ eV

$_{35}Br \rightarrow {_{35}Br^{3-}}$: $-4.934 - 8.1147 = -13.0487$ eV

Observe the following about Bromine: electrically, it is not very different from Fluorine. Its reason for presenting as a liquid is probably not related to ionic configuration. That leaves the mass as the probable determining parameter. The halogen before Bromine is lighter, and presents as a gas. The halogen after Bromine is heavier, and presents as a crystalline solid. The uniqueness of Bromine may just reflect the rarity of the liquid state throughout Nature, which was noted in the last Chapter.

4. METALS

Typical metals are solid, but Mercury is not. Mercury is the one other element, besides Bromine, that is liquid so far as humans are concerned. And Gallium is close to it. Can

Algebraic Chemistry shed light on why Mercury is a liquid, and why Gallium is almost a liquid?

4.1. Mercury

If you Google the element 'Mercury', you will find a narrative acknowledging first of all that a complete explanation of its behavior requires a deep excursion into quantum physics. The narrative then proceeds without much math. When the mathematical analysis of a question is burdensomely complicated, readers and authors alike revert a more narrative style of discourse. It is to be hoped that Algebraic Chemistry can tell the story with a more comfortable and appropriate level of math.

The summarized explanation given on the web refers to the filled sets of single-electron states, especially the set with angular momentum $l=3$. The narrative refers to the common belief in an association between the various sets of single-electron states and electron shells enclosing and shielding the nucleus. Completely filled sub-shells would mean better shielding of the nucleus, and hence weaker bonding of the outermost electrons.

One problem with the narrative is that we really do not know, not by any direct observation, that sets of single-electron states actually make closed (topologically equivalent to spherical) shells, or that electron shells actually enclose the nucleus, or that weaker bonding between the nucleus and the outer electrons makes for lower melting point.

What we really *do* know is the data that Algebraic Chemistry develops and uses. So let us see what that tells us. The relevant information about Mercury is this:

$$_{80}Hg \rightarrow {_{80}Hg^+}: 6.2427 \text{ eV}$$

$$_{80}Hg \rightarrow {_{80}Hg^{2+}}: 6.2427 + 6.2898 = 12.5325 \text{ eV}$$

$$_{80}Hg \rightarrow {_{80}Hg^{3+}}: 12.5325 + 6.1014 = 18.6339 \text{ eV}$$

$$_{80}Hg \rightarrow {_{80}Hg^{4+}}: 18.6339 + 5.0893 = 23.7237 \text{ eV}$$

$$_{80}Hg \rightarrow {_{80}Hg^{5+}}: 23.7237 + 5.0856 = 28.8093 \text{ eV}$$

$$_{80}Hg \rightarrow {_{80}Hg^{6+}}: 28.8093 + 5.4671 = 34.2764 \text{ eV}$$

From this we infer:

$$2Hg \rightarrow Hg^+ + Hg^-: 6.2427 - 1.7810 = 4.4617 \text{ eV}$$

$$2\text{Hg} \to \text{Hg}^{2+} + \text{Hg}^{2-} : 12.5325 - 8.8487 = 3.6838 \text{ eV}$$

$$2\text{Hg} \to \text{Hg}^{3+} + \text{Hg}^{3-} : 18.6339 - 16.2367 = 2.3972 \text{ eV}$$

$$2\text{Hg} \to \text{Hg}^{4+} + \text{Hg}^{4-} : 23.7237 - 22.1080 = 1.6157 \text{ eV}$$

$$2\text{Hg} \to \text{Hg}^{5+} + \text{Hg}^{5-} : 28.8093 - 29.9321 = -1.1228 \text{ eV}$$

$$2\text{Hg} \to \text{Hg}^{6+} + \text{Hg}^{6-} : 34.2764 - 38.2104 = -3.934 \text{ eV}$$

The first four of these numbers are all positive. Other metals examined to this same level of detail for this Book are not like this. You will see in subsequent Chapters that Mercury's neighbor, Gold, is totally unlike this, and Platinum is unlike this, and Palladium is unlike this.

The string of four positive energy values says that Mercury is largely unable to ionize with itself. So it cannot readily make the sort of crystal lattice that is typical in a solid. A sample of Mercury is something like a pile of neutral billiard balls. So Mercury spreads out, like any liquid would do.

Mercury does have some small surface tension (4.76 dyne / cm at 20°C), which does limit the spread. The existence of some surface tension suggests the existence of some small fraction of ion pairs, as the concept of the Boltzmann factor would indeed predict.

Basically, Mercury can make enough ion pairs to form a two-dimensional external containment net, but not enough to make a three-dimensional internal support scaffold. So when Mercury does eventually appear to freeze (at temperature -38.87°C), it may actually become a glass-like, highly viscous liquid, instead of a true crystalline solid. If so, the seemingly solid Mercury would offer little in the way of structural strength. This is a prediction, and it should be possible to test it in a proper laboratory.

4.2. Gallium

Gallium has the lowest melting point of any metal (unless one counts Hydrogen as a metal). Gallium melts a little below human body temperature. It is the source of the practical joke that provides the title for Sam Kean's book [1].

The relevant information about Gallium is the following:

$$_{31}\text{Ga} \to {}_{31}\text{Ga}^{+}: 4.6191 \text{ eV}$$

$$_{31}\text{Ga} \to {}_{31}\text{Ga}^{2+}: 4.6191 + 6.6210 = 11.2401 \text{ eV}$$

$_{31}Ga \to {}_{31}Ga^{3+}$: $11.2401 + 6.4803 = 17.7204$ eV

$_{31}Ga \to {}_{31}Ga^{4+}$: $17.7204 + 6.3394 = 24.0598$ eV

$_{31}Ga \to {}_{31}Ga^{5+}$: $24.0598 + 6.1989 = 30.2587$ eV

$_{31}Ga \to {}_{31}Ga^{-}$: -7.9855 eV

$_{31}Ga \to {}_{31}Ga^{2-}$: $-7.9855 - 8.5515 = -16.5370$ eV

$_{31}Ga \to {}_{31}Ga^{3-}$: $-16.5370 - 6.7466 = -23.2836$ eV

$_{31}Ga \to {}_{31}Ga^{4-}$: $-23.2836 - 9.4080 = -32.6916$ eV

$_{31}Ga \to {}_{31}Ga^{5-}$: $-32.6916 - 5.6159 = -38.3075$ eV

From this information we infer that Gallium glides easily through all listed states of self-ionization:

$2Ga \to Ga^+ + Ga^-$ takes $4.6191 - 7.9855 = -3.3664$ eV

$2Ga \to Ga^{2+} + Ga^{2-}$ takes $11.2401 - 16.5370 = -5.2969$ eV

$2Ga \to Ga^{3+} + Ga^{3-}$ takes $17.7204 - 23.2836 = -5.5632$ eV

$2Ga \to Ga^{4+} + Ga^{4-}$ takes $24.0598 - 32.6916 = -8.6318$ eV

$2Ga \to Ga^{5+} + Ga^{5-}$ takes $30.2587 - 38.3075 = -8.0488$ eV

At any given temperature, all these ionization states will be present in proportion to their Boltzmann factors. Since they are so similar in energy, it takes a rather low temperature to freeze out any of them. That makes the freezing (*i.e.* melting) point of this metal unusually low.

Conclusion

This stories about the several singular elements discussed in this Chapter are stories featuring numbers. The stories illustrate the application of Algebraic Chemistry as a productive approach for understanding macroscopic states of matter, and much more.

The most endlessly fascinating state that matter can exhibit is the state of being Chemically Bound. By this comment, I mean to introduce a deeper study of molecules.

And just as the study of ionization states illuminates the workings of the singular elements, it also illuminates the construction of many typical molecules. That exercise forms the next Chapter.

A Project for Readers

Some of you may be studying rare earths, or trans-uranic elements, or dangerous elements. You may have interesting problems to pose. They can be featured in subsequent editions of this Book, with a citation to you, or a by-line for you. Please let us know about them.

Reference

[1] Kean, S. *The Disappearing Spoon,* Little Brown and Co., Hachette Book Group, New York, 2011.

Chapter 5

TYPICAL MOLECULES

ABSTRACT

This Chapter illustrates the techniques of Algebraic Chemistry as applied to modeling issues involving molecules in one particularly important family. The example family consists of various light hydrocarbons, and a few other molecules that are consumed by, or generated by, or otherwise involved in, reactions that burn the hydrocarbons. The issues include the ionic configurations of the individual molecules and the reactions involving the molecules. These illustrations show how ionization states of their constituent atoms relate to the energy content of molecules. Most molecules present more than one possible ionic configuration, and hence they have more than one energy state. One of the ionic configurations will have lowest total ionic energy, and hence be preferred on energy grounds. But all of the ionic configurations can be present, in proportion to their so-called 'Boltzmann factor'. Sometimes, the state with lowest ionic energy is actually disqualified on grounds of geometry, at least under normal conditions of temperature and pressure. In such cases, the ionic configuration next lowest in energy becomes dominant.

INTRODUCTION

With full energy information about the possible ionic configurations of molecules, one is in a good position to make quantitative assessments of possible reactions among them. In the case of light hydrocarbons, the first interesting reactions are combustion reactions. Combustion invokes plasma, and for the family of hydrocarbons completes the association between ionic configurations and the macroscopic states of matter: solid, liquid, gas, plasma, discussed in the last Chapter. Combustion reactions also invite a breakdown into stages, and the idea of reaction stages leads to a subject for the Chapter next after this one: catalysis.

1. GENERAL INFORMATION

Before we start on particular hydrocarbons, there is some general information that is relevant to all of them. The elements Hydrogen, Carbon, Nitrogen, and Oxygen are involved throughout the discussions of the various hydrocarbons. This Section collects that information from Chapter 2.

To start with, recall from the last Chapter that the relevant information about Hydrogen is:

$$_1H \rightarrow {_1H^+}: 14.1369 \text{ eV and}$$

$$_1H \rightarrow {_1H^-}: -90.6769 \text{ eV}$$

Carbon is the jack-of-all-trades, involved in all sorts of organic, *i.e.* biological, molecules, including our own. Recall from the last Chapter that the relevant information about Carbon is:

$$_6C \rightarrow {_6C^+}: 9.6074 \text{ eV}$$

$$_6C \rightarrow {_6C^{2+}}: 9.6074 + 4.6431 = 14.2505 \text{ eV}$$

$$_6C \rightarrow {_6C^{3+}}: 14.2505 + 9.2799 = 23.5304 \text{ eV}$$

$$_6C \rightarrow {_6C^{4+}}: 23.5304 - 1.3434 = 22.187 \text{ eV}$$

$$_6C \rightarrow {_6C^-}: -11.6267 \text{ eV}$$
$$_6C \rightarrow {_6C^{2-}}: -11.6267 - 9.3046 = -20.9313 \text{ eV}$$

$$_6C \rightarrow {_6C^{3-}}: -20.9313 - 15.2156 = -36.1469 \text{ eV}$$

$$_6C \rightarrow {_6C^{4-}}: -36.1469 - 18.4835 = -54.6304 \text{ eV}$$

Carbon is like Hydrogen in that the positive energy requirements for removing electrons are of lesser magnitude than the negative energies for adding electrons. This means a population of neutral Carbon atoms does not want to remain neutral. Carbon atoms will either give electrons or take electrons, and then form molecules - all sorts of molecules. Carbon is extremely versatile that way.

Nitrogen forms a variety of compounds, for which the relevant information appears to be limited to:

$$_7N \rightarrow {_7N^+}: 10.498 \text{ eV}$$

$_7N \to {_7N^{2+}}$: $10.498 + 8.7277 = 19.2257$ eV

$_7N \to {_7N^{3+}}$: $19.2257 + 4.4278 = 23.6535$ eV

$_7N \to {_7N^{-}}$: -8.5432 eV

$_7N \to {_7N^{2-}}$: $-8.5432 - 13.646 = -22.1892$ eV

$_7N \to {_7N^{3-}}$: $-22.1892 - 25.4814 = -47.6706$ eV

Oxygen too forms a variety of compounds, and the relevant information extends to:

$_8O \to {_8O^{+}}$: 7.9399 eV

$_8O \to {_8O^{2+}}$: $7.9399 + 9.6214 = 17.5613$ eV

$_8O \to {_8O^{3+}}$: $17.5613 + 8.0394 = 26.6007$ eV

$_8O \to {_8O^{4+}}$: $25.6007 + 4.2404 = 29.8411$ eV

$_8O \to {_8O^{-}}$: -12.4354 eV

$_8O \to {_8O^{2-}}$: $-12.4354 - 14.9434 = -27.3788$ eV

$_8O \to {_8O^{3-}}$: $-27.3788 + 12.3116 = -15.0672$ eV

$_8O \to {_8O^{4-}}$: $-15.0672 - 12.7597 = -27.8269$ eV

2. Diatomic Molecules

Hydrogen forms the molecule H_2 with energy $14.1369 - 90.6769 = -76.54$ eV. That number is very negative, which means the Hydrogen molecule forms quickly, even explosively. Isolated neutral Hydrogen atoms are rare in Nature. Even at very low density, in deep space, Hydrogen atoms would rather form molecules, or form plasma, than remain as neutral atoms.

Nitrogen has a diatomic molecule N_2 that allows more than one ionic configuration, but one is especially comfortable: $_7N^{3+} + _7N^{3-}$. The information about Nitrogen tells us the $_7N^{3+} + _7N^{3-}$ ionic configuration takes $23.6535 - 47.6706 = -24.0171$ eV. That makes for a very stable molecule. Basically, our life on planet Earth is feasible because the Nitrogen in the atmosphere controls the Oxygen there.

Oxygen has a diatomic molecule O_2 that allows two likely ionic configurations, $_8O^+ + _8O^-$ and $_8O^{2+} + _8O^{2-}$. Forming $_8O^+ + _8O^-$ takes $7.9399 - 12.4354 = -4.4955$ eV, and forming $_8O^{2+} + _8O^{2-}$ takes $17.5613 - 27.3788 = -9.8175$ eV. Although the second ionic configuration is better in terms of energy, the two are close. In situations like this, the relative proportions of different ionic configurations are determined by the concept mentioned in the last Chapter, the Boltzmann factor, $\exp(-E/kT)$, where E is the energy of the condition or state considered, k is Boltzmann's constant, and T is temperature. The fact that the two ionic configurations are so close in energy means that transitions between them are easy. This can make O_2 an absorber and emitter of low energy photons; *i.e.* infrared photons, *i.e.* heat. This can in turn make Oxygen act as a so-called 'greenhouse gas'; *i.e.*, a contributor to atmospheric warming. But as O_2-consuming animals, we never speak of it in that derogatory way, especially since some other gasses are far more significant.

In the case of O_2, neither of the likely ionic configurations permits both atoms to have a noble-gas electron count, like 2 or 10, or even a lesser good count, like 4. The first ionic configuration has 7 and 9, and the second has 6 and 10. The 10 is good, but the 6 is not, and the 7 and 9 are just frustrating. So the Oxygen atoms have to time-share electrons to achieve some comfortable number of electrons at least some of the time. This is covalent bonding. It is not a happy situation. For any Oxygen atom, just about *any* fate is better than being covalently bound to another Oxygen atom. This makes the O_2 molecule very reactive.

Carbon monoxide CO illustrates the possibility of several ionic configurations for a molecule. We could be looking at $C^+ + O^-$ or $C^- + O^+$ or $C^{2+} + O^{2-}$ or $C^{2-} + O^{2+}$. The energy requirements are $9.6074 - 12.4354 = -2.828$ eV or $-11.6267 + 7.9399 = -3.6868$ eV or $14.2505 - 27.3788 = -13.1283$ eV or $-20.9313 + 17.5613 = -3.37$ eV. All are feasible, but the likely one is $C^{2+} + O^{2-}$ at -13.1283 eV.

3. TRIATOMIC MOLECULES

Ozone O_3 is an even more interesting form of Oxygen. It has at least four possible ionic configurations, $_8O^{4+} + 2\,_8O^{2-}$, $_8O^{2-} + 2\,_8O^{+}$, $_8O^{2+} + 2\,_8O^{-}$, and $_8O^{4-} + 2\,_8O^{2+}$, with energy requirements:

$_8O^{4+} + 2\,_8O^{2-}$:
$29.8411 + 2 \times (-27.3788) = 29.8411 - 54.7576 = -24.9165$ eV

$_8O^{2-} + 2\,_8O^{+}$:
$-27.3788 + 2 \times 7.9399 = -27.3788 + 15.8798 = -11.499$ eV

$_8O^{2+} + 2\,_8O^{-}$:
$17.5613 + 2 \times (-12.4354) = 17.5613 - 24.8708 = -7.3095$ eV

$_8O^{4-} + 2\,_8O^{2+}$:
$-27.8269 + 2 \times 17.5613 = -27.8269 + 35.1226 = 7.2957$ eV

Three of these energies are negative, and hence these ionic configurations are likely to occur. Transitions between those ionic configurations involve energy increments that range up to $24.9165 - 7.3095 = 17.607$ eV. So it takes energetic photons to provoke such transitions. For this reason, ozone in the high atmosphere serves as shield against ultra violet radiation from space.

Carbon dioxide CO_2 again illustrates the possibility of several ionic configurations for a molecule. CO_2 has at least four plausible ionization configurations: $C^{4+} + 2O^{2-}$, $C^{4-} + 2O^{2+}$, $C^{2+} + 2O^{-}$, and $C^{2-} + 2O^{+}$. The energy requirements to make them from the neutral atoms are:

$C^{4+} + 2O^{2-}$:
$22.187 + 2 \times (-27.3788) = 22.187 - 54.7576 = -32.5706$ eV

$C^{4-} + 2O^{2+}$:
$-54.6304 + 2 \times 17.5613 = -54.6304 + 35.1226 = -19.5078$ eV

$C^{2+} + 2O^-$:
$14.2505 + 2 \times (-12.4354) = 14.2505 - 24.8708 = -10.6203$ eV

$C^{2-} + 2O^+$:
$-20.9313 + 2 \times 7.9399 = -20.9313 + 15.8798 = -5.0515$ eV

The first one listed is the one favored electrically, but the others may also occur, all in proportion to their Boltzmann factors.

With so many ionic configurations available, all so close together, and so allowing low-energy transitions between them, CO_2 can be an even more effective greenhouse gas than O_2 can be. As CO_2-generating animals, living by CO_2-generating technologies, we rightly worry about this fact.

Water H_2O yet again illustrates the possibility of more than one ionic configuration for a molecule. Water is known to dissociate into the naked proton H^+ and the hydroxyl radical OH^-. In turn, the hydroxyl radical has to be the combination of ions $O^{2-} + H^+$; there is not an alternative form using H^-, because then O would have to be neutral. So it might well be imagined that H_2O has ionic configuration $2H^+ + O^{2-}$. But the formation of that ionic configuration takes $2 \times 14.1369 - 27.3788 = +0.8950$ eV, a slightly positive energy. That can't be right for the common water molecule covering our planet. So in fact, common water must not live in the ionic configuration to which it dissociates, and thereby dies.

Therefore, consider the alternative ionic configuration $2H^- + O^{2+}$. This one requires $2 \times (-90.6769) + 17.5613 = -163.7925$ eV. This is a decidedly negative energy, and so is believable for a decidedly stable molecule.

Water in the normal $2H^- + O^{2+}$ ionic configuration has to form a tetrahedron, with two H^+ naked protons on two vertices and two $2e^-$ electron pairs on and the other two vertices. That is why the water molecule we know has a bend to it. Viewing the Hydrogen nuclei as lying on arms originating from the Oxygen nucleus, the angle of the bend is the angle characteristic of arms from the center to two corners of a regular tetrahedron – on the order of $109.5°$.

The tetrahedral shape of the normal water molecule relates to an interesting behavior that water exhibits. Unlike just about anything else, water expands upon freezing. Why? Consider that the electron pairs have to orbit their respective protons, and being identical, they have to orbit in synchrony. That makes the whole tetrahedron spin about an axis that is the tetrahedron edge connecting the two protons. A tetrahedron spinning on an axis that is an edge is a lop-sided occupier of space. It sweeps out a volume that is larger than its instantaneous self. The swept shape resembles two 'Hershey kisses' candies, melted bottom-to-bottom. When the water is in liquid state, the spinning tetrahedrons can tolerate other spinning tetrahedrons temporarily invading their temporarily vacant space: they can bounce, or otherwise adjust. But when water is in solid state, such adjustments are not possible: each spinning tetrahedron needs to have sole ownership of the volume of space in which it spins.

The 'Hershey kisses' image helps explain another interesting feature of freezing water: the development of snowflakes. Imagine one Hershey kiss sitting bottom down on a horizontal mirror, so that one sees the volume swept by the tetrahedron. Obviously, the most efficient packing arrangement for such volumes is one of them at the center surrounded by six on the outside; *i.e.*, hexagonal. Note too that the center of mass of the H_2O molecule is near the oxygen nucleus, and the two protons are outboard. So not only must the tetrahedron spin, but also it must translate: the spin axis must traverse a path that sweeps out a cylinder around the molecular center of mass. This means the polarization vector of the molecule spins in the plane of the Hershey kisses, *i.e.* the horizontal mirror surface. It is always in that plane. That is why a snowflake nearly always turns out planar.

The other ionic configuration for water, $2H^+ + O^{2-}$, is the charge mirror image of the commonly known one. But it is a completely different shape: not tetrahedral, but instead linear. It just looks like $H \bullet O \bullet H$, where the dots mean 'chemical bond'. This form of water apparently does exist, but only in a very un-natural circumstance. There exists an electrochemically created substance known as 'Brown's gas' that has occasioned some impossibly wild claims about energy generation, but has also been investigated quite legitimately for applications in welding. A linear isomer of water is thought to be the active ingredient in Brown's gas.

Brown's gas acts a lot like methane: it seems to burn. But if the linear $2H^+ + O^{2-}$ is indeed the active ingredient Brown's gas, then Brown's gas does not actually consume any oxygen, as methane does. Instead, the linear $2H^+ + O^{2-}$ molecule just reverts to the normal $2H^- + O^{2+}$ molecule, and dumps energy in so doing. We can tell this by comparing the energies required to make the two ionic configurations for the two water molecules, computed above.

The story of water tells us that even the most familiar of compounds in our world can have some very interesting isomers. The conclusion to be drawn is that *any* molecule with three or more atoms can have isomers that differ, certainly in ionic configuration, but probably also in molecular shape, and in resultant chemical properties.

4. HYDROCARBONS

The very negative -163.7925 eV result for forming the $2H^- + O^{2+}$ ionic configuration of normal water is what makes the burning of all the hydrocarbons to follow so worthwhile as energy sources.

The slightly positive $+0.8950$ eV requirement for forming the Brown's gas ionic configuration $2H^+ + O^{2-}$ is not very different from what we find with the first of two forms of methane, the first of the hydrocarbons following.

4.1. Methane CH_4

The methane molecule can be drawn as:

$$\begin{matrix} H\cdot & & \cdot H \\ & C & \\ H\cdot & & \cdot H \end{matrix}$$

Of course, it is not really flat, as drawn on the page. Because of its symmetry, we should expect a tetrahedral shape, and hence bond angles on the order of $109.5°$.

Methane admits two ionic configurations that are mirror image to each other; *i.e.* with plus and minus signs exchanged. They are $C^{4-} + 4H^+$ and $C^{4-} + 4H^+$. The energy requirements to form them from the neutral atoms are:

$C^{4-} + 4H^+$:

$-54.6304 + 4 \times 14.1369 = -54.6304 + 56.5476 = +1.9172$ eV,

and

$C^{4+} + 4H^-$:

$22.187 + 4 \times (-90.6769) = 22.187 - 362.7076 = -340.5206$ eV.

Note that one ionic configuration is at a positive energy, and the other is at a negative energy. The negative-energy ionic configuration is favored electrically, but it is not the fuel we are familiar with. Apparently, it is not so favorable for some other reason. That other reason appears to be simple geometry: the H^- ions involve two electrons, and so are very large compared to the naked proton H^+ ions. So it is difficult to fit four of them onto a tetrahedral molecule. So the $C^{4+} + 4H^-$ form of methane may exist only under high pressure. Such a condition is produced artificially, to make liquefied natural gas (LNG). It also exists naturally, for example at the bottom of the seas. And methane is indeed known to exist there, in some form. So is $C^{4+} + 4H^-$ the form of methane found under the seas? If so, what evidence would we see? When such a form of methane is disturbed, a lapse of steady pressure can start a chain of events such that the molecule can revert to its normal ionic configuration by taking on heat energy from its environment. That means it can appear to be accompanied by lots of water ice crystals. So it naturally acquires the name 'methane hydrate'. So the observed existence in our seas of something we call 'methane hydrate' is good evidence that $C^{4+} + 4H^-$ is the form of methane that exists under the extreme pressure condition that prevails under the seas. *In situ*, that methane is likely to be in liquid, or even solid, state.

The other ionic configuration for methane, $C^{4-} + 4H^+$, is favored geometrically, but being at positive energy, it is highly reactive; in particular, it is readily burnable in Oxygen. The combustion reaction goes:

$$CH_4 + 2O_2 \rightarrow CO_2 + 2H_2O$$

Assuming most likely ionic configurations of all participants, the left side of the combustion reaction takes:

$$1.9172 - 2 \times 9.8175 = 1.9172 - 19.635 = -17.7178 \text{ eV}$$

and the right side of the combustion reaction takes:

$$-32.5706 + 2(-163.7925) = -32.5706 - 327.585 = -360.1556 \text{ eV}$$

so the combustion reaction delivers:

$$-17.7178 + 360.1556 = 342.4378 \text{ eV}.$$

Observe that the most numerically significant part of the combustion reaction is actually the water. It is the generation of normal water that makes the burning of all hydrocarbons worthwhile as a source of energy.

The positive-energy Methane is common as an atmospheric gas, but atmospheric Nitrogen keeps it from burning there. It poses a more subtle risk as a greenhouse gas. That is because its very small positive energy invites transitions between that ionic configuration and the dissociated neutral atoms. That can mean absorption and re-emission of infrared radiation that might otherwise quickly escape the planet.

The story of methane tells us that, like water, hydrocarbons too can occur with charge mirror-image ionic configurations, like the C^{4-} with H^+'s and the C^{4+}'s with H^-'s for methane. And like water, the charge mirror-image molecules may behave in profoundly different ways. One ionic configuration may be inert, while the other may react with oxygen molecules to produce normal water and energy.

A similar situation is true of all odd-numbered hydrocarbon chains. They admit two ionic configurations, one at higher energy (although not generally positive), and so relatively reactive, and the other at quite negative energy, and so relatively inert.

For odd-numbered hydrocarbon chains, the ionic configuration that we see under the conditions that humans typically experience is the higher-energy one. For this reason, the odd-numbered linear hydrocarbon chains make good fuels.

Even numbered hydrocarbon chains are a little different, as we see next.

4.2. Ethane C_2H_6

The ethane molecule is usually drawn as:

$$\begin{array}{cc} H & H \\ \cdot & \cdot \\ H \cdot C \cdot C \cdot H \\ \cdot & \cdot \\ H & H \end{array}$$

But again, all the bond angles are really tetrahedral.

Ethane does not admit two ionic configurations that are mirror images, because the mirror images are not different from each other. Ethane has one ionic configuration, $C^{4-} + 3H^+ + C^{4+} + 3H^-$. That ionic configuration takes energy

$$-54.6304 + 3 \times 14.1369 + 22.187 + 3 \times (-90.6769) =$$
$$-54.6304 + 42.4107 + 22.187 - 272.0307 = -262.0634 \text{ eV}$$

This energy is negative. Ethane is stable, except in the presence of a spark to initiate burning. The combustion reaction is:

$$2C_2H_6 + 7O_2 \rightarrow 4CO_2 + 6H_2O \ .$$

Again assuming most likely ionic configurations for all participant molecules, the left side of the combustion reaction takes:

$$2 \times (-262.0634) + 7 \times (-5.322) =$$
$$-524.1268 - 37.254 = -561.3808 \text{ eV}$$

and the right side takes:

$$4 \times (-32.5706) + 6 \times (-163.7925) =$$
$$-130.2824 - 982.755 = -1113.0374 \text{ eV}$$

So, although the left side is negative, the right side is more negative, and, once initiated, the combustion reaction proceeds. Indeed, ethane can be explosive.

Ethane can work as a fuel, and has historically been a constituent of natural gas. But today, ethane is considered less valuable as a fuel than as a petrochemical feed stock. Ethane is a polarized molecule, and that makes it a candidate for polymerization, and hence commercial utilization for purposes other than fuel.

Let us look at ethylene and then polyethylene. Ethylene is C_2H_4, or $2(CH_2)$, or a pair of two interior units from a hydrocarbon chain. The ionic configuration is $C^{4-} + 2H^+ + C^{4+} + 2H^-$, and that takes:

$$-54.6304 + 2 \times 14.1369 + 22.187 + 2 \times (-90.6769) =$$
$$-54.6304 + 28.2738 + 22.187 - 181.3538 = -185.5234 \text{ eV}$$

which is quite negative. So ethylene is quite stable. It is a natural plant hormone involved in growth, ripening, and decay of fruits and vegetables. We eat it.

Next look at the ethylene combustion reaction. It is:

$$C_2H_4 + 3O_2 \rightarrow 2CO_2 + 2H_2O \;.$$

The left side of the reaction likely takes:

$$-185.5234 + 3 \times (-5.322) = -185.5234 - 15.966 = -201.4894 \text{ eV},$$

and the right side of the reaction likely takes:

$$2 \times (-32.5706) + 2 \times (-163.7925) =$$
$$-65.1412 - 327.585 = -392.7262 \text{ eV}.$$

So ethylene combustion is robust, but nothing like that of ethane.

Polyethylene is mainly a big association of ethylene pairs. It can involve long chains, branches, and, especially, cross-links. It can be made, in all its various forms, starting from ethane. Ethane is a constituent of natural gas, and can be extracted through cryogenic cooling. This process works, in part, because ethane freezes before methane. That behavior is consistent with the negative energy of ethane, -262.0634 eV, since normal methane has (slightly) positive energy $+1.9172$ eV. But what of propane? Will ethane also freeze before propane?

4.3. Propane C_3H_8

The propane molecule can be drawn:

$$\begin{array}{ccc} H & H & H \\ \cdot & \cdot & \cdot \\ H\cdot C\cdot C\cdot C\cdot H \\ \cdot & \cdot & \cdot \\ H & H & H \end{array}$$

Because propane is an odd-numbered hydrocarbon chain, it admits two ionic configurations. One way to think of each is as two methyl ends and one ethyl middle:

$$2(C^{4-} + 3H^+) + 1(C^{4+} + 2H^-)$$

and

$$2(C^{4+} + 3H^-) + 1(C^{4-} + 2H^+).$$

Viewed in this way, the energy requirement are:

$$2 \times (-54.6304 + 3 \times 14.1369) + [22.187 + 2 \times (-90.6769)] =$$
$$2 \times (-54.6304 + 42.4107) + 22.187 - 181.3538 = -183.6062 \text{ eV}$$

and

$$2 \times [22.187 - 3 \times (-90.6769)] + [-54.6304 + 2 \times (14.1369)] =$$
$$2 \times (22.187 - 272.0307) - 54.6304 + 28.2738 = -526.044 \text{ eV}$$

But anticipating subsequent hydrocarbon chains, another way to think of these ionic configurations is as one methane molecule with an ethylene pair inserted. Viewed that way, the ionic configurations are:

$$(C^{4-} + 4H^+) + (C^{4+} + 2H^- + C^{4-} + 2H^+)$$

and

$$(C^{4+} + 4H^-) + (C^{4+} + 2H^- + C^{4-} + 2H^+)$$

and the energy requirements are:

$$1.9172 - 185.5234 = -183.6062 \text{ eV}$$

and

$-340.5206 - 185.5234 = -526.044$ eV

Viewed either way, the energy requirements are the same because the ions listed are the same. But the second way is a little simpler. And its pattern will be repeated in all subsequent linear-chain hydrocarbons. All those with odd number n will be like methane ($+1.9172$ eV) with $(n-1)/2$ ethylene pairs added (-185.5234 eV each). All those with even number n will be like ethane (-262.0634 eV) with $(n-2)/2$ ethylene pairs added (-185.5234 eV each). Ethylene pairs always require -185.5234 eV, so that is the increment by which both odd and even hydrocarbon-chain energy requirements decrease linearly with length.

In the case of all odd-numbered hydrocarbon chains, there are charge mirror-image ionic configurations, one at higher energy and the other at lower energy. In the case of propane, the higher-energy one is the fuel as used under normal pressure conditions, and the lower-energy one is the commodity in storage or transport under high pressure, as in LNG.

We earlier posed a question about the cryogenic-cooling method for separating ethane from natural gas. Would ethane freeze before propane? Now we see how energies of the two propane ionic configurations, at -183.6062 eV and -526.044 eV, bracket that of ethane, at -262.0634 eV. So yes, the ethane would surely freeze before the higher-energy form of propane.

The propane total combustion reaction can be written:

$$C_3H_8 + 5O_2 \rightarrow 3CO_2 + 4H_2O,$$

but it may be more informative to write it as:

$$3C_3H_8 + 15O_2 \rightarrow 9CO_2 + 12H_2O,$$

because this form helps reveal the developing pattern for all such total combustion reactions. We have:

methane: $CH_4 + 2O_2 \rightarrow CO_2 + 2H_2O$,

ethane: $2C_2H_6 + 7O_2 \rightarrow 4CO_2 + 6H_2O$,

propane : $3C_3H_8 + 15O_2 \rightarrow 9CO_2 + 12H_2O$,

and together these three cases suggest that the general case is:

$$NC_NH_{2N+2} + \left[N^2 + (N^2+N)/2\right]O_2 \rightarrow$$
$$N^2CO_2 + (N^2+N)H_2O$$

We see that, while the number of fuel molecules consumed grows as N, the number of water molecules produced grows as $N^2 + N$, which is faster, so the energy produced grows faster than the number of fuel molecules consumed. So the hydrocarbon chain fuels do get more energy-efficient with increasing chain length. This is one reason why the propane tanks are located on the outsides of summer cottages.

The propane chain molecule is also long enough that one can imagine dropping two Hydrogen atoms and making a ring. This is cyclo propane. It is written as C_3H_6, and can be drawn as:

$$H_2C \quad \bullet \quad CH_2$$
$$\bullet \quad \bullet$$
$$H\,C\,H$$

But this molecule is not likely to be as symmetrical as it is drawn, because if it were, the bond angles between the Carbon atoms would be $60°$ – quite far from the desired $109.5°$.

What might be the ionic configuration of cyclo propane? Unlike the chain propane molecule, the cyclo propane molecule cannot have Carbon atoms of alternating charge, because the number of Carbon atoms is not even. So for cyclo propane, we have to consider ionic configurations that use neutral units, like $(C^{2-} + 2H^+)$ or $(C^{2+} + 2H^-)$. A molecule made of such neutral units can hang together, but only with covalent bonding between Carbon atoms. We already know from the case of the bonding between Oxygen atoms in O_2 that covalent bonding is fraught with risk.

The unit $(C^{2-} + 2H^+)$ takes $-20.9313 + 2 \times 14.1369 = 7.3425$ eV, while the unit $(C^{2+} + 2H^-)$ takes $14.2505 + 2 \times (-90.6769) = -167.1033$ eV. The molecules take three times the unit energies: $3 \times 7.3425 = 22.0275$ eV, and $3 \times (-167.1033) = -501.3099$ eV. Again, the electronically better choice is the one using H^- ions, and for this molecule, two of them per Carbon is probably geometrically acceptable. That makes cyclo propane very stable. But the cyclo propane molecule is actually problematic, both because of the stressed geometry and because of the covalent bonding keeping it together. All this makes the molecule highly reactive, even explosive. It is also bioactive, as an anesthetic. Despite its dangers, it was used that way back in the 20th century.

It is true of all longer hydrocarbon chains that they admit cyclic structures. And like cyclo propane, the cyclic molecules derived from all odd numbered hydrocarbon chains have to have neutral units, and so have to rely on covalent bonding to hold them together.

Even numbered cyclic molecules are a little different, as we learn next.

4.4. Butane C_4H_{10}

There is no need to draw this; it is like ethane with an ethylene pair inserted. There is only one ionic configuration, and there is no need to detail the computation of its energy; it is that of the ethane (-262.0634 eV) and the ethylene pair (-185.5234 eV), which gives -447.5868 eV. This lies intermediate between the energies of the two propane ionic configuration, -183.6062 eV and -526.044 eV. It is a liquid, and it is a reliable and safe fuel for uses such as in cigarette lighters (especially in comparison to the danger of the cigarettes themselves).

As in the case of propane, there is a corresponding cyclic molecule. Cyclo butane C_4H_8 can be drawn as:

$$H_2C \cdot CH_2$$
$$\cdot \quad \cdot$$
$$H_2C \cdot CH_2$$

Like linear butane, and unlike cyclo propane, cyclo butane allows an ionic configuration involving charged units. That ionic configuration is $2\left[(C^{4-}+2H^+)+(C^{4+}+2H^-)\right]$, and it takes:

$$2\left[-54.6304 + 2 \times 14.1369 + 22.187 + 2 \times (-90.6769)\right] =$$
$$2(-54.6304 + 28.2738 + 22.187 - 181.3538) = -371.0468 \text{ eV}$$

There is no covalent bonding involved in this ionic configuration, so this cyclo butane is biologically safe. This assessment is confirmed by the fact that cyclo butane occurs naturally in biological systems.

Cyclo butane also admits structures like those of cyclo propane, with the neutral units, $C^{2-}+2H^+$ and / or $C^{2+}+2H^-$. With four units, cyclo butane can alternate between these possibilities (which was not possible with cyclo propane), and so avoid reliance on covalent bonding. This ionic configuration is $2(C^{2-}+2H^+)+2(C^{2+}+2H^-)$, and it takes:

$$2(-20.9313 + 28.2738) + 2(14.2505 - 181.3538) = -319.5216 \text{ eV}$$

This form can occur along with the form that has charged units, in its Boltzmann-appropriate proportion.

Butane is also large enough to permit a branching structure. This structure can be drawn:

$$\begin{array}{c} CH_3 \\ \cdot \\ H_3C \cdot C \cdot CH_3 \\ \cdot \\ H \end{array}$$

One could call it 'methyl propane'. Unlike linear butane, this form of butane does admit two ionic configurations. One of them is:

$$C^{4-} + H^+ + 3(C^{4+} + 3H^-),$$

and the other is:

$$C^{4+} + H^- + 3(C^{4-} + 3H^+).$$

In the first ionic configuration, the one $C^{4-} + H^+$, takes $-54.6304 + 14.1369 = -40.2614$ eV, and each of the three $C^{4+} + 3H^-$ takes $22.187 + 3 \times (-90.6769) = -249.8437$ eV, so the total energy requirement is:

$$C^{4-} + H^+ + 3(C^{4+} + 3H^-): -40.2614 + 3 \times (-249.8437) = -789.7925 \text{ eV}.$$

This energy being so negative, this branching form of butane is favored energetically. But it is probably geometrically awkward, and so may not occur without high pressure.

In the second ionic configuration, the $C^{4+} + H^-$ takes $22.187 - 90.6769 = -68.4899$ eV, and each $C^{4-} + 3H^+$ takes $-54.6304 + 3 \times 14.1369 = -12.2197$ eV, so the total energy requirement is

$$-68.4899 + 3 \times (-12.2197) = -105.149 \text{ eV}.$$

This energy is higher than the -789.7925 eV of the other ionic configuration for branched butane. It is also higher than the -447.5868 eV for linear butane. This energy is so high mainly because this ionic configuration has only one H^- ion in it.

Being at higher energy should make this branched butane molecule burn faster than the linear butane molecule. And it is indeed generally true that all subsequent linear hydrocarbons have branched variants that burn faster. Algebraic Chemistry is here revealing a quantitative 'reason why' for that fact.

4.5. Pentane C_5H_{12}

Linear pentane is like propane with an ethylene inserted. There are two ionic configurations,

$$\begin{array}{ccc} H^- & H^+ & H^- \\ \bullet & \bullet & \bullet \end{array}$$
$$3H^+C^{4-} \bullet C^{4+} \bullet C^{4-} \bullet C^{4+} \bullet C^{4-}3H^+$$
$$\begin{array}{ccc} \bullet & \bullet & \bullet \\ H^- & H^+ & H^- \end{array}$$

and

$$\begin{array}{ccc} H^+ & H^- & H^+ \\ \bullet & \bullet & \bullet \end{array}$$
$$3H^-C^{4+} \bullet C^{4-} \bullet C^{4+} \bullet C^{4-} \bullet C^{4+}3H^-$$
$$\begin{array}{ccc} \bullet & \bullet & \bullet \\ H^+ & H^- & H^+ \end{array}$$

with ion counts $3C^{4-}, 2C^{4+}, 8H^+, 4H^-$, and $2C^{4-}, 3C^{4+}, 4H^+, 8H^-$.

There is no need to detail the computation of their energies; they are those of propane (-183.6062 eV and -526.044 eV) and with another ethylene added (-185.5234 eV), or -369.1296 eV and -711.5674 eV. The higher-energy one is preferred functionally as a fuel, and, based on the pattern from propane and methane, also preferred geometrically.

The corresponding cyclic molecule, cyclo pentane C_5H_{10}, can be drawn as:

$$\begin{array}{cc} \text{H C H} & \\ \bullet & \bullet \\ H_2C & CH_2 \\ \bullet & \bullet \\ H_2C \bullet & CH_2 \end{array}$$

There are two possible ionic configurations, $5(C^{2+}+2H^-)$ and $5(C^{2-}+2H^+)$. The first one takes:

$$5\left[14.2505+2\times(-90.6769)\right]=5(14.2505-181.3538)=-835.5165 \text{ eV}$$

and the second one takes

$$5(-9.3046 + 2 \times 14.1369) = 5(-9.3046 + 28.2738) = 94.846 \text{ eV}$$

Only the first one looks realistically feasible.

Like butane, pentane also admits branching. One such structure is very symmetric:

$$\begin{array}{ccc} & CH_3 \quad CH_3 & \\ & \cdot \quad \cdot & \\ & C & \\ & \cdot \quad \cdot & \\ & CH_3 \quad CH_3 & \end{array}$$

It can reasonably be called 'dimethyl propane'. There are two mirror image ionic configurations, but the one more likely on geometry grounds is $C^{4+} + 4(C^{4-} + 3H^+)$. It takes

$$22.187 + 4 \times (-54.6304) + 12 \times 14.1369 =$$
$$22.187 - 218.5216 + 169.6428 = -26.6918 \text{ eV}$$

Compare this to the linear pentane molecules at -369.1296 eV and -711.5674 eV. This branched pentane is at much higher energy than either linear pentane, which means it burns much faster.

Another branched structure for pentane is:

$$\begin{array}{ccc} H \quad CH_3 & & H \quad CH_3 \\ \cdot \quad \cdot & & \cdot \quad \cdot \\ H_3C \cdot C \cdot C \cdot CH_3 \text{, or, equivalently, } & H_3C \cdot C \cdot C \cdot H \\ \cdot \quad \cdot & & \cdot \quad \cdot \\ H \quad H & & H \quad CH_3 \end{array}$$

Drawn the first way, it invites the name 'methyl butane'. Drawn the second way, it emphasizes perfect antisymmetry in placement of methyl groups and Hydrogen atoms.

For methyl butane, there are two mirror image ionic configurations:

$$\begin{array}{c} H^- \quad C^{4+} 3H^- \\ \cdot \quad \cdot \\ 3H^+C^{4-} \cdot C^{4+} \cdot C^{4-} \cdot C^{4+} 3H^- \\ \cdot \quad \cdot \\ H^- \quad H^+ \end{array}$$

and

$$\begin{array}{ccc} H^+ & C^{4-} & 3H^+ \\ \cdot & \cdot & \\ 3H^-C^{4+} \cdot C^{4+} \cdot C^{4-} \cdot & C^{4-}3H^+ \\ \cdot & \cdot & \\ H^+ & H^- & \end{array}$$

To evaluate these, it is sufficient to look at the ion counts. The first one has $3\,C^{4-}$, $2\,C^{4+}$, $4\,H^+$, $8\,H^-$, three identical to the first linear pentane, at $-369.1296\,\text{eV}$. The second one has $2\,C^{4-}$, $3\,C^{4+}$, eight $8\,H^+$, $4\,H^-$, identical to the second linear pentane at -711.5674 eV. So methyl butane is an example of branching without benefit.

4.6. Hexane C_6H_{14}

Linear hexane is like ethane (-262.0634 eV) with two ethylene pairs inserted ($2 \times (-185.5234) = -371.0468$ eV), which gives -633.1102 eV. Linear hexane is reliable but unremarkable as a fuel. All the even-numbered hydrocarbons are like that.

There are two hexane isomers that have branching that can be described as 'methyl pentane'. They are

$$\begin{array}{ccc} H_3C & H & H \\ \cdot & \cdot & \cdot \\ H_3C \cdot C \cdot C \cdot C \cdot CH_3 \\ \cdot & \cdot & \cdot \\ H & H & H \end{array} \quad \text{and} \quad \begin{array}{ccc} H & CH_3 & H \\ \cdot & \cdot & \cdot \\ H_3C \cdot C \cdot C \cdot C \cdot CH_3 \\ \cdot & \cdot & \cdot \\ H & H & H \end{array}$$

The first one has likely ionic configuration

$$\begin{array}{cccc} 3H^+C^{4-} & H^+ & H^- \\ \cdot & \cdot & \cdot \\ 3H^+C^{4-} \cdot C^{4+} \cdot C^{4-} \cdot C^{4+} \cdot C^{4-}3H^+ \\ \cdot & \cdot & \cdot \\ H^- & H^+ & H^- \end{array}$$

with ion count $4\,C^{4-}$, $2\,C^{4+}$, $11\,H^+$, $3\,H^-$. This implies energy

$$4 \times (-54.6304) + 2 \times 22.187 + 11 \times 14.1369 + 3 \times (-90.6769) =$$
$$-218.5216 + 44.374 + 155.5059 - 272.0307 = -290.6724 \text{ eV}$$

This is an improvement over linear hexane at -633.1102 eV. So it is unlike the case of methyl butane branching, which was without benefit.

The second branched hexane isomer has ionic configuration

$$H^- \quad C^{4+}3H^- \quad H^-$$
$$\cdot \quad \cdot \quad \cdot$$
$$3H^+C^{4-} \cdot C^{4+} \cdot C^{4-} \cdot C^{4+} \cdot C^{4-}3H^+$$
$$\cdot \quad \cdot \quad \cdot$$
$$H^- \quad H^+ \quad H^-$$

with ion count $3C^{4-}$, $3C^{4+}$, $7H^+$, $7H^-$. This is the same as linear hexane, and again shows the phenomenon of branching without benefit. Clearly, the benefit of branching depends on where the branching is located.

The cyclic molecule corresponding to hexane, C_6H_{12}, can be drawn as:

$$H_2C \quad \cdot \quad CH_2$$
$$\cdot \qquad\qquad \cdot$$
$$H_2C \qquad\qquad CH_2$$
$$\cdot \qquad\qquad \cdot$$
$$H_2C \quad \cdot \quad CH_2$$

Of the cyclic molecules drawn so far, the Carbon frame for this one comes closest to being actually flat, since the hexagon angle $120°$ is close to the tetrahedral angle $109.5°$. Cyclo hexane is a comfortable molecule, though of course a flammable one. Its ionic configuration must be similar to that of cyclo butane, with units of alternating charge. Being a liquid, cyclo hexane makes a useful solvent.

Cyclo hexane has an interesting molecular cousin, with only one Hydrogen atom, instead of two, on each Carbon atom. That cousin is the seriously toxic molecule benzene. What makes the big difference? Consider the units on the rings. Cyclo hexane can have units of alternating charge, like cyclo butane can have: $(C^{4-} + 2H^+)$ and $(C^{4+} + 2H^-)$. But for benzene, each unit can have only one Hydrogen atom. It turns out that the energetically best ionic configuration for benzene has six neutral units $(C^+ + H^-)$, each of which takes

$$9.6074 + (-90.6769) = -81.0695 \text{ eV}$$

At $6 \times (-81.0695) = -486.417$ eV, the benzene molecule looks stable. But with neutral units, the benzene molecule is entirely dependent on covalent Carbon-Carbon bonding. The Carbon atoms are not at all happy here. So, like the O_2 oxygen molecule, the benzene molecule wants to react in any way that it can.

4.7. Septane, Octane, and Beyond

Septanes have been the perfect fuel for passenger vehicles throughout the twentieth century. Linear septane is an odd-numbered hydrocarbon, all of which are good as fuels. And linear septane is in liquid state under most temperature / pressure conditions that occur on the surface of this planet.

Linear septane is like methane with three ethylene pairs added. Therefore, the energy of its likely ionic configuration is $1.9172 - 3 \times 185.5234 = -554.6530$ eV.

To look at linear septane in more detail, consider its first ionic configuration

$$
\begin{array}{ccccc}
H^- & H^+ & H^- & H^+ & H^- \\
\cdot & \cdot & \cdot & \cdot & \cdot \\
3H^+C^{4-} \cdot C^{4+} \cdot C^{4-} \cdot C^{4+} \cdot C^{4-} \cdot C^{4+} \cdot C^{4-} 3H^+ \\
\cdot & \cdot & \cdot & \cdot & \cdot \\
H^- & H^+ & H^- & H^+ & H^-
\end{array}
$$

with ion count $4 C^{4-}, 3 C^{4+}, 10 H^+, 6 H^-$. This leads to the energy

$$4 \times (-54.6304) + 3 \times 22.187 + 10 \times 14.1369 + 6 \times (-90.6769) =$$
$$-218.5216 + 66.561 + 141.369 - 544.0614 = -554.6530 \text{ eV}$$

Linear septane in this first ionic configuration is the desirable fuel, and it is geometrically comfortable. For comparison, the charge mirror image ionic configuration, with ion count $3 C^{4-}, 4 C^{4+}, 6 H^+, 10 H^-$, is geometrically uncomfortable, and it leads to a poorer energy result

$$4 \times 22.187 + 3 \times (-54.6304) + 10 \times (-90.6769) + 6 \times 14.1369 =$$
$$88.748 - 163.8912 - 906.769 + 84.8214 = -897.0908 \text{ eV}$$

But it is impossible to produce a gasoline product that is completely free of all molecules other than the desired linear septane molecules. One can expect, for example, that the product will contain some linear hexane molecules, and some linear octane molecules, and some branched molecules of various sorts. The penalty for having a variety of molecules is having a variety of burn rates, possibly leading to timing problems in the combustion cycle. If some molecules burn slower than others, then there will be engine knock.

So getting an automotive engine to work smoothly on a fuel that is feasible to produce appears to require some kind of intervention. We use fuel additives to improve engine performance. In the twentieth century, we relied on tetraethyl lead, but that left a residue of toxic lead in the environment, and has been banned. Today, the legal fuel additive is called 'trimethyl pentane'. It is one of many forms of octane, which has 18 structural isomers, or 24 isomers counting the steroisomers, or even more, if we distinguish charge mirror image ionic configurations.

Plain linear octane is like ethane with three ethylene pairs added. Therefore it has energy

$$-262.0634 - 3 \times 185.5234 = -262.0634 - 556.5702 = -818.6336 \text{ eV}$$

That is notably lower than linear septane at -554.6530 eV, and doesn't look very helpful. But as we have seen, branching can make a difference. Trimethy pentane is octane with lots of branches. Where linear octane has only two methyl groups, trimethyl pentane has five methyl groups, meaning three branch points have been created.

But the number of branches is not enough information to specify what the desirable trimethyl pentane really is. The name admits three isomers:

$$\begin{array}{ccc} H_3C & H & CH_3 \\ \bullet & \bullet & \bullet \\ H_3C \bullet C \bullet C \bullet & C \bullet CH_3 \\ \bullet & \bullet & \bullet \\ H_3C & H & H \end{array},$$

and

$$\begin{array}{ccc} H_3C & CH_3 & H \\ \bullet & \bullet & \bullet \\ H_3C \bullet C \bullet C \bullet & C \bullet CH_3 \\ \bullet & \bullet & \bullet \\ H_3C & H & H \end{array},$$

and

$$\begin{array}{ccc} H_3C & CH_3 & H \\ \bullet & \bullet & \bullet \\ H_3C \bullet C \bullet C \bullet & C \bullet CH_3 \\ \bullet & \bullet & \bullet \\ H & H_3C & H \end{array}$$

And each of those admits two ionic configurations. Algebraic Chemistry allows one to evaluate all of these possibilities, and quickly see which one is the useful fuel additive.

The first trimethyl pentane molecules listed admits as its first ionic configuration

$$3H^+C^{4-} \quad H^+ \quad C^{4-}3H^+$$
$$\cdot \quad \cdot \quad \cdot$$
$$3H^+C^{4-} \cdot C^{4+} \cdot C^{4-} \cdot C^{4+} \cdot C^{4-}3H^+$$
$$\cdot \quad \cdot \quad \cdot$$
$$3H^+C^{4-} \quad H^+ \quad H^-$$

with ion count $6\,C^{4-}$, $2\,C^{4+}$, $17\,H^+$, $1\,H^-$. That leads to energy

$$6 \times (-54.6304) + 2 \times 22.187 + 17 \times 14.1369 + 1 \times (-90.6769) =$$
$$-327.7824 + 44.374 + 240.3273 - 90.6769 = -133.758 \text{ eV}$$

This is the best of the six possibilities allowed by the name 'trimethyl pentane'. It is both appropriate as a fuel, and comfortable geometrically.

The effectiveness of trimethyl pentane as a fuel additive is generally attributed to the speed of burning that comes with all the branching. But Algebraic Chemistry can suggest that there is more to the story. Consider the possibility that one reason for the existence engine knock is the presence within the fuel, which is mainly linear septane, of a contaminant: traces of slower-burning linear hexane. Maybe the trimethyl pentane does something to the linear hexane.

What can happen if a molecule of trimethyl pentane encounters a molecule of linear hexane somewhere in the system, way before the combustion chamber – say, in the fuel tank? Can a reaction occur? Consider the possibility trimethyl pentane + linear hexane → 2 linear septanes .

We already know that the trimethyl pentane has energy -133.758 eV, and that linear hexane has energy -633.1102 eV. So the left side of this reaction represents

$$-133.758 - 633.1102 = -766.8682 \text{ eV}.$$

We also already know that linear septane has energy is -554.6530 eV. This means the right side of the proposed reaction involves

$$2 \times (-554.6530) = -1109.306 \text{ eV}.$$

This result is more negative than the result for the left side of the proposed reaction. So clearly, if these reactants are both present, the proposed reaction ensues. It releases energy

$$1109.306 - 7668682 = 342.4378 \text{ eV}.$$

And, more importantly, it would clean up the linear hexane, and leaves two good linear-septane fuel molecules.

But to complete the picture, we should also ask what happens when the trimethyl pentane molecule encounters, not a rare linear hexane molecule, but rather an ordinary linear septane molecule. Those two might, for example, reform into a methyl hexane molecule and a methyl septane molecule:

Trimethyl pentane + linear septane → methyl hexane + methyl septane.

First consider the methyl hexane product. It is a singly branched septane, and presents two structural isomers,

$$
\begin{array}{c}
\text{CH}_3 \quad \text{H} \quad \text{H} \quad \text{H} \\
\cdot \quad \cdot \quad \cdot \quad \cdot \\
\text{H}_3\text{C} \cdot \text{C} \cdot \text{C} \cdot \text{C} \cdot \text{C} \cdot \text{CH}_3 \\
\cdot \quad \cdot \quad \cdot \quad \cdot \\
\text{H} \quad \text{H} \quad \text{H} \quad \text{H}
\end{array}
\quad \text{and} \quad
\begin{array}{c}
\text{H} \quad \text{CH}_3 \quad \text{H} \quad \text{H} \\
\cdot \quad \cdot \quad \cdot \quad \cdot \\
\text{H}_3\text{C} \cdot \text{C} \cdot \text{C} \cdot \text{C} \cdot \text{C} \cdot \text{CH}_3 \\
\cdot \quad \cdot \quad \cdot \quad \cdot \\
\text{H} \quad \text{H} \quad \text{H} \quad \text{H}
\end{array}
$$

The first one has likely ionic configuration

$$
\begin{array}{c}
\text{C}^{4-}3\text{H}^+ \quad \text{H}^+ \quad \text{H}^- \quad \text{H}^+ \\
\cdot \quad \cdot \quad \cdot \quad \cdot \\
3\text{H}^+\text{C}^{4-} \cdot \text{C}^{4+} \cdot \text{C}^{4-} \cdot \text{C}^{4+} \cdot \text{C}^{4-} \cdot \text{C}^{4+}3\text{H}^- \\
\cdot \quad \cdot \quad \cdot \quad \cdot \\
\text{H}^- \quad \text{H}^+ \quad \text{H}^- \quad \text{H}^+
\end{array}
$$

with ion count 4C^{4-}, 3C^{4+}, 10H^+, 6H^-. This is the same as the favored linear septane, and leads to energy -554.6530 eV. The situation here is the same as it was with methyl butane as a pentane: branching with no benefit.

So to try for a more interesting example, consider the second isomer of methyl hexane, which has first ionic configuration

$$
\begin{array}{c}
\text{H}^- \quad \text{C}^{4+}3\text{H}^- \quad \text{H}^- \quad \text{H}^+ \\
\cdot \quad \cdot \quad \cdot \quad \cdot \\
3\text{H}^+\text{C}^{4-} \cdot \text{C}^{4+} \cdot \text{C}^{4-} \cdot \text{C}^{4+} \cdot \text{C}^{4-} \cdot \text{C}^{4+}3\text{H}^- \\
\cdot \quad \cdot \quad \cdot \quad \cdot \\
\text{H}^- \quad \text{H}^+ \quad \text{H}^- \quad \text{H}^+
\end{array}.
$$

This has ion count 3C^{4-}, 4C^{4+}, 6H^+, 10H^-, the same as the geometrically unfavorable linear septane, which makes for energy -897.0908 eV.

Of course the charge mirror image ionic configuration

$$H^+ \quad C^{4-}3H^+ \quad H^+ \quad H^-$$
$$\cdot \qquad \cdot \qquad \cdot \qquad \cdot$$
$$3H^-C^{4+} \cdot C^{4-} \cdot C^{4+} \cdot C^{4-} \cdot C^{4+} \cdot C^{4-}3H^+$$
$$\cdot \qquad \cdot \qquad \cdot \qquad \cdot$$
$$H^+ \quad H^- \quad H^+ \quad H^-$$

has ion count $4C^{4-}$, $3C^{4+}$, $10H^+$, $6H^-$, the same as the favored linear septane, and leads to energy -554.6530 eV. In short, the phenomenon of branching with no benefit always occurs with single branching of septane as methyl hexane, and by extension, probably occurs with single branching of all odd-numbered hydrocarbons.

Now consider the methyl septane. It is a singly branched octane. There are three structural isomers:

$$\begin{array}{c} CH_3 \quad H \quad H \quad H \quad H \\ \cdot \quad \cdot \quad \cdot \quad \cdot \quad \cdot \\ H_3C \cdot C \cdot C \cdot C \cdot C \cdot C \cdot CH_3 \\ \cdot \quad \cdot \quad \cdot \quad \cdot \quad \cdot \\ H \quad H \quad H \quad H \quad H \end{array},$$

and

$$\begin{array}{c} H \quad CH_3 \quad H \quad H \quad H \\ \cdot \quad \cdot \quad \cdot \quad \cdot \quad \cdot \\ H_3C \cdot C \cdot C \cdot C \cdot C \cdot C \cdot CH_3 \\ \cdot \quad \cdot \quad \cdot \quad \cdot \quad \cdot \\ H \quad H \quad H \quad H \quad H \end{array},$$

and

$$\begin{array}{c} H \quad H \quad CH_3 \quad H \quad H \\ \cdot \quad \cdot \quad \cdot \quad \cdot \quad \cdot \\ H_3C \cdot C \cdot C \cdot C \cdot C \cdot C \cdot CH_3 \\ \cdot \quad \cdot \quad \cdot \quad \cdot \quad \cdot \\ H \quad H \quad H \quad H \quad H \end{array}.$$

The first methyl septane has likely ionic configuration

$$3H^+C^{4-} \quad H^+ \quad H^- \quad H^+ \quad H^-$$
$$\cdot \qquad \cdot \qquad \cdot \qquad \cdot \qquad \cdot$$
$$3H^+C^{4-} \cdot C^{4+} \cdot C^{4-} \cdot C^{4+} \cdot C^{4-} \cdot C^{4+} \cdot C^{4-}3H^+$$
$$\cdot \qquad \cdot \qquad \cdot \qquad \cdot \qquad \cdot$$
$$H^- \quad H^+ \quad H^- \quad H^+ \quad H^-$$

with ion count $5\,C^{4-}$, $3\,C^{4+}$, $13\,H^{+}$, $5\,H^{-}$, which implies energy

$$5 \times (-54.6304) + 3 \times 22.187 + 13 \times 14.1369 + 5 \times (-90.6769) =$$
$$-273.1520 + 66.561 + 183.779 - 453.3845 = -476.1965 \text{ eV}$$

This is an improvement over linear septane at -554.6530 eV, and a big improvement over linear octane at -818.6336 eV.

The charge mirror image ion count $3\,C^{4-}$, $5\,C^{4+}$, $5\,H^{+}$, $13\,H^{-}$ implies energy

$$3 \times (-54.6304) + 5 \times 22.187 + 5 \times 14.1369 + 13 \times (-90.6769) =$$
$$-163.8912 + 110.935 + 70.6845 - 1178.7997 = -1161.0714 \text{ eV}$$

which is pretty useless.

The second methyl septane has ionic configuration is

$$
\begin{array}{ccccc}
H^- & 3H^-C^{4+} & H^- & H^+ & H^- \\
\bullet & \bullet & \bullet & \bullet & \bullet \\
3H^+C^{4-}\bullet C^{4+}\bullet C^{4-}\bullet C^{4+}\bullet C^{4-}\bullet C^{4+}\bullet C^{4-}3H^+ \\
\bullet & \bullet & \bullet & \bullet & \bullet \\
H^- & H^+ & H^- & H^+ & H^-
\end{array}
$$

with ion count $4\,C^{4-}$, $4\,C^{4+}$, $9\,H^{+}$, $9\,H^{-}$, which implies energy

$$4 \times (-54.6304) + 4 \times 22.187 + 9 \times 14.1369 + 9 \times (-90.6769) =$$
$$-218.5216 + 88.748 + 127.2321 - 816.0921 = -818.6336 \text{ eV}$$

and of course its charge mirror image is the same, and not useful.

The third methyl septane has ionic configuration is

$$
\begin{array}{ccccc}
H^- & H^+ & C^{4-}3H^+ & H^+ & H^- \\
\bullet & \bullet & \bullet & \bullet & \bullet \\
3H^+C^{4-}\bullet C^{4+}\bullet C^{4-}\bullet C^{4+}\bullet C^{4-}\bullet C^{4+}\bullet C^{4-}3H^+ \\
\bullet & \bullet & \bullet & \bullet & \bullet \\
H^- & H^+ & H^- & H^+ & H^-
\end{array}
$$

with ion count $5\,C^{4-}$, $3\,C^{4+}$, $13\,H^{+}$, $5\,H^{-}$, like the first methyl septane, which again implies energy -476.1965 eV.

Thus for the reaction proposed,

trimethyl pentane + linear septane → methyl hexane + methyl septane,

the left side of the reaction, trimethyl pentane + linear septane, takes

$$-133.758 \text{ eV} - 554.6530 \text{ eV} = -688.411 \text{ eV}$$

and the right side of the reaction, methyl hexane + methyl septane, takes

$$-554.6530 \text{ eV} - 476.1965 \text{ eV} = -1030.8495 \text{ eV}.$$

So this reaction will, indeed, transpire. It releases

$$1030.8495 - 688.411 = 342.4385 \text{ eV}.$$

This release is comparable to the 342.4378 eV from the earlier analysis of trimethyl pentane + linear hexane → two linear septanes.

The numbers are almost identical, and again the difference between them arises from round-off in the arithmetic. A detailed examination of the ionization states involved shows that both reactions cause the same profile of changes: $-1C^{4-}$, $+1C^{4+}$, $-4H^+$, $+4H^-$, and hence the same energy release.

This all means that trimethyl pentane as fuel additive for a mainly septane fuel with a trace of linear hexane will quickly disappear, before interacting with any of the linear hexane. It will produce two good fuel molecules - methyl hexane and methyl septane, but it won't reduce the linear hexane.

Fortunately, the methyl septane can go on to interact with the linear hexane. Consider the reaction

methyl septane + linear hexane → two linear septanes

The left side of this reaction embodies $-476.1965 - 633.1102 = -1109.3067$ eV, and the right side of the reaction embodies $2 \times (-554.6530) = -1109.306$ eV. These numbers are so close that one may suspect that they differ only on account of round off in the arithmetic. A quick check of the ionic configurations involved confirms that this is indeed the case:

methyl septane has $5 C^{4-}$, $3 C^{4+}$, $13 H^+$, $5 H^-$,

linear hexane has $3 C^{4-}$, $3 C^{4+}$, $7 H^+$, $7 H^-$,

left side has sum $8 C^{4-}$, $6 C^{4+}$, $20 H^+$, $12 H^-$;

linear septane has $4\,C^{4-}$, $3\,C^{4+}$, $10\,H^+$, $6\,H^-$,

right side has double $8\,C^{4-}$, $6\,C^{4+}$, $20\,H^+$, $12\,H^-$.

This means the reaction really takes a double arrow:

methyl septane + linear hexane ↔ two linear septanes

So the cleanup looks a bit half-hearted.

This analysis suggests that trimethyl pentane may not survive long enough to speed the action in the combustion chamber. But it may somewhat indirectly improve the mix of fuel molecules back in the fuel tank.

Trimethyl pentane is certainly considered expensive to use. That must mean it is expensive to produce. What can be said of the production process? Surely it is better to find it occurring naturally among normal petroleum refining products than to synthesize it specially. So it is useful to have an idea of where to look for it. If trimethyl pentane exists at all among normal petroleum refining products, its high energy means it probably exists in the gas state, and therefore within the natural-gas bubble that forms in the fractionation column. It would probably be found near the bottom of the bubble. This assessment follows from the correlation between ionic configurations and macroscopic states of matter proposed in the preceding Chapter.

But can something other than trimethyl pentane do the needed job just as well, but more economically? Consider this fact: the scenario described above is using trimethyl pentane to produce both methyl septane and methyl hexane. But what is the benefit of making any methyl hexane? It has the same ionic configuration as linear septane does, $4\,C^{4-}$, $3\,C^{4+}$, $10\,H^+$, $6\,H^-$, and so carries the same energy, -554.6530 eV. So there is no obvious benefit to creating the methyl hexane. And there is a cost for the process that makes it: we have to provide the trimethyl pentane. We could instead make a gasoline additive out of just methyl septane itself, and it might be much cheaper to make.

4.8. A Brief Revisit to Hydrocarbons and States of Matter

The family of hydrocarbons does a good job of illustrating the correlation between ionic configurations and states of matter that was proposed in the preceding Chapter.

- The lowest-energy ionic configurations occur with the heaviest hydrocarbons. Those are found in solid coal.
- The intermediate-energy ionic configurations occur with the liquid fuels –gasoline, kerosene, oil. Those are all liquids.
- The highest-energy ionic configurations occur with the lightest hydrocarbons: methane, ethane, and propane. Under normal pressure conditions, those are gasses.

- The flame that they all exhibit when provided with Oxygen and a spark is certainly the plasma state.

So, does hydrocarbon combustion support the correlation between plasma state and higher-energy ionization states that was proposed in the last Chapter? Consider for example the methane combustion reaction

$$CH_4 + 2O_2 \rightarrow CO_2 + 2H_2O$$

Expressed in terms of the likeliest ionic configurations, the reaction goes

$$(C^{4-} + 4H^+) + 2(O^{2+} + O^{2-}) \rightarrow (C^{4+} + 2O^{2-}) + 2(2H^- + O^{2+})$$

Observe that the set of Oxygen ionic configurations does not change; it remains $2(O^{2+} + O^{2-})$. Only the Carbon and Hydrogen ionic configurations change. Those changes are:

$$C^{4-} \rightarrow C^{4+}, \text{ or } -54.6304 \rightarrow 22.187 \text{ eV}$$

and

$$4H^+ \rightarrow 4H^-, \; 4 \times 14.1369 \rightarrow 4 \times (-90.6769)$$

or

$$56.5476 \rightarrow -362.7076 \text{ eV}$$

The Carbon ion energies average to

$$(-54.6304 + 22.187)/2 = -16.2217 \text{ eV},$$

and the Hydrogen ion energies average to

$$(56.5476 - 362.7076)/2 = -153.08 \text{ eV},$$

so the Hydrocarbon ions overall averages to

$$-16.2217 - 153.08 = -169.3017 \text{ eV}.$$

The Oxygen ion energies remain steady at

$$2 \times (17.5613 - 27.3788) = -19.635 \text{ eV}.$$

This means the presence of Oxygen is a big factor in raising the average energy of ions in the mix. So high ionic-configuration energy can indeed be said to correlate with the combustion scene being the flaming plasma that it is.

CONCLUSION

The light hydrocarbons exhibit a number of interesting behaviors. In this Chapter, Algebraic Chemistry has been shown to comport with those behaviors, and to offer a numerical accounting for them in terms of ionic configurations. We learn why methane has both a familiar fuel version and an exotic methane hydrate version: methane has two possible ionic configurations. And so do all odd-numbered linear hydrocarbons. We learn that even-numbered linear hydrocarbons don't have this feature, and as a result their performance as fuels is different.

The combustion of hydrocarbons is a major topic that deserves much further discussion. This subject is explored in the next Chapter. So far, combustion reactions have been presented as if they occurred in one step, but in actuality, they occur in stages. A rich story flows from that.

A PROJECT FOR READERS

Many of you have on your shelves some Chemistry textbooks that you have studied from. Try re-reading them, this time armed with the techniques of Algebraic Chemistry. You may see arguments that can be made more rigorous by the inclusion of numerical analysis. Those would be good topics for student term papers, thesis chapters, *etc*.

Chapter 6

IMPORTANT REACTIONS

ABSTRACT

This Chapter examines hydrocarbon combustion and associated technologies as an example class of reactions that are economically important. The purpose is to illustrate the application of the analysis technique that Algebraic Chemistry provides, and to develop some insights about hydrocarbon combustion itself, especially for the case of hydrocarbon fuels for inernal combustion engines in vehicles. The analysis technique is based on the possible ionic configurations of molecules, and the evaluation of the energies that the ionic configurations imply. The associated issues addressed include the use gasoline additives to address engine knock, and the use of catalytic converters to clean up noxious constituents in combustion exhaust. The particular additive evaluated is the methyl septane, which was discussed in the last Chapter. The particular catalysts evaluated are representative ones that also turn up in other problems addressed later in this Book.

INTRODUCTION

Section 1 is about hydrocarbon combustion in general, and how we write combustion equations about it for the purposes of analyses. There is much history [1], some of it to be overcome. People have often written combustion reactions with attention to proportions, which sometimes led to presentations with fractional molecules, which cannot correspond to physical reality. With fractional molecules disallowed, another problem emerges; namely, that odd-numbered hydrocarbons look very different from even-numbered ones, as to whether combustion requires one molecule or two molecules. We recommend that two molecules be presumed for all cases, and find that choice vindicated in Section 2.

Section 2 is about the four steps of hydrocarbon combustion. The model hydrocarbon is methane. The first step of combustion is the only one that is particular to methane. So that step is developed and detailed for the several vehicle fuel hydrocarbons in Section 3. Then in Section 4, the combustion-cycle timing implications are discussed. Section 5 is about cleaning up the exhaust that hydrocarbon combustion leaves. The action of several example catalysts is spelled out. Section 6 collects conclusions.

1. Hydrocarbon Combustion

To enable comparisons between different chemical reactions that are related to each other, it is generally customary to present them all in some common format. For the studies to follow, it is important to do this for the total combustion reactions of hydrocarbon chain molecules. We need to see them all in one format that fits them all.

Furthermore, we need that common format to correspond to real physical events. That means: without a notation involving fractional molecules. The use of fractions is common in discussions about relative proportions of reactants, but it isn't good for discussing actual chemical reactions since there are not actual fractional molecules.

Let us begin with expressions using only integer molecules. The pattern that emerges is the following:

The methane total combustion reaction looks like

$$CH_4 + 2O_2 \rightarrow CO_2 + 2H_2O ,$$

but the ethane total combustion reaction looks like

$$2C_2H_6 + 7O_2 \rightarrow 4CO_2 + 6H_2O ;$$

The propane total combustion reaction looks like

$$C_3H_8 + 5O_2 \rightarrow 3CO_2 + 4H_2O ,$$

but the butane total combustion reaction looks like

$$2C_4H_{10} + 13O_2 \rightarrow 8CO_2 + 10H_2O ;$$

The pentane total combustion reaction looks like

$$C_5H_{12} + 8O_2 \rightarrow 5CO_2 + 6H_2O ,$$

but the hexane total combustion reaction looks like

$$2C_6H_{14} + 19O_2 \rightarrow 12CO_2 + 14H_2O ;$$

The septane total combustion reaction looks like

$$C_7H_{16} + 11O_2 \rightarrow 7CO_2 + 8H_2O ,$$

but the octane total combustion reaction looks like

$$2C_8H_{18} + 25O_2 \rightarrow 16CO_2 + 18H_2O\ .$$

The pattern here is that the total combustion reactions for odd-numbered hydrocarbons look as if they involve a single hydrocarbon molecule, but the total combustion reactions for even-numbered hydrocarbons demand two hydrocarbon molecules. That pattern is a warning that the odd-numbered hydrocarbons probably require two molecules too, even if they don't look that way at first.

Heeding that warning, the set of total combustion reactions of interest here goes:

$$2CH_4 + 4O_2 \rightarrow 2CO_2 + 4H_2O$$

$$2C_2H_6 + 7O_2 \rightarrow 4CO_2 + 6H_2O$$

$$2C_3H_8 + 10O_2 \rightarrow 6CO_2 + 8H_2O$$
$$2C_4H_{10} + 13O_2 \rightarrow 8CO_2 + 10H_2O$$

$$2C_5H_{12} + 16O_2 \rightarrow 10CO_2 + 12H_2O$$

$$2C_6H_{14} + 19O_2 \rightarrow 12CO_2 + 14H_2O$$

$$2C_7H_{16} + 22O_2 \rightarrow 14CO_2 + 16H_2O$$

$$2C_8H_{18} + 25O_2 \rightarrow 16CO_2 + 18H_2O$$

The general case can be written

$$2C_NH_{2N+2} + \left[4 + 3(N-1)\right]O_2 \rightarrow 2NCO_2 + 2(N+1)H_2O\ .$$

2. Hydrocarbon Combustion in Steps

In the last Section, hydrocarbon combustion reactions were all summarized to a single total reaction. But combustion reactions actually occur in a number of steps. Many other reactions are like this too, so combustion reactions make good examples to illustrate ideas that are of general utility. In the case of combustion, the focus on reaction stages has practical important because any interruption of the staged process can leave partial combustion products that are toxic. When that happens, it is a serious problem.

2.1. Steps in Methane Combustion

Methane combustion illustrates the staging of general hydrocarbon combustion. The total methane combustion reaction goes

$$2CH_4 + 4O_2 \rightarrow 2CO_2 + 4H_2O \ .$$

Observe that for every Carbon atom processed, two oxygen molecules go in, and two water molecules go out.
The first step of methane combustion goes

$$2CH_4 + 2O_2 \rightarrow 2H_2O + 2H_2CO \ .$$

where H_2CO is formaldehyde. Observe that two oxygen molecules go in, and two water molecules go out in this step.

The second step gives the formyl radical HCO. This can step be written:

$$2H_2CO \rightarrow 2HCO + H_2 \ .$$

Observe that a hydrogen molecule comes out. It will be used next.
The third step then gives carbon monoxide, CO, and water H_2O.

$$2HCO + H_2 + O_2 \rightarrow 2CO + 2H_2O \ .$$

Observe that another oxygen molecule goes in, and two water molecules come out.
A fourth and last step oxidizes the carbon monoxide

$$2CO + O_2 \rightarrow 2CO_2 \ .$$

Observe that another oxygen molecule is used in this last step.

Overall, four oxygen molecules have been consumed, and four water molecules have been produced, so the stated overall reaction $2CH_4 + 4O_2 \rightarrow 2CO_2 + 4H_2O$ has been fulfilled.

But the individual steps exist, and can become desynchronized, so the overall process can stop short of completion. This can leave interim products, the H_2CO formaldehyde, the HCO formyl radical, and the CO carbon monoxide. They are toxic. So we need to study the individual reaction steps. Algebraic Chemistry is a tool for this.

Algebraic Chemistry generally deals with energies. The relative speed of each reaction step should be correlated with the energy released at that step. Generally, one expects an inverse relationship: a large energy release makes for a speedy reaction step. This expectation comes from Quantum Mechanics, which gives an uncertainty relationship $\Delta e \Delta t \geq h$, where Δ means uncertainty, e means energy, t means time, and h is Planck's constant.

2.2. Energies from Steps in Methane Combustion

Details about energy releases from the steps of methane combustion can be quantified from ionic configurations of the participant molecules. With that approach, we can evaluate each reaction step within the overall combustion reaction $2CH_4 + 4O_2 \rightarrow 2CO_2 + 4H_2O$, and establish an energy release for each reaction step.

Consider the first reaction step, $2CH_4 + 2O_2 \rightarrow 2H_2O + 2H_2CO$. On the left side of the reaction, each methane molecule has likely ionic configuration $C^{4-} + 4H^+$, with energy +1.9172 eV. Each oxygen molecule allows two believable ionic configurations,

$O^+ + O^-$ at $7.9399 - 12.4354 = -4.4955$ eV,

and

$O^{2+} + O^{2-}$ at $17.5613 - 27.3788 = -9.8175$ eV.

Both of these ionic configurations are probably present in Nature, but the lower-energy one will be more plentiful. Assuming that ionic configuration to be dominant, the left side of the reaction involves energy:

$2 \times (1.9172 - 9.8175) = -15.8006$ eV.

On the right side of the reaction, each water molecule has likely ionic configuration $2H^- + O^{2+}$ at

$2 \times (-90.6769) + 17.5613 = -181.3538 + 17.5613 = -163.7925$ eV.

Each formaldehyde molecule H_2CO has two possible ionic configurations, $2H^+C^{4-}O^{2+}$ and $2H^-C^{4+}O^{2-}$. The first one, $2H^+C^{4-}O^{2+}$, has energy

$2 \times 14.1369 + (-54.6304) + 17.5613 = -8.7953$ eV.

The second one, $2H^-C^{4+}O^{2-}$, has energy

$$2\times(-90.6769)+22.187-27.3788=$$
$$-181.3538+22.187-27.3788=-186.5456 \text{ eV}$$

This second ionic configuration looks more likely. The right side of the reaction then embodies total energy

$$2\times(-163.7925-186.5456)=-700.6762 \text{ eV}.$$

The first reaction step then delivers energy

$$700.6762-15.8006=684.8756 \text{ eV}.$$

This is robust, meaning this reaction step will proceed quickly.

Now consider the second reaction step, $2H_2CO \rightarrow 2HCO + H_2$. On the left side of this reaction, the two formaldehyde molecules have energy $2\times(-186.5456)=-373.0912$ eV, and that energy represents the left side of the reaction. On the right side of the reaction, a formyl radical HCO has two possible ionic configurations, $H^+C^{2-}O^+$ and $H^-C^{2+}O^-$. The $H^+C^{2-}O^+$ ionic configuration has energy

$$14.1369+(-20.9313)+7.9399=+1.1455 \text{ eV},$$

and the $H^-C^{2+}O^-$ ionic configuration has energy

$$-90.6769+14.2505-27.3788=-103.8052 \text{ eV}.$$

This second ionic configuration, $H^-C^{2+}O^-$, is the more likely one. The hydrogen molecule has to have ionic configuration H^+H^-, with energy

$$14.1369+(-90.6769)=-76.54 \text{ eV}$$

So the right side of the reaction likely has energy $-103.8052-76.54=-180.3452$ eV. So the second reaction step likely releases

$$180.3452-17.5906=162.7546 \text{ eV}.$$

This is much less energy than the first reaction step released, 684.8756 eV. So this reaction step likely takes more time than the first reaction step did. So the H_2CO may temporarily build up.

Now consider the third reaction step,

$$2HCO + H_2 + O_2 \rightarrow 2CO + 2H_2O$$

On the left side of the reaction, the two formyl radicals embody $2.291\,eV$, the hydrogen molecule embodies $-76.54\,eV$, and the oxygen molecule embodies $-9.8175\,eV$. So the left side of the reaction embodies

$$2.291 - 76.54 - 9.8175 = -84.0665\,eV.$$

On the right side of the reaction, each carbon monoxide CO molecule has likely ionic configuration $C^{2+}O^{2-}$ with energies $14.2505\,eV$ for C^{2+} and $-27.3788\,eV$ for O^{2-}, which makes

$$14.2505 - 27.3788 = -13.1283\,eV$$

for each CO molecule. Each water molecule embodies $-163.7925\,eV$. So, altogether, the right side of the reaction embodies

$$2 \times (-13.1283) + 2 \times (-163.7925) =$$
$$-26.2566 - 327.585 = -353.8416\,eV$$

So the reaction delivers $353.8416 - 84.0665 = 269.7751\,eV$. That is robust, and makes for speedy completion of this third reaction step.

Now consider the fourth and final reaction step: $2CO + O_2 \rightarrow 2CO_2$. On the left side, two carbon monoxide CO molecules embody $-26.2566\,eV$ and the oxygen O_2 molecule embodies $-9.8175\,eV$, so the left side of the reaction embodies

$$-26.2566 - 9.8175 = -36.0741\,eV.$$

On the right side of the reaction, each carbon dioxide CO_2 molecule has possible ionic configurations $C^{4+} + 2O^{2-}$ and $C^{4-} + 2O^{2+}$, with energies

$$22.187 + 2 \times (-27.3788) = 22.187 - 54.7576 = -32.5706\,eV,$$

and

$$-54.6304 + 2 \times 17.5613 = -54.6304 + 35.1226 = -19.5078\,eV.$$

The first of these ionic configurations is the more likely one, and assuming it to be the dominant one, the right side of the reaction embodies $-65.1412\,eV$. The reaction then releases $65.1412 - 36.0741 = 29.0671\,eV$. This is really modest compared to the third reaction step, at $269.7751\,\text{eV}$. This fourth reaction step will go slow in comparison. This circumstance could result in build up of carbon monoxide that will then appear in the combustion exhaust.

From the methane example, we see how Algebraic Chemistry can give a numerical quantification for any toxicity worries associated with hydrocarbon combustion.

2.3. Focus on the First Step of Methane Combustion

Observe that in the methane combustion reaction

$$2CH_4 + 2O_2 \rightarrow 2H_2O + 2H_2CO,$$

the first reaction step $2CH_4 + 2O_2 \rightarrow 2H_2O + 2H_2CO$ is the only one that actually reveals which hydrocarbon is being burned. So to compare the all of the hydrocarbons, we only need to write the corresponding first reaction step for the general case, and compare its implication in the several particular cases. The general first reaction step can be written:

$$2C_N H_{2N+2} + (N+1)O_2 \rightarrow 2H_2O + 2NH_2CO ,$$

Notice again the factor of 2. In order to accomplish the first step of combustion, we always need *two* hydrocarbon molecules.

3. First Step of Combustion for Other Hydrocarbons

We can evaluate the first step of hydrocarbon combustion for several cases of interest from the last Chapter: linear hexane (a possibly undesirable trace constituent of gasoline), linear septane (the ideal gasoline), methyl septane (an octane, proposed as a candidate additive for gasoline in the last Chapter).

3.1. Hexane Combustion, First Step

For any hexane, $N = 6$, and the first step of the combustion reaction reads

$$2C_6H_{14} + 7O_2 \rightarrow 2H_2O + 12H_2CO$$

On the left side of the reaction, each linear hexane molecule with ionic configuration

$$\begin{array}{cccc} H^- & H^+ & H^- & H^+ \\ \cdot & \cdot & \cdot & \cdot \end{array}$$

$$3H^+C^{4-} \cdot C^{4+} \cdot C^{4-} \cdot C^{4+} \cdot C^{4-} \cdot C^{4+}3H^-$$

$$\begin{array}{cccc} \cdot & \cdot & \cdot & \cdot \\ H^- & H^+ & H^- & H^+ \end{array}$$

has ion count $3\,C^{4-}$, $3\,C^{4+}$, $7\,H^+$, $7\,H^-$, and each oxygen molecule has likely ion count $1O^{2+}$, $1O^{2-}$. So the left side of the reaction embodies the energy for Carbon,

$$3 \times 2 \times (-54.6304) + 3 \times 2 \times 22.187 =$$
$$-327.7824 + 133.122 = -194.6604 \text{ eV}$$

and the energy for Hydrogen,

$$7 \times 2 \times 14.1369 + 7 \times 2 \times (-90.6769) =$$
$$197.9166 - 1269.4766 = -1071.56 \text{ eV}$$

and the energy for Oxygen,

$$1 \times 7 \times (17.5613) + 1 \times 7 \times (-27.3788) =$$
$$122.9291 - 191.6516 = -68.7225 \text{ eV}$$

for a total of $-194.6604 - 1071.56 - 68.7225 = -1334.9429$ eV.

On the right side of the reaction, each water molecule embodies -163.7925 eV, and each formaldehyde molecule embodies energy -186.5456 eV. So the right side of the reaction embodies total energy

$$2 \times (-163.7925) + 12 \times (-186.5456) =$$
$$-327.585 - 2238.5472 = -2566.1322 \text{ eV}$$

The reaction delivers energy

$$2566.1322 - 1334.9429 = 1231.1893 \text{ eV}.$$

On a per Carbon atom basis, that comes to $1231.1893 / 12 = 102.5991$ eV.

3.2. Septane Combustion, First Step

For any septane, $N = 7$, and the first step of the combustion reaction

$$2C_NH_{2N+2} + (N+1)O_2 \rightarrow 2H_2O + 2NH_2CO$$

reads

$$2C_7H_{16} + 8O_2 \rightarrow 2H_2O + 14H_2CO$$

Linear septane with the ionic configuration

$$\begin{array}{ccccc} H^- & H^+ & C^{4-}3H^+ & H^+ & H^- \\ \bullet & \bullet & \bullet & \bullet & \bullet \\ 3H^+C^{4-} \bullet C^{4+} \bullet C^{4-} \bullet C^{4+} \bullet C^{4-} \bullet C^{4+} \bullet C^{4-}3H^+ \\ \bullet & \bullet & \bullet & \bullet & \bullet \\ H^- & H^+ & H^- & H^+ & H^- \end{array}$$

has ion count has ion count $4C^{4-}$, $3C^{4+}$, 10 H^+, 6 H^-, and each oxygen molecule has likely ion count $1O^{2+}$, $1O^{2-}$. So the left side of the reaction embodies the energy for Carbon,

$$4 \times 2 \times (-54.6304) + 3 \times 2 \times 22.187 =$$
$$-437.0432 + 133.122 = -303.9212 \text{ eV}$$

and the energy for Hydrogen,

$$10 \times 2 \times 14.1369 + 6 \times 2 \times (-90.6769) =$$
$$282.738 - 1088.1228 = -805.3848 \text{ eV}$$

and the energy for Oxygen,

$$1 \times 8 \times (17.5613) + 1 \times 8 \times (-27.3788) =$$
$$140.4904 - 219.0304 = -78.54 \text{ eV}$$

for a total of $-303.9212 - 805.3848 - 78.54 = -1187.846$ eV.
On the right side of the reaction, each water molecule embodies energy -163.7925 eV, and each formaldehyde molecule embodies energy -186.5456 eV.
So the right side of the reaction embodies total energy

$$2 \times (-163.7925) + 14 \times (-186.5456) =$$
$$-327.585 - 2611.6384 = -2939.2234 \text{ eV}$$

This reaction delivers energy $2939.2234 - 1187.846 = 1751.3774$ eV. On a per Carbon atom basis, that is $1751.3774/14 = 125.0984$ eV. This is better than the 102.5991 eV for linear hexane.

For future reference, recall that methyl hexane tells exactly the same energy story as linear septane does.

3.3. Octane Combustion, First Step

For any octane $N = 8$, and the first step of the combustion reaction

$$2C_NH_{2N+2} + (N+1)O_2 \to 2H_2O + 2NH_2CO$$

reads

$$2C_8H_{18} + 9O_2 \to 2H_2O + 16H_2CO .$$

For the particular octane called methyl septane, the good isomers have ionic configurations

$$\begin{array}{ccccccc}
& 3H^+C^{4-} & H^+ & H^- & H^+ & H^- & \\
& \cdot & \cdot & \cdot & \cdot & \cdot & \\
3H^+C^{4-} \cdot & C^{4+} \cdot & C^{4-} \cdot & C^{4+} \cdot & C^{4-} \cdot & C^{4+} \cdot & C^{4-}3H^+ \\
& \cdot & \cdot & \cdot & \cdot & \cdot & \\
& H^- & H^+ & H^- & H^+ & H^- &
\end{array}$$

and

$$\begin{array}{ccccccc}
& H^- & H^+ & C^{4-}3H^+ & H^+ & H^- & \\
& \cdot & \cdot & \cdot & \cdot & \cdot & \\
3H^+C^{4-} \cdot & C^{4+} \cdot & C^{4-} \cdot & C^{4+} \cdot & C^{4-} \cdot & C^{4+} \cdot & C^{4-}3H^+ \\
& \cdot & \cdot & \cdot & \cdot & \cdot & \\
& H^- & H^+ & H^- & H^+ & H^- &
\end{array}$$

with ion count $5C^{4-}$, $3C^{4+}$, $13H^+$, $5H^-$. Each oxygen molecule has likely ion count $1O^{2+}$, $1O^{2-}$. So the left side of the reaction embodies the energy for Carbon,

$$5 \times 2 \times (-54.6304) + 3 \times 2 \times 22.187 =$$
$$-546.304 + 133.122 = -413.182 \text{ eV}$$

and the energy for Hydrogen

$$13 \times 2 \times 14.1369 + 5 \times 2 \times (-90.6769) =$$
$$367.5594 - 906.769 = -539.2096 \text{ eV}$$

and the energy for Oxygen

$$1 \times 9 \times (17.5613) + 1 \times 9 \times (-27.3788) =$$
$$158.0517 - 246.4092 = -88.3575 \text{ eV}$$

for a total of

$$-413.182 - 539.2096 - 88.3575 = -1040.7491 \text{ eV}.$$

On the right side of the reaction, each water molecule embodies energy -163.7925 eV, and each formaldehyde molecule embodies energy -186.5456 eV. So the right side of the reaction embodies total energy

$$2 \times (-163.7925) + 16 \times (-186.5456) =$$
$$-327.585 - 2984.7296 = -3312.3146 \text{ eV}$$

This reaction delivers $3312.3146 - 1040.7491 = 2271.5655$ eV. On a per-Carbon-atom basis, that is $2271.5655 / 16 = 141.9728$ eV. That is again a bigger number.

3.4. Real Combustion, First Step

Remember that factor of 2 that occurred in the general first step combustion reaction:

$$2C_N H_{2N+2} + (N+1)O_2 \rightarrow 2H_2O + 2NH_2CO.$$

The factor of 2 was necessary to write a reaction equation that does not use a non-existent half oxygen molecule. But consider that linear hexane is no more than trace constituent of gasoline. Getting two molecules of linear hexane together is a decidedly rare event. So trace linear hexane in gasoline might never burn at all. That would be a big problem. Similarly, as a trace additive, the octane methyl septane also might never burn at all.

One remedy for the problem is the half-hearted clean-up reaction analyzed in the last Chapter:

linear hexane + methyl septane ↔ two linear septanes

When this reaction occurs, the two linear septane molecules are then ready for the first combustion step. But even if the reaction doesn't occur, or occurs in the wrong direction, proximity between the linear hexane molecule and the methyl septane molecule is enough to let the first combustion step proceed:

$$C_6H_{14} + C_8H_{18} + 8O_2 \rightarrow 2H_2O + 14H_2CO .$$

The basic pattern here is that the general hydrocarbon combustion reaction Step 1 requires *two* hydrocarbon fuel molecules. Combustion cannot commence without two hydrocarbon molecules. The molecules do not, however, have to be the same. We can have:

$$C_NH_{2N+2} + C_MH_{2M+2} + \frac{1}{2}(N+M+2)O_2 \rightarrow$$
$$2H_2O + (N+M+2)H_2CO$$

The corresponding total combustion reaction then goes

$$C_NH_{2N+2} + C_MH_{2M+2} + \left[4 + \frac{3}{2}(N+M-2)\right]O_2 \rightarrow$$
$$(N+M)CO_2 + (N+M+2)H_2O$$

Observe that, while N and M do not have to be the same, they do have to be compatible: either both odd, or else both even. Only then can they add to an even number and so produce an integer coefficient for O_2 in both Step 1 of the combustion reaction, and in the total combustion reaction.

4. First Step of Combustion for a Fuel Mix

Let us collect and compare the energies delivered by the first step of combustion for the several fuel molecules investigated above:

Linear hexane delivers 1231.1893 eV,

Linear septane, and methyl hexane, and linear hexane with methyl septane,

all deliver 1751.3774 eV,

Methyl septane delivers 2271.5655 eV.

We can see that the energy produced per reaction varies, so that if all these reactions were triggered at the same instant, they would take distinctly different times to complete:

Linear hexane takes a long time $\propto 1/1231.1893 \approx 0.00081$,

Linear septane, and methyl hexane, and inear hexane with methyl septane,

all take a medium time $\propto 1/1751.3774 \approx 0.00057$,

Methyl septane takes a short time $\propto 1/2271.5655 = 0.00044$.

The range of time requirements is almost a factor of 2. What could this much disparity mean to engine function? Certainly it could mean three sharp peaks in energy delivery: a large main peak for septane and its cohorts, a small late peak for linear hexane, and a small early peak for the octane methyl septane. (The amplitudes of the peaks would depend on the relative concentrations of the different constituents. If A and B are the small fractions of linear hexane and methyl septane, then the hexane peak will be proportional to A^2, the methyl septane peak will be proportional to B^2, and the main peak will be proportional to $(1-A-B)^2 + AB$.

But in reality, these reactions are not triggered all at once, but rather as one or several triggered initially, typically by a spark, and then several more, triggered by the energy generated by the first ones, and then some number more, triggered successively in a cascade. So peaks won't be so sharp. However, having a late one for linear hexane is not good. But observe that having the additive methyl septane in the mix will help reduce the size of the linear hexane peak. It does so by providing non-hexane even-number hydrocarbon molecules for the linear hexanes to partner with in combustion. So the energy produced is moved out of the hexane peak and into the main peak.

If the concentration of methyl septane equals the concentration of linear hexane, then the troublesome hexane peak will be cut in half. If the methyl septane concentration is more, then the hexane peak will be cut more. Of course, the methyl septane peak will grow accordingly. But that is not a problem. Fast is fine; slow is the condition that is capable of creating engine knock.

5. THE NECESSARY POST SCRIPT TO HYDROCARBON COMBUSTION

If completed perfectly, the combustion steps discussed above leave only water and carbon dioxide to exhaust. But in reality, there will always be some unwanted products: un-combusted hydrocarbon fuel vapors, and un-oxidized carbon monoxide molecules. These things are toxic at best, and lethal at worst. In addition, internal combustion engines use air as their source of O_2 oxygen molecules, which means that they take in many more N_2 nitrogen molecules than oxygen molecules. This leads to some production of nitrogen compounds. The exhaust contains nitrogen oxides, NO and NO_2 (referred to jointly as NO_x), which in turn lead to uncomfortable ozone and ugly smog in the atmosphere we breathe.

The water and the carbon dioxide are inevitable, but the rest of these products are all targets for reduction. A variety of approaches are currently in use. Algebraic Chemistry can reveal a few new facts about the problems, and so help suggest possible improvements for the solution approaches.

About the un-unburned fuel molecules: we learned that the first step of hydrocarbon combustion cannot occur without two fuel molecules that are both odd-numbered, or else both even-numbered. Gasoline is mostly septanes, odd-numbered, but it can have traces of residual hexanes, and of remedial octanes, both of which are even-numbered. These rare even-numbered molecules may not find partner even-numbered molecules with which to enter into combustion, and so may go un-burned. This means the un-burned fuel molecules that reach the catalytic converter will be mostly linear hexane and the octane methyl septane, both of which are even-numbered. Gathered together, they will burn. So there is an opportunity here for some type of secondary energy harvest, not presently exploited. One could envision exploiting it to help keep the vehicle battery charged up, just for one example.

About the un-oxidized carbon monoxide molecules: CO is produced in the third combustion step, and is supposed to be consumed in the fourth combustion step. That fourth step requires two CO molecules. As happened with the fuel molecules in the first combustion step, the lack of a partner CO molecule prevents a lone CO molecule from entering the fourth combustion step. If gathered together, and given oxygen, the CO molecules can complete their combustion step.

About the NO_x molecules: there are two of them routinely discussed, NO and NO_2, and a third one, NO_3, routinely disregarded. Algebraic Chemistry shows that NO and NO_2 are not equally problematic, and NO_3 won't be a problem at all.

The NO molecule reduces according to the reaction

$$2NO \rightarrow N_2 + O_2 \ .$$

Let us analyze this reaction. NO has many possible ionic configurations. Consider:

$N^+ + O^-$ at $10.498 - 12.4354 = -1.9374$ eV

$N^- + O^+$ at $-8.5432 + 7.9399 = -0.6033$ eV

$N^{2+} + O^{2-}$ at $19.2257 - 27.3788 = -8.1531$ eV

$N^{2-} + O^{2+}$ at $-22.1892 + 17.5613 = -4.6279$ eV

$N^{3+} + O^{3-}$ at $23.6535 - 15.0672 = 8.5863$ eV

$N^{3-} + O^{3+}$ at $-47.6706 + 26.6007 = -21.0699$ eV

The most likely ionic configuration for NO is the one at the lowest energy, $N^{3-} + O^{3+}$, at -21.0699 eV. So the left side of the reaction embodies $2 \times (-21.0699) = -42.1398$ eV. It is of some interest that the second best ionic configuration is $N^{2+} + O^{2-}$ at -8.1531 eV. Observe that N^{3-} and O^{2-} are both examples of electron count 10, a noble gas number.

The nitrogen molecule N_2 has only three plausible ionic configurations:

$N^{+} + N^{-}$ at $10.498 - 8.5432 = 1.9548$ eV

$N^{2+} + N^{2-}$ at eV $19.2257 - 22.1892 = -2.9635$ eV

$N^{3+} + N^{3-}$ at $23.6535 - 47.6706 = -24.0171$ eV

The likeliest ionic configuration for N_2 therefore $N^{3+} + N^{3-}$, at -24.0171 eV.

We already know that O_2 has likely ionic configuration

$O^{2+} + O^{2-}$ at $17.5613 - 27.3788 = -9.8175$ eV.

So the right side of the reaction $2NO \rightarrow N_2 + O_2$ embodies

$-24.0171 - 9.8175 = -33.8346$ eV.

So the reaction overall requires $-33.8346 + 42.1398 = 8.3053$ eV. This is modest.

The other common NO_x molecule, NO_2, reduces according to the reaction

$2NO_2 \rightarrow N_2 + 2O_2$.

NO_2 has four plausible ionic configurations:

$N^{2+} + 2O^{-}$ at:

$19.2257 + 2 \times (-12.4354) = 19.2257 - 24.8708 = -5.6451$ eV,

$N^{2-} + 2O^+$ at:

$$-22.1892 + 2 \times (7.9399) = -22.1892 + 15.8798 = -6.3094 \text{ eV},$$

$N^{3+} + O^- + O^{2-}$ at:

$$23.6535 - 12.4354 - 27.3788 = -16.1607 \text{ eV},$$

$N^{3-} + O^+ + O^{2+}$ at:

$$-47.6706 + 7.9399 + 17.5613 = -22.1694 \text{ eV}.$$

The likeliest ionic configuration is $N^{3-} + O^+ + O^{2+}$ at -22.1694 eV. So the left side of the reaction embodies $2 \times (-22.1694) = -44.3388$ eV. The right side of the reaction embodies

$$-24.0171 + 2 \times (-9.8175) = -43.6521 \text{ eV}.$$

So the reaction requires $44.3388 - 43.6521 = 0.6867$ eV. This is practically nothing.

The last nitrogen oxide mentioned, NO_3, reduces according to the reaction

$$2NO_3 \rightarrow N_2 + 3O_2 \ .$$

NO_3 has likely ionic configuration $N^{3-} + 3O^+$ at

$$-47.6706 + 3 \times 7.9399 = -23.8509 \text{ eV}.$$

So the left side of the reaction embodies energy

$$2 \times (-23.8509) = -47.7018 \text{ eV}.$$

The right side of the reaction embodies

$$-24.0171 + 3 \times (-9.8175) = -53.4696 \text{ eV}.$$

So the reaction requires no energy at all; instead, it actually delivers $53.4696 - 47.7018 = 5.7678$ eV. It will occur spontaneously. This particular result is symptomatic of NO_3 being highly reactive in general.

Modern vehicles use catalytic converters to reduce their exhaust content of un-burned fuel, carbon monoxide, and the two nitrous oxides. The job to be done includes oxidation of hydrocarbon fuel molecules, and of CO molecules, and reduction of NO and NO_2 molecules. The oxidation of unburned fuel is not really a matter for catalysis; the fuel that arrives unburned consists mostly of hexane and octane molecules that did not find each other amidst all the now departed septane molecules. Given the chance that renewed proximity offers, they will burn if given a spark.

The carbon monoxide and nitrous oxide molecules are the ones that really need catalysis for disposal. The materials at hand include various metals and metal oxides that catalyze the reactions needed. The metals include: Aluminum $_{13}Al$, Manganese $_{25}Mn$, Iron $_{26}Fe$, Nickel $_{28}Ni$, Rhodium $_{45}Rh$, Palladium $_{46}Pd$, Cerium $_{58}Ce$, and Platinum $_{78}Pt$. The metal oxides include: aluminum oxide, titanium dioxide, silicon dioxide, and mixes of silica and alumina. Algebraic Chemistry can help explain what any of these materials can do. The following Subsections look at three of the metals: Rhodium, Palladium, and Platinum.

5.1. Rhodium

It is claimed that Rhodium $_{45}Rh$ helps with reduction reactions, such as those needed to process nitrogen oxide NO_x molecules. The relevant information about Rhodium is:

$$_{45}Rh \rightarrow {_{45}Rh^+} : 6.5461 \text{ eV}$$

$$_{45}Rh \rightarrow {_{45}Rh^{2+}} : 6.5461 + 7.6975 = 14.2436 \text{ eV}$$

$$_{45}Rh \rightarrow {_{45}Rh^{3+}} : 14.2436 + 6.3741 = 20.6177 \text{ eV}$$

$$_{45}Rh \rightarrow {_{45}Rh^{4+}} : 20.6177 + 6.2784 = 26.8961 \text{ eV}$$

$$_{45}Rh \rightarrow {_{45}Rh^-} : -6.6418 \text{ eV}$$

$$_{45}Rh \rightarrow {_{45}Rh^{2-}} : -6.6418 - 6.7390 = -13.3808 \text{ eV}$$

$$_{45}Rh \rightarrow {_{45}Rh^{3-}} : -13.3808 - 6.8365 = -20.2173 \text{ eV}$$

$$_{45}\text{Rh} \rightarrow {}_{45}\text{Rh}^{4-}: -20.2173 - 4.6400 = -24.8573 \text{ eV}$$

When neutral Rhodium ionizes without consuming or creating free electrons, the reactions are:

$$2\text{Rh}^0 \rightarrow \text{Rh}^+ + \text{Rh}^-: 6.5461 - 6.6418 = -0.0957 \text{ eV}$$

$$2\text{Rh}^0 \rightarrow \text{Rh}^{2+} + \text{Rh}^{2-}: 14.2436 - 13.3808 = 0.8628 \text{ eV}$$

$$2\text{Rh}^0 \rightarrow \text{Rh}^{3+} + \text{Rh}^{3-}: 20.6177 - 20.2173 = 0.4004 \text{ eV}$$

$$2\text{Rh}^0 \rightarrow \text{Rh}^{4+} + \text{Rh}^{4-}: 26.8961 - 24.8573 = 2.0388 \text{ eV}$$

Only the first of these is a negative energy. So Rhodium naturally ionizes only to one unit of charge. So for Rhodium, we may really have only singly ionized ions. However, because Rhodium is a metal, we do also have some free electrons, e^-, and so we have more Rh^+ ions than Rh^- ions.

How might these facts assist the reduction reaction

$$2\text{NO}_2 \rightarrow \text{N}_2 + 2\text{O}_2,$$

which needs 0.6867 eV? The remedy offered by a Rhodium ion is to change the reaction to

$$2\text{NO}_2 + \text{Rh}^+ + e^- \rightarrow \text{N}_2 + 2\text{O}_2 + \text{Rh}^0.$$

This puts an extra 6.5461 eV on the left side of the reaction, more than enough to make the reaction transpire.

And how might the facts assist the reduction reaction $2\text{NO} \rightarrow \text{N}_2 + \text{O}_2$, which needs a larger energy increment, 8.3053 eV? For this case, Rhodium ions can change the reaction to

$$2\text{NO} + 2\text{Rh}^+ + 2e^- \rightarrow \text{N}_2 + \text{O}_2 + 2\text{Rh}^0$$

This puts an extra $2 \times 6.5461 = 13.0922$ eV on the left side of the reaction, again sufficient to make the reaction transpire.

For the sake of completeness, let us ask how Rhodium might perform for an oxidation reaction. Carbon monoxide is supposed to oxidize according to the reaction

$$2\text{CO} + \text{O}_2 \rightarrow 2\text{CO}_2.$$

But that reaction can fail, for example if two CO molecules are not simultaneously at hand. How might Rhodium help out? It could trap CO molecules as they arrive at a surface in the catalytic converter, keeping them there until another CO and an O_2 arrive in close proximity. One candidate trapping reaction could make rhodium carbide and rhodium oxide:

$$CO + 2Rh \rightarrow RhC + RhO ,$$

or in complete detail

$$C^{2+}O^{2-} + 2Rh^0 \rightarrow Rh^{2-}C^{2+} + Rh^{2+}O^{2-}$$

But the right side of this reaction has an additional energy for $Rh^{2+} + Rh^{2-}$ that embodies $14.2436 - 13.3808 = 0.8628$ eV, which is positive. So this is no help.

Evidently, Rhodium cannot trap the Carbon and Oxygen ions for a subsequent oxidation reaction with another CO molecule and an O_2 molecule. That is a good reason why Rhodium was not recommended for this job.

5.2. Palladium

It is claimed that Palladium $_{46}Pd$ helps with oxidation, so can it help dispose of un-combusted fuel molecules and carbon monoxide CO molecules? The relevant information about Palladium is the following:

$$_{46}Pd \rightarrow {_{46}Pd^+} : 6.4899 \text{ eV}$$

$$_{46}Pd \rightarrow {_{46}Pd^{2+}} : 6.4899 + 6.3964 = 12.8863 \text{ eV}$$

$$_{46}Pd \rightarrow {_{46}Pd^{3+}} : 12.8863 + 6.2286 = 19.1149 \text{ eV}$$

$$_{46}Pd \rightarrow {_{46}Pd^{4+}} : 19.1149 + 6.0363 = 25.1512 \text{ eV}$$

$$_{46}Pd \rightarrow {_{46}Pd^-} : -6.5845 \text{ eV}$$

$$_{46}Pd \rightarrow {_{46}Pd^{2-}} : -6.5845 - 6.6794 = -13.2639 \text{ eV}$$

$$_{46}Pd \rightarrow {}_{46}Pd^{3-}: -13.2639 - 4.5562 = -17.8201 \text{ eV}$$

$$_{46}Pd \rightarrow {}_{46}Pd^{4-}: -17.8201 - 8.0056 = -25.8257 \text{ eV}$$

When neutral Palladium ionizes without consuming or creating free electrons, the reactions are:

$$2Pd^0 \rightarrow Pd^+ + Pd^-: 6.4899 - 6.5845 = -0.0946 \text{ eV},$$

$$2Pd^0 \rightarrow Pd^{2+} + Pd^{2-}: 12.8863 - 13.2639 = -0.3776 \text{ eV},$$

$$2Pd^0 \rightarrow Pd^{3+} + Pd^{3-}: 19.1149 - 17.8201 = 1.2948 \text{ eV},$$

$$2Pd^0 \rightarrow Pd^{4+} + Pd^{4-}: 25.1512 - 25.8257 = -0.6745 \text{ eV}.$$

All but one of these reactions involves a negative energy. So Palladium provides a much greater selection of catalytic ions than Rhodium does. How might that help with the needed trapping reaction? The reaction is

$$CO + 2Pd \rightarrow PdC + PdO,$$

or, in complete detail,

$$C^{2+}O^{2-} + 2Pd^0 \rightarrow Pd^{2-}C^{2+} + Pd^{2+}O^{2-}.$$

The right side of the reaction has an additional energyy for $Pd^{2+} + Pd^{2-}$, which embodies $-13.2639 + 12.8863 = -0.3776 \text{ eV}$. So yes, the Palladium can trap the Carbon and Oxygen ions for a subsequent oxidation reaction with another CO molecule and an O_2 molecule. That reaction goes

$$PdC + PdO + CO + O_2 \rightarrow 2CO_2 + 2Pd,$$

with ionic configurations

$$(Pd^{2-}C^{2+}) + (Pd^{2+}O^{2-}) + (C^{2+}O^{2-}) + (O^{2+} + O^{2-}) \rightarrow$$

$$2(C^{4+} + 2O^{2-}) + 2Pd,$$

which evaluate to

$$(-13.2639+14.2505)+(12.8863-27.3788)+$$
$$(14.2505-27.3788)+(17.5613-27.3788)=$$
$$0.9866-14.4925-13.1283-9.8175=-36.4517 \text{ eV}$$

and

$$2[22.187+2\times(-27.3788)]+2\times 0 =$$
$$2\times(22.187-54.7576)=2\times(-32.5706)=-65.1412 \text{ eV}$$

so that the needed trapping reaction transpires without difficulty.

For the sake of completeness, we should ask how Palladium does in comparison to Rhodium in assisting the reduction reaction

$$2NO_2 \rightarrow N_2 + 2O_2,$$

which needs 0.6867 eV. The remedy offered by a palladium ion is to change the reaction to

$$2NO_2 + Pd^+ + e^- \rightarrow N_2 + 2O_2 + Pd^0.$$

This puts an extra 6.4899 eV on the left side of the reaction, again more than enough to make the reaction transpire. So Palladium too could have been recommended for this job.

5.3. Platinium

It is claimed that Platinum $_{78}Pt$ helps with both oxidation and reduction. The relevant information about Platinum is:

$$_{78}Pt \rightarrow {}_{78}Pt^+ : 6.1995 \text{ eV}$$

$$_{78}Pt \rightarrow {}_{78}Pt^{2+} : 6.1995 + 6.1293 = 12.3288 \text{ eV}$$

$$_{78}Pt \rightarrow {}_{78}Pt^{3+} : 12.3288 + 5.6626 = 17.9914 \text{ eV}$$

$$_{78}Pt \rightarrow {}_{78}Pt^{4+} : 17.9914 + 6.0187 = 24.0101 \text{ eV}$$

$$_{78}\text{Pt} \to {}_{78}\text{Pt}^- : -6.2703 \text{ eV}$$

$$_{78}\text{Pt} \to {}_{78}\text{Pt}^{2-} : -6.2703 - 6.3439 = -12.6142 \text{ eV}$$

$$_{78}\text{Pt} \to {}_{78}\text{Pt}^{3-} : -12.6142 - 1.7567 = -14.3709 \text{ eV}$$

$$_{78}\text{Pt} \to {}_{78}\text{Pt}^{4-} : -14.3709 - 7.1908 = -21.5617 \text{ eV}$$

When neutral Platinum ionizes without consuming or creating free electrons, the reactions are:

$$2\text{Pt}^0 \to \text{Pt}^+ + \text{Pt}^- : 6.1995 - 6.2703 = -0.0708 \text{ eV},$$

$$2\text{Pt}^0 \to \text{Pt}^{2+} + \text{Pt}^{2-} : 12.3288 - 12.6142 = -0.2854 \text{ eV},$$

$$2\text{Pt}^0 \to \text{Pt}^{3+} + \text{Pt}^{3-} : 17.9914 - 14.3709 = 3.6205 \text{ eV},$$

$$2\text{Pt}^0 \to \text{Pt}^{4+} + \text{Pt}^{4-} : 24.0101 - 21.5617 = 2.4484 \text{ eV}.$$

As it was with Palladium, the third reaction $2\text{Pt}^0 \to \text{Pt}^{3+} + \text{Pt}^{3-}$, takes a positive energy, 3.6205 eV. But unlike Palladium, the fourth reaction $2\text{Pt}^0 \to \text{Pt}^{4+} + \text{Pt}^{4-}$ also takes a positive energy: 2.4484 eV. So Platinum is less versatile than Palladium was. But it is certainly more versatile than Rhodium was.

So could Platinum do both of the jobs that were illustrated for Rhodium and Palladium? Consider the corresponding reactions.

The reduction reaction is $2\text{NO}_2 \to \text{N}_2 + 2\text{O}_2$, which needs 0.6867 eV, becomes

$$2\text{NO} + 2\text{Pt}^+ + 2e^- \to \text{N}_2 + \text{O}_2 + 2\text{Pt}^0.$$

The Platinum adds 6.1995 eV on the left side of this reaction, so, yes, it helps enough to make the reduction reaction transpire.

The trapping reaction to allow a subsequent oxidation

$$2\text{CO} + \text{O}_2 \to 2\text{CO}_2$$

becomes

$$C^{2+}O^{2-} + 2Pt^0 \rightarrow Pt^{2-}C^{2+} + Pt^{2+}O^{2-}.$$

the right side of the reaction has an additional energy for $Pd^{2+} + Pd^{2-}$, which embodies $-12.6142 + 12.3288 = -0.2854$ eV. So yes, the Palladium can trap Carbon and Oxygen ions for subsequent reaction with another CO and an O_2 – although not as well as Palladium does it. This fact is of economic interest, since Platinum is quite expensive.

CONCLUSION

This Chapter has examined hydrocarbon combustion enough to highlight a few important points. Firstly, hydrocarbon combustion always takes two input fuel molecules. So we ought not obscure the fact by factoring a redundant-looking 2 out of reaction equations, which we may be sorely tempted to do in the case of even-numbered hydrocarbons.

Secondly, combustion occurs in steps, and only the first step depends on which hydrocarbon we are burning. So we can understand the timing problems that occur in internal combustion engines by looking at that first combustion step for the several hydrocarbons that may be present. The requirement for two input fuel molecules for every combustion event may be obscured in total combustion reactions for even-numbered hydrocarbons when written in the way that they are often written. But it is clear when we look at that crucial first reaction step of combustion.

Thirdly, we can be aware of the existence of first-step combustion where the two fuel molecules involved are not the same. It is clear that in general the two molecules required to initiate combustion need not be the same; they need only mutually compatible; *i.e.* add to an even number of Carbon atoms. Combinations that fail that test will not combust. The resulting un-burned fuel is one of three problems that a catalytic converter has to deal with. There is an apparently unexploited opportunity to harvest some more energy here, since the fuel molecules that have gone unburned will be mostly hexanes and octanes, all compatible with each other.

Finally, there are economic ramifications for ideas generated out of Algebraic Chemistry. The last Chapter proposed a potentially more economical fuel additive to help suppress engine knock, and the present Chapter has identified a potentially more economical metal to use in catalytic converters, plus a potential source of additional combustion energy available to harvest: combustion of trace even-numbered hydrocarbons collected at the catalytic converter.

A PROJECT FOR READERS

Those of you who do have web access can follow any evolving information on fuel additives. For whatever you see discussed, you can use Algebraic Chemistry to make some numerical evaluation. Use that capability to decide what makes sense, and what does not.

REFERENCES

[1] If you have web access, Google the term "hydrocarbon combustion". You will find a wealth of Wikipedia items covering every aspect of the phenomenon and related technologies.

Chapter 7

CATALYSIS OF CHEMICAL REACTIONS

ABSTRACT

This Chapter illustrates the application of Algebraic Chemistry (AC) to the problem of catalysis. It focuses on a particular example reaction from which the reader can then generalize to others. The example reaction comes from a college level textbook. The reaction will not transpire spontaneously, and AC documents the reason why. The reaction will transpire with the addition of Silver ions, which are not at all consumed to make the final product. The textbook explanation for that phenomenon was not very numerical in style, but AC supplies the corresponding numbers. The textbook explanation for this catalysis scenario does not pass the AC numerical test. So another explanation for this example catalysis scenario is generated for testing, and it passes the AC numerical test. So the exercise as a whole demonstrates that AC can help us understand catalysis a little better.

INTRODUCTION

What do we do when a chemical reaction in aqueous solution runs too slowly? We can shake, we can stir, we can supply heat, and we can supply salt. All this has been known for tens of millennia. It is called 'cooking'. In modern Chemistry, the salt ions just become the catalytic ions. So at some level, humans have always employed catalysis. But like cooking, the subject of catalysis has always seemed a bit mysterious. This has been the case mainly because discussions of catalysis have tended to be more qualitative than quantitative. But Algebraic Chemistry now offers an opportunity to replace a qualitative narrative account with a quantitative numerical account.

This Chapter examines an example problem from the Chemistry textbook [1]. There, a reluctant reaction transpires with added catalytic ions. I used this example reaction in an early paper about Algebraic Chemistry [2], and here update that work with everything learned since. The example reaction fits in with the themes developing here, which consist of various issues surrounding hydrocarbons, combustion, and vehicles. This reaction involves Chromium, the metal used to make enhancing trim on vehicles, and also a constituent in the steel of the vehicle body. The subject reaction also involves Sulfur, an element involved in acid rain that attacks vehicle surfaces. Sulfur arises from combustion of coal, a mostly heavy hydrocarbon fuel with trace contaminants, like the Sulfur. The example reaction uses Silver ions as the catalyst. And, of course, the scenario that develops involves lots of Oxygen, Hydrogen, and water. Section 1 revisits the reaction from [1], and updates the re-analysis from [2] with notation since found to keep track electron balance more visibly. The original reaction did not transpire by itself, but did transpire in the presence of Silver ions. The mechanism presumed in [1] involved two reaction steps, the first of which uses, and the second of which returns, certain catalytic ions. Numerical analysis indicates that the proposed first reaction step indeed transpires, but the second proposed reaction step still does not work.

The implication is that the original problem statement was over-constrained in a way that Nature does not actually require. Section 2 removes a likely over-constraint. There then exists a solution whereby two reaction steps can indeed both transpire. But it is not a very believable solution. Section 3 then explores the possibility that the overall reaction that actually transpires is *not* the one originally proposed, but another similar one. The overall reaction originally proposed used water as a reagent, but the overall reaction now proposed uses oxygen molecules dissolved in the water, instead of the water itself, as the likely reagent. That finally works.

Section 4 points out that there is nothing unique about Silver ions. Much more mundane metal ions will do the same catalysis job. For example, Iron can do the job. So wanted or not, the reactions described here will occur naturally at points of surface damage on your vehicle.

1. THE EXAMPLE REACTION

The reaction of interest as an example came from a very old textbook, and was written in the notation of that time, and for the purposes of that work. It went

$$2Cr^{+++} + 3S_2O_8^= + 7H_2O = Cr_2O_7^= + 6SO_4^= + 14H^+$$

The in-line equal sign meant 'reaction'. The numerical coefficients and subscripts made for equal atom counts of all element on both sides of the reaction. The superscript equal signs meant double negatives.

The reaction was chosen for discussion because it did not to transpire without catalysis. The introduction of Silver ions as catalysts was said to cause it to transpire in two steps:

$$S_2O_8^= + Ag^+ = 2SO_4^= + Ag^{+++} ,$$

followed by

$$2Cr^{+++} + 3Ag^{+++} + 7H_2O = Cr_2O_7^= + 3Ag^+ + 14H^+ .$$

The notation used originally for the total reaction, and for the two catalyzed step reactions, was a bit difficult to read. For example, the superscript equal signs were difficult to resolve visually. Rewritten with the notation of the present work, and for the purposes of the present work, the full reaction looks like

$$2Cr^{3+} + 6(SO_4)^- + 7H_2O \to ? \to (Cr_2O_7)^{2-} + 6(SO_4)^{2-} + 14H^+$$

Exponents like $3+$ and $2-$ are meant to be easier to read than their counterparts $+++$ and $=$. The $6(SO_4)^-$ in place of $3S_2O_8^=$ expresses the fact that $S_2O_8^=$ appears to fit together with a central linking that is not a typical ionic chemical bond, but instead a rather shaky covalent bond. The symbol $\to ? \to$ warns of the fact that the overall reaction does not actually transpire spontaneously.

Expressed in the notation of the present work, the presumed steps of the reaction catalyzed by Silver ions look like

$$3\left[2(SO_4)^- + Ag^+ \to 2(SO_4)^{2-} + Ag^{3+} \right]$$

and

$$2Cr^{3+} + 3Ag^{3+} + 7H_2O \to (Cr_2O_7)^{2-} + 3Ag^+ + 14H^+ .$$

Observe how the Silver ions start as Ag^+, transform to Ag^{3+}, and then revert back to Ag^+. This cycling back and forth is the standard idea of catalysis.

My earliest analysis of the original total reaction supported the textbook conclusion that the total reaction would not transpire spontaneously. As for the pair of catalyzed reactions, my analysis indicated that the first one would occur, but the second one would not. That meant there was a conflict between textbook claims and Algebraic Chemistry calculations.

The following Sub-Sections revisit those analyses and conclusions with the new and improved notation. The conclusions remain the same.

1.1. The Full Reaction

The full reaction was written above as:

$$2Cr^{3+} + 6(SO_4)^- + 7H_2O \to ? \to (Cr_2O_7)^{2-} + 6(SO_4)^{2-} + 14H^+ ,$$

Rewritten in detail sufficient for a numerical evaluation of ionic configuration energies, it becomes

$$2Cr^{3+} + 6(S^{4+} + 3O^- + O^{2-}) + 7(2H^- + O^{2+}) \to ? \to$$

$$\left[2Cr^{4+} + 4O^- + 3O^{2-}\right] + 6\left[S^{4+} + 2O^- + 2O^{2-}\right] + 14H^+ .$$

Drawn with detail sufficient to explain the proposed ionic configurations, the reaction becomes

$$2Cr^{3+} + 6 \times \begin{bmatrix} O^- & _G & _G O^{2-} \\ & S^{4+} & \\ O^- & _G & _G O^- \end{bmatrix} + 7 \times (2H^- + O^{2+}) \to ? \to$$

$$\begin{bmatrix} O^- \cdot & & \cdot O^- \\ O^{2-} \cdot Cr^{4+} \cdot O^{2-} \cdot Cr^{4+} \cdot O^{2-} \\ O^- \cdot & & \cdot O^- \end{bmatrix} + 6 \times \begin{bmatrix} O^- \cdot & \cdot O^- \\ & S^{4+} & \\ O^{2-} \cdot & \cdot O^{2-} \end{bmatrix} + 14H^+$$

In further, but superfluous, detail, the non-typical linking that could make one $(S_2O_8)^{2-}$ from two $(SO_4)^-$ would look like

$$\begin{bmatrix} & O^- & \\ O^{2-} & S^{4+} & O^- \\ & O^- & \end{bmatrix} \begin{bmatrix} & O^- & \\ O^- & S^{4+} & O^{2-} \\ & O^- & \end{bmatrix},$$

where the two O^- near each other would together make an electron count of $9 + 9 = 18$, a noble gas number, so that the two O^- could make a covalent bond instead of repelling each other.

Now let us evaluate all these ionic configurations. The relevant information for Hydrogen, Oxygen, and water was already developed. The relevant information for Chromium is:

$_{24}Cr \to {}_{24}Cr^+$: 6.8695 eV

$_{24}Cr \to {}_{24}Cr^{2+}$: $6.8695 + 6.6951 = 13.5646$ eV

$_{24}Cr \to {}_{24}Cr^{3+}$: $13.5646 + 6.5199 = 20.0845$ eV

$_{24}Cr \to {}_{24}Cr^{4+}$: $20.0845 + 6.1297 = 26.2142$ eV

$_{24}Cr \to {}_{24}Cr^{-}$: -7.0435 eV

$_{24}Cr \to {}_{24}Cr^{2-}$: $-7.0435 - 6.8707 = -13.9142$ eV

$_{24}Cr \to {}_{24}Cr^{3-}$: $-13.9142 - 7.3601 = -21.2743$ eV

$_{24}Cr \to {}_{24}Cr^{4-}$: $-21.2743 - 7.5307 = -28.805$ eV

The relevant information for Sulfur is:

$_{16}S \to {}_{16}S^{+}$: 7.4103 eV

$_{16}S \to {}_{16}S^{2+}$: $7.4103 + 9.3663 = 16.7766$ eV

$_{16}S \to {}_{16}S^{3+}$: $16.7766 + 8.3973 = 25.1739$ eV

$_{16}S \to {}_{16}S^{4+}$: $25.1739 + 5.8156 = 30.9895$ eV

$_{16}S \to {}_{16}S^{-}$: -3.6315 eV

$_{16}S \to {}_{16}S^{2-}$: $-3.6315 - 20.5253 = -24.1568$ eV

$_{16}S \to {}_{16}S^{3-}$: $-24.1568 + 8.4753 = -15.6815$ eV

$_{16}S \to {}_{16}S^{4-}$: $-15.6815 - 9.9928 = -25.6743$ eV

The left side of the reaction, $2Cr^{3+} + 6(SO_4)^{-} + 7H_2O$, expands to

$$2Cr^{3+} + 6(S^{4+} + 3O^- + O^{2-}) + 7(2H^- + O^{2+})$$

and evaluates to

$$2 \times 20.0845 + 6 \times [30.9895 + 3 \times (-12.4354) + (-27.3788)]$$
$$+ 7 \times (-163.7925)$$

or

$$40.169 + 6 \times (30.9895 - 37.3062 - 27.3788) - 1146.5475 \text{ eV}$$

or

$$40.169 + 6 \times (-33.6955) - 1146.5475 =$$
$$40.169 - 202.173 - 1146.5475 = -1308.5515 \text{ eV}$$

The right side of the reaction, $(Cr_2O_7)^{2-} + 6(SO_4)^{2-} + 14H^+$, expands to

$$\left[2Cr^{4+} + 4O^- + 3O^{2-} \right] + 6\left[S^{4+} + 2O^- + 2O^{2-} \right] + 14H^+$$

and evaluates to

$$\left[2 \times 26.2142 + 4 \times (-12.4354) + 3 \times (-27.3788) \right] +$$
$$6 \times \left[30.9895 + 2 \times (-12.4354) + 2 \times (-27.3788) \right] + 14 \times 14.1369 =$$
$$\left[52.4284 - 49.7416 + -82.1364 \right] +$$
$$6 \times \left[30.9895 - 24.8708 - 54.7576 \right] + 197.9166 =$$
$$-79.4496 + 6 \times (-48.6389) + 197.9166 =$$
$$-79.4496 - 291.8334 + 197.9166 = -173.3664 \text{ eV}$$

Clearly, the right side of the reaction,

$$(Cr_2O_7)^{2-} + 6(SO_4)^{2-} + 14H^+ \text{, at } -173.3664 \text{ eV,}$$

represents higher energy than the left side,

$$2Cr^{3+} + 6(SO_4)^- + 7H_2O \text{, at } -1308.5515 \text{ eV.}$$

So this reaction indeed does *not* transpire spontaneously. To transpire, it would need an energy input of

$$-173.3664 + 1308.5515 = 1135.1851 \text{ eV}$$

from somewhere.

1.2. The Textbook Catalyzed Reaction Steps

The relevant information about Silver is

$$_{47}\text{Ag} \rightarrow {}_{47}\text{Ag}^+ : 6.5626 \text{ eV}$$

$$_{47}\text{Ag} \rightarrow {}_{47}\text{Ag}^{2+} : 6.5626 + 6.4684 = 13.031 \text{ eV}$$

$$_{47}\text{Ag} \rightarrow {}_{47}\text{Ag}^{3+} : 13.031 + 6.3756 = 19.4066 \text{ eV}$$

$$_{47}\text{Ag} \rightarrow {}_{47}\text{Ag}^- : -6.6569 \text{ eV}$$

$$_{47}\text{Ag} \rightarrow {}_{47}\text{Ag}^{2-} : -6.6569 - 4.5629 = -11.2198 \text{ eV}$$

$$_{47}\text{Ag} \rightarrow {}_{47}\text{Ag}^{3-} : -11.2198 - 7.9666 = -19.1864 \text{ eV}$$

The first catalyzed reaction step,

$$3\left[2(\text{SO}_4)^- + \text{Ag}^+ \rightarrow 2(\text{SO}_4)^{2-} + \text{Ag}^{3+}\right]$$

focuses to

$$2(\text{SO}_4)^- + \text{Ag}^+ \rightarrow 2(\text{SO}_4)^{2-} + \text{Ag}^{3+} ,$$

and evaluates to

$$2 \times (-33.6955) + 6.5626 \rightarrow 2 \times (-60.8763) + 19.4066$$

or $\quad -67.391 + 6.5626 \rightarrow -121.7526 + 19.4066$

or $\quad -60.8284 \rightarrow -102.346$ eV

So this reaction step can indeed transpire. It yields energy $102.346 - 60.8284 = 41.5176$ eV, or $3 \times 41.5176 = 124.5528$ eV for the three of them that occur.

The second catalyzed reaction step was written as:

$$2Cr^{3+} + 3Ag^{3+} + 7H_2O \rightarrow ? \rightarrow (Cr_2O_7)^{2-} + 3Ag^+ + 14H^+ .$$

It deserves the question mark because it evaluates to:

$$2 \times 20.0845 + 3 \times 19.4066 + 7 \times (-163.7925) \rightarrow ? \rightarrow$$
$$-79.4496 + 3 \times 6.5626 + 197.9166$$

or
$$40.169 + 58.2198 - 1146.5475 \rightarrow ? \rightarrow$$
$$-79.4496 + 19.6878 + 197.9166$$

or
$$-1048.1587 \text{ eV} \rightarrow ? \rightarrow +138.1548 \text{ eV}$$

This reaction will not transpire spontaneously. It demands an energy input of

$$1048.1587 + 138.1548 = 1186.3135 \text{ eV},$$

which is even greater than the energy demanded by the original un-catalyzed total reaction, 1135.1851 eV. Thus the historical understanding of the mechanism of catalysis in this example problem did *not at all* explain what actually happened.

1.3. Why the Textbook Story Didn't Work

With any total reaction that will not transpire on account of energy demands, and with any catalysis scheme that is limited to simply cycling back and forth between ionization states of catalytic ions, failure is inevitable. The reason why is that such cycling provides no net energy.

2. A NEW ATTACK ON THE PROBLEM

Despite the failure demonstrated for the specific catalysis scheme discussed in the last Section, catalysis can still work, because strict cycling is not actually required. The laws of Statistical Mechanics imply that all possible ionization states of catalytic ions will be populated in proportion to their Boltzmann factors, and if our catalyzed reactions unbalance that population, then Nature will restore the proper balance. The catalytic ions will interact with each other, giving and taking electrons as necessary, until the entropy of their population distribution is again at its maximum.

Historically, this way of speaking about catalytic ions was not considered because there was an implicit assumption that the subject of discourse, the reaction vessel, was a 'closed system'. The concept of 'closed system' is common in mathematical physics, starting from Newtonian mechanics. But it is only a concept, not a reality. Nature does not actually have closed systems. For the problem at hand, the reaction vessel is embedded in an environment. The chemical scenario inside can always take some heat energy from the environment, or give some heat energy to it, to restore its appropriate balance of catalytic ions.

Given these liberating ideas, a variety of catalyzed reactions that don't use strict cycling might do the trick for this total reaction. In fact, the variety can be so great that it amounts to a riotous profusion of possibilities. The following Subsections give numerical details about a few of the possible catalysis scenarios for the subject reaction.

2.1. The First Catalyzed Reaction Step

Let us generalize the first catalyzed reaction step

$$3\left[2(SO_4)^- + Ag^+ \rightarrow 2(SO_4)^{2-} + Ag^{3+}\right]$$

to

$$3\left[2(SO_4)^- + Ag^{n+} \rightarrow 2(SO_4)^{2-} + Ag^{(n+2)+}\right]$$

where we do not specify what n is. Indeed, n could even be negative, in which case the meaning of Ag^{n+} is $Ag^{|n|-}$. In fact, even $n+2$ could be negative, in which case the meaning of $Ag^{(n+2)+}$ is $Ag^{|n+2|-}$. The energy spacing between any Ag^n and Ag^{n+2} for $-3 < n < +3$ is fairly similar, so any such reaction is likely to work. This situation greatly expands the number of catalyzing possibilities available.

2.2. The Second Catalyzed Reaction Step

How might the second catalyzed reaction step

$$2Cr^{3+} + 3Ag^{3+} + 7H_2O \rightarrow ? \rightarrow (Cr_2O_7)^{2-} + 3Ag^+ + 14H^+ ,$$

be similarly generalized? The basic problem is that a really large amount of energy is needed. The only way I have found to provide so much energy is to use many more Silver ions. And that requires the use of some free electrons to maintain charge balance. For example, we can use

$$2Cr^{3+} + 33Ag^{3+} + 7H_2O + 192e^- \rightarrow ? \rightarrow$$
$$(Cr_2O_7)^{2-} + 33Ag^{3-} + 14H^+$$

which evaluates to

$$2 \times 20.0845 + 33 \times 19.4066 + 7 \times (-163.7925) \rightarrow ? \rightarrow$$
$$-79.4496 + 33 \times (-19.1864) + 14 \times 14.1369$$

or

$$40.169 + 640.4178 - 1146.5475 \rightarrow ? \rightarrow$$
$$-79.4496 - 633.1512 + 197.9166$$

or

$$-465.9607 \rightarrow ? \rightarrow -514.6842 \text{ eV} .$$

Thus we see that, if given enough catalytic ions, and if given enough free electrons, both catalyzed reaction steps for the subject reaction can be made to transpire as needed. But this is only an existence proof, not a plausibly realistic solution. The use of 33 Silver ions makes the reaction much too complicated.

2.3. Define More Reaction Steps?

The situation with the subject reaction,

$$2Cr^{3+} + 6(SO_4)^- + 7H_2O \rightarrow ? \rightarrow (Cr_2O_7)^{2-} + 6(SO_4)^{2-} + 14H^+ ,$$

and especially its purported second step,

$$2Cr^{3+} + 3Ag^{3+} + 7H_2O \to ? \to (Cr_2O_7)^{2-} + 3Ag^+ + 14H^+ ,$$

is not unlike that of the hydrocarbon combustion reactions studied in the last Chapter. Recall the geneeral-case hydrocarbon total combustion reaction:

$$2C_NH_{2N+2} + [4 + 3(N-1)]O_2 \to 2NCO_2 + 2(N+1)H_2O$$

The number of molecules on the left side is $2 + 4 + 3(N-1)$, and the number of molecules on the right side is $2N + 2(N+1)$. For cxample, at $N = 8$, there are 27 molecules on the left. The gathering of 27 molecules for a reaction sounds like a hugely complicated proposition.

But the complexity is much ameliorated by the fact that hydrocarbon combustion actually proceeds in four separate steps. The first step is:

$$2C_NH_{2N+2} + (N+1)O_2 \to 2H_2O + 2NH_2CO .$$

There are $2(N+1)$ molecules on the left side. For $N = 8$, that comes to 11 molecules on the left. That is a big improvement over 27.

The remaining three hydrocarbon combustion reaction steps are all independent of N, and extremely simple. The second one,

$$2H_2CO \to 2HCO + H_2 ,$$

has just two molecules on the left side. The third one,

$$2HCO + H_2 + O_2 \to 2CO + 2H_2O ,$$

has just four molecules on the left side. The last one,

$$2CO + O_2 \to 2CO_2 ,$$

has just three molecules on the left side.

Can any such expansion of our offending second catalyzed reaction step help? Observe that this second step

$$2Cr^{3+} + 3Ag^{3+} + 7H_2O \to ? \to (Cr_2O_7)^{2-} + 3Ag^+ + 14H^+ ,$$

has 5 ions and 7 molecules – 12 entities total - on its left side. That is a lot.

Now consider expanding that second reaction step to

$$7\left[2H^- + O^{2+} + 3Ag^{3+} + 18e \rightarrow 2H^+ + O^{2-} + 3Ag^{3-}\right]$$

and

$$2\left[Cr^{3+} + 6Ag^{3-} \rightarrow Cr^{4+} + 6Ag^{3+} + 37e\right] .$$

The first constituent reaction has 6 ions (and 18 electrons) on its left side. The second constituent reaction has 7 ions on its left side. Assuming that the use of 37 electrons is not a problem, that probably is an improved situation.

But I just don't really believe it.

3. QUESTIONING THE ASSUMED REACTION

We should remember that the subject total reaction is really a reaction between neutral molecules:

$$2Cr(SO_4)_3 + 7H_2O \rightarrow ? \rightarrow (Cr_2O_7)H_2 + 6H_2(SO_4) .$$

With the reaction written this way, we can see clearly that it is using water molecules as a source for Oxygen atoms. That fact reflects the source of the stated problem: a book Chapter about reactions in aqueous solutions. But the water molecule is very stable. Remember, its formation was the main energy-producing process in hydrocarbon combustion. So the water molecule is very difficult to mine for Oxygen atoms. So it seems unlikely that the total reaction that occurs dissociates water molecules to obtain Oxygen atoms. But remember that this planet is nearly covered with water, and the water is full of fish and other oxygen-consuming organisms. So water must generally contain dissolved molecular oxygen. So the reaction that actually occurs most likely uses the oxygen molecules that are dissolved in the water.

So, for example, consider the following alternative reaction:

$$4Cr(SO_4)_3 + 7O_2 + 28Ag \rightarrow ? \rightarrow 2(Cr_2O_7)Ag_2 + 12(SO_4)Ag_2$$

This one expands to:

$$4\left[Cr^{3+} + 3(SO_4)^-\right] + 7\left[O^{2+} + O^{2-}\right] + 28Ag \rightarrow ? \rightarrow$$
$$2\left[(2Cr^{3+} + 4O^- + 3O^{2-}) + 2Ag^+\right] + 12\left[(SO_4)^{2-} + 2Ag^+\right]$$

or more fully to:

$$4\left[Cr^{3+} + 3(S^{4+} + 3O^- + O^{2-})\right] + 7\left[O^{2+} + O^{2-}\right] + 28Ag \rightarrow ? \rightarrow$$
$$2\left[(2Cr^{3+} + 4O^- + 3O^{2-}) + 2Ag^+\right] + 12\left[(S^{4+} + 2O^- + 2O^{2-}) + 2Ag^+\right]$$

and evaluates to:

$$4\{20.0845 + 3[30.9895 + 3(-12.4354) + (-27.3788)]\} +$$
$$7[17.5613 + (-27.3788)] + 28 \times 0 \rightarrow ? \rightarrow$$
$$2\{[2 \times 20.0845 + 4(-12.4354) + 3 \times (-27.3788)] + 2 \times 6.5626\}$$
$$+ 12[(30.9895 - 24.8708 - 54.7576) + 13.1252]$$

or

$$4[20.0845 + 3(30.9895 - 37.3062 - 27.3788)] +$$
$$7[17.5613 - 27.3788] + 0 \rightarrow ? \rightarrow$$
$$2[(40.169 - 49.7416 - 82.1364) + 13.1252] +$$
$$12[(30.9895 - 24.8708 - 54.7576) + 13.1252]$$

or

$$4[20.0845 + 3 \times (-33.6955)] + 7 \times (-9.8175) \rightarrow ? \rightarrow$$
$$2[(40.169 + -49.7416 - 82.1364) + 13.1252]$$
$$+ 12[(-48.6389) + 13.1252]$$

or

$$4 \times (20.0845 - 101.0865) - 68.7225 \rightarrow ? \rightarrow$$
$$2 \times (-78.5838) + 12 \times (-35.5137)$$

or

$$4 \times (-81.002) - 68.7225 \rightarrow ? \rightarrow 2 \times (-78.5838) + 12 \times (-35.5137)$$
$$-324.008 - 68.7225 \rightarrow ? \rightarrow -157.1676 - 426.1644$$
$$-392.7305 \text{ eV} \rightarrow ? \rightarrow -583.332 \text{ eV}$$

So this alternative reaction will transpire without difficulty. Extra energy is not required, and multiple reaction steps are probably not required.

Could this alternative reaction fit the circumstances that the original subject reaction was meant to fit? Certainly, the alternative reaction cannot transpire without Silver being present.

But unlike the original reaction, the alternative reaction does not produce any H^+ ions. Such ions indicate an acid condition. The aqueous solution must have tested acidic. But the reason for that acidity cannot lie within this alternative reaction. To provide the acidity, there has to be some subsequent reaction involving a reaction product. Consider that water always has a small proportion of its molecules dissociated into H^+ and $(OH)^-$ ions. When the $(SO_4)Ag_2$ product encounters an $(OH)^-$ ion, a no-cost rearrangement of ions can occur, from

$$(S^{4+} + 2O^- + 2O^{2-} + 2Ag^+) + (H^+ + O^{2-})$$

to

$$(2Ag^+ + O^{2-}) + (S^{4+} + 2O^- + 2O^{2-}) + H^+$$

4. Natural Catalysis

The catalysis discussed so far has used Silver ions. That works, but does not seem relevant to the theme so far developing in the present book, which features hydrocarbons, combustion, and vehicles. So let us look into catalysis using the element Iron, which is, of course, the main constituent of steel, and so would be present in any vehicle, maybe right under plating of Chromium atoms, or right beside some Chromium atoms included in the recipe for steel. In this scenario, one wants to prevent, not assist, any reaction that may occur, because it would damage the beauty or integrity of the vehicle. But alas, Nature is inexorable!

The relevant information about Iron is:

$$_{26}Fe \rightarrow {}_{26}Fe^+ : 6.6575 \text{ eV}$$

$$_{26}Fe^+ \rightarrow {}_{26}Fe^{2+} : 6.6575 + 6.8131 = 13.4706 \text{ eV}$$

$$_{26}Fe^{2+} \rightarrow {}_{26}Fe^{3+} : \text{eV } 13.4706 + 6.6462 = 20.1168 \text{ eV}$$

$$_{26}Fe^{3+} \rightarrow {}_{26}Fe^{4+} : 20.1168 + 6.4785 = 26.5953 \text{ eV}$$

$$_{26}Fe \rightarrow {}_{26}Fe^- : -7.1180 \text{ eV}$$

$$_{26}Fe^- \rightarrow {}_{26}Fe^{2-} : -7.1180 - 7.2818 = -14.3998 \text{ eV}$$

$$_{26}Fe^{2-} \rightarrow {}_{26}Fe^{3-} : -14.3998 - 7.4468 = -21.8466 \text{ eV}$$

$$_{26}Fe^{3-} \rightarrow {}_{26}Fe^{4-} : -21.8466 - 7.6111 = -29.4577 \text{ eV}$$

The information about Iron is barely distinguishable from the information about Silver. So reactions similar to those discussed for Silver ions may transpire. Now consider your vehicle. Chrome is present, and Iron is present, but what about the Sulfur? Consider this: the whole industrialized world burns coal, and coal generally contains some compounds of Sulfur, leading to some sulfuric acid, $H_2(SO_4)$, in the rain that falls everywhere. So eventually, similar reactions will happen on your vehicle.

Conclusion

We learn here that Algebraic Chemistry can help us critique a mostly verbal argument and formulate a replacement argument that is more numerical in character. We can figure out if a verbal argument is true or false, and we can feel more comfortable with a replacement argument that is supported numerically.

A Project for Readers

Again, many of you have on your shelves some Chemistry textbooks that you have studied from, and can now try re-reading them with the techniques of Algebraic Chemistry in mind. You may actually see some arguments that can be seriously challenged through application of numerical analysis. Those could be of interest for prompting new development both in theory and in technology.

Acknowledgments

I thank the International Journal of Chemical Modeling for providing a forum for model development in this area.

References

[1] Noyes, A.A., and Sherrill, M.S. *A Course of Study in Chemical Principles*, second edition, re-written, MacMillan Co., New York, 1934.
[2] Whitney, C.K. On the Algebraic Chemistry of Catalysis, *International Journal of Chemical Modeling*, Vol. 1, pp. 245-297, 2008.

Chapter 8

ELECTRO-CHEMISTRY IN POWER GENERATION

ABSTRACT

This Chapter examines a controversial new technology from the viewpoint of Algebraic Chemistry. The subject technology is so-called Cold Fusion. There are good reasons why this technology is controversial, and they are addressed here. The most important one is lack of a credible Theory to describe what is going on. That is where Algebraic Chemistry comes in. It offers a quantitative Model that tracks purported events. This task is the mirror image to the task undertaken in the last Chapter. There, an accepted narrative was challenged, and a numerically based one was offered. Here a generally disbelieved narrative is disregarded, and a numerically based one is offered.

INTRODUCTION

This Book has a strong theme about hydrocarbon combustion and vehicles. But some day your vehicle will operate entirely on electricity. One development that could help hasten the arrival of that day would be commercialization of electric power production based on nuclear fusion. The term 'nuclear fusion' implies forcing nuclei of lighter elements together to form heavier elements, which releases tremendous energy. It is easier said than done.

The requirements for producing nuclear fusion can be broken down into the following:

1) Removal of shielding electrons from the subject nuclei.
2) Confinement of the naked nuclei within the intended fusion volume.
3) Defeat of natural Coulomb repulsion forces between naked nuclei.
4) Handover to nuclear attraction forces between naked nuclei.

To meet these requirements, two concepts are presently being pursued. They are called 'Hot Fusion' (HF) and 'Cold Fusion' (CF).

The term 'Hot Fusion' refers to thermonuclear fusion, the kind of process that occurs in the Sun and other stars. In HF, plasma consisting of Hydrogen and its heavier isotopes, Deuterium and Tritium, is confined magnetically. The various nuclei fuse into Helium. In the stars, the Helium eventually fuses into successively heavier elements.

But here on Earth, just producing Helium would be quite enough. A huge amount of energy is released in making Helium. We know it can support a useless weapon. We hope it can support peaceful power production. But there has been a very long, but not yet very successful, pursuit of HF as a commercial energy technology.

The term 'Cold Fusion' refers to nuclear fusion without the many difficult features of Hot Fusion: the high temperature, the plasma state, the intense magnetic confinement, and the generally big hardware. CF is meant to be small in scale, for local, maybe even domestic, use. But CF is not at all well understood. That is why it is defined by what it is not, namely not HF, rather than by what it is. Indeed, there exists considerable doubt that the term 'Cold Fusion' is matched by any real physical phenomenon.

Here I wish to comment on some beliefs about CF that are commonly articulated, and might be wrong, and some other ideas that haven't been articulated much, and may deserve more study. Section 1 comments on the origins of controversy over the possible reality of CF. Section 2 collects the information needed for a proper numerical study of the CF electrolytic cell. Section 3 a studies events in the electrolyte. Section 4 studies events at the cathode.

1. THE ORIGINS OF CONTROVERSY

There are three main problem areas: 1) unfulfilled expectations that neutrons should be among reaction products, 2) lack of predictability in excess heat production, and 3) lack of any credible theoretical explanation of what actually transpires.

1.1. Lack of Neutrons

Those who are skeptical about CF have often pointed to its meager generation of neutrons, which are common byproducts of many of the more familiar nuclear processes that involve fission. But there is a flaw in this sort of objection. When we study the Periodic Table, we see that the proportion of neutrons in stable isotopes increases with atomic number. That means fission reactions typically start with elements that have more neutrons than the daughter elements will need for stability. So the excess neutrons are liberated in the fission reaction. By contrast, fusion reactions do not occur between the abundant isotopes of input elements because those isotopes do not have enough neutrons to make a stable isotope of the product element. Fusion reactions require some unusual heavy isotope(s) as input. And even then, any neutrons generated may be completely used up in stabilizing the product element. So none of them may be liberated.

So the absence of neutrons in the near environs of a purported CF demonstration does not mean there is no fusion occurring.

1.2. Variability of Excess Heat

It is not the presence of neutrons, but rather the presence of excess heat that can reveal that fusion can indeed be occurring. And excess heat has been difficult to produce reliably. The lack of repeatability has always been cited as an argument against the possibility of CF.

But lack of repeatability does not mean nothing happens; it only means we do not know what it is that happens.

Over many years, the repeatability situation has improved a lot. CF technology has been nurtured and developed over several decades with support from the US government, administered through the US Navy. Knowledge about the conditions required has been gathered, and whatever it is that happens, that thing can be made to happen pretty reliably

But we still do not know in detail what it is that happens.

1.3. Lack of Credible Theory

What is really missing for CF is a candidate explanation that can be derived from a credible theory and reduced to testable numbers. Without that, people can only generate patent nonsense, argue fruitlessly about it, any never convince each other of anything.

Algebraic Chemistry offers a candidate approach for analyzing aspects of CF in a numerical way. What emerges is at the very least a credible numerical Model. And if you read to the end of the book, you may begin to think it is the beginning of a credible Theory. The following Sections detail the AC analysis of CF.

2. The Numerical Information Needed

The relevant information about Hydrogen that we have used previously in this book represents a natural mix of all the isotopes, with average mass number $M = 1.008$. Now we need that information separated out for pure normal Hydrogen, for Deuterium, and for Tritium. The information for pure Hydrogen uses mass $M = 1$, the information for Deuterium uses $M = 2$, and that for Tritium uses $M = 3$. So we have:

$_1H \rightarrow {_1H^+}$: 14.250 eV

$_1H \rightarrow {_1H^-}$: $-20.1525 - 71.2500 + 0 = -91.4025$ eV

$_1D \rightarrow {_1D^+}$: 7.125 eV

$_1D \rightarrow {_1D^-}$: $-10.07625 - 35.625 = -45.70125$ eV

$_1T \rightarrow {_1T^+}$: 4.750 eV

$_1T \rightarrow {_1T^-}$: $-6.71745 - 23.750 + 0 = -30.46745$ eV

Observe that for the positive ions of the heavier isotopes are easier to make than the positive ion of Hydrogen. That means the heavier nuclei are easier to strip naked, and thus easier to prepare for fusion. Observe too that for the heavier isotopes, the negative energies of negative ions are less negative than the negative energy of the negative Hydrogen ion. That means heavier water molecules can more easily be driven out of their typical tetrahedral isomer and into their rather exotic linear isomer. The linear isomers expose the heavy nuclei without even destroying the molecules they are in.

An important part in the CF story concerns several metals that can act as electrodes and/or catalysts. One such metal is Palladium. It figures in the aqueous solution in which shielding electrons are stripped from heavy nuclei, and in the cathode where the cold fusion is said to occur.

Recall from the Chapter on Important Reactions that we had for Palladium:

$$_{46}Pd \rightarrow {}_{46}Pd^+: 6.4899 \text{ eV}$$

$$_{46}Pd \rightarrow {}_{46}Pd^{2+}: 6.4899 + 6.3964 = 12.8863 \text{ eV}$$

$$_{46}Pd \rightarrow {}_{46}Pd^{3+}: 12.8863 + 6.2286 = 19.1149 \text{ eV}$$

$$_{46}Pd \rightarrow {}_{46}Pd^{4+}: 19.1149 + 6.0363 = 25.1512 \text{ eV}$$

$$_{46}Pd \rightarrow {}_{46}Pd^-: -6.5845 \text{ eV}$$

$$_{46}Pd \rightarrow {}_{46}Pd^{2-}: -6.5845 - 6.6794 = -13.2639 \text{ eV}$$

$$_{46}Pd \rightarrow {}_{46}Pd^{3-}: -13.2639 - 4.5562 = -17.8201 \text{ eV}$$

$$_{46}Pd \rightarrow {}_{46}Pd^{4-}: -17.8201 - 8.0056 = -25.8257 \text{ eV}$$

and from that we have

$$2Pd^0 \rightarrow Pd^+ + Pd^-: 6.4899 - 6.5845 = -0.0946 \text{ eV}$$

$$2Pd^0 \rightarrow Pd^{2+} + Pd^{2-}: 12.8863 - 13.2639 = -0.3776 \text{ eV}$$

$$2Pd^0 \rightarrow Pd^{3+} + Pd^{3-}: 19.1149 - 17.8201 = 1.2948 \text{ eV}$$

$$2\text{Pd}^0 \to \text{Pd}^{4+} + \text{Pd}^{4-} : 25.1512 - 25.8257 = -0.6745 \text{ eV}$$

The importance of these numerical results in the CF problem will be that ionization states involving both two charges and four charges will be available, because the reactions $2\text{Pd}^0 \to \text{Pd}^{2+} + \text{Pd}^{2-}$ and $2\text{Pd}^0 \to \text{Pd}^{4+} + \text{Pd}^{4-}$ both take negative energies; namely -0.3776 eV and -0.6745 eV.

Another metal important in the CF problem is Platinum, since the anode in the CF electrolytic cell is usually Platinum. Recall that for Platinum we had:

$$_{78}\text{Pt} \to {}_{78}\text{Pt}^+ : 6.1995 \text{ eV}$$

$$_{78}\text{Pt} \to {}_{78}\text{Pt}^{2+} : 6.1995 + 6.1293 = 12.3288 \text{ eV}$$

$$_{78}\text{Pt} \to {}_{78}\text{Pt}^{3+} : 12.3288 + 5.6626 = 17.9914 \text{ eV}$$

$$_{78}\text{Pt} \to {}_{78}\text{Pt}^{4+} : 17.9914 + 6.0187 = 24.0101 \text{ eV}$$

$$_{78}\text{Pt} \to {}_{78}\text{Pt}^- : -6.2703 \text{ eV}$$

$$_{78}\text{Pt} \to {}_{78}\text{Pt}^{2-} : -6.2703 - 6.3439 = -12.6142 \text{ eV}$$

$$_{78}\text{Pt} \to {}_{78}\text{Pt}^{3-} : -12.6142 - 1.7567 = -14.3709 \text{ eV}$$

$$_{78}\text{Pt} \to {}_{78}\text{Pt}^{4-} : -14.3709 - 7.1908 = -21.5617 \text{ eV}$$

and from that we have

$$2\text{Pt}^0 \to \text{Pt}^+ + \text{Pt}^- : 6.1995 - 6.2703 = -0.0708 \text{ eV}$$

$$2\text{Pt}^0 \to \text{Pt}^{2+} + \text{Pt}^{2-} : 12.3288 - 12.6142 = -0.2854 \text{ eV}$$

$$2\text{Pt}^0 \to \text{Pt}^{3+} + \text{Pt}^{3-} : 17.9914 - 14.3709 = 3.6205 \text{ eV}$$

$$2\text{Pt}^0 \to \text{Pt}^{4+} + \text{Pt}^{4-} : 24.0101 - 21.5617 = 2.4484 \text{ eV}$$

The significance of these numerical results is that Platinum does not easily self ionize beyond its second ionization state. This reluctance to ionize may be desirable because the anode in a CF cell is not meant to do anything dramatic that might compete with the action at the cathode.

One may wonder if any other metal might work even better for the anode, and possibly be more economical too. For example, what about Gold? As we know, Gold is the most noble of noble metals: it hardly forms compounds at all. The relevant information about Gold is goes to at least 6 charges:

$_{79}Au \rightarrow {}_{79}Au^{+}$: 6.2464 eV

$_{79}Au \rightarrow {}_{79}Au^{2+}$: 6.2464 + 6.1759 = 12.4223 eV

$_{79}Au \rightarrow {}_{79}Au^{3+}$: 12.4223 + 6.1062 = 18.5285 eV

$_{79}Au \rightarrow {}_{79}Au^{4+}$: 18.5285 + 5.6438 = 24.1723 eV

$_{79}Au \rightarrow {}_{79}Au^{5+}$: 24.1723 + 5.9304 = 30.1027 eV

$_{79}Au \rightarrow {}_{79}Au^{6+}$: 30.1027 + 5.9963 = 36.0991 eV

$_{79}Au \rightarrow {}_{79}Au^{-}$: −6.3194 eV

$_{79}Au \rightarrow {}_{79}Au^{2-}$: −6.3194 − 7.6368 = −13.9562 eV

$_{79}Au \rightarrow {}_{79}Au^{3-}$: −13.9562 − 7.1586 = −21.1148 eV

$_{79}Au \rightarrow {}_{79}Au^{4-}$: −21.1148 − 7.4847 = −28.5995 eV

$_{79}Au \rightarrow {}_{79}Au^{5-}$: −28.5995 − 5.9405 = −34.5400 eV

$_{79}Au \rightarrow {}_{79}Au^{6-}$: −34.5400 − 7.9281 = −42.4681 eV

and from that we have

$2Au^0 \rightarrow Au^+ + Au^-$: $6.2464 - 6.3194 = -0.077$ eV

$2Au^0 \rightarrow Au^{2+} + Au^{2-}$: $12.4223 - 13.9562 = -1.5339$ eV

$2Au^0 \rightarrow Au^{3+} + Au^{3-}$: $18.5285 - 21.1148 = -2.5863$ eV

$2Au^0 \rightarrow Au^{4+} + Au^{4-}$: $24.1723 - 28.5995 = -4.4272$ eV

$2Au^0 \rightarrow Au^{5+} + Au^{5-}$: $30.1027 - 34.5400 = -4.4373$ eV

$2Au^0 \rightarrow Au^{6+} + Au^{6-}$: $36.0991 - 42.4681 = -6.3690$ eV

Observe that *all* of these numbers are negative. So all the ionization states mentioned are accessible; that is, for every value of n from 1 to 6, and very likely beyond, the reaction $2Au^0 \rightarrow Au^{n+} + Au^{n-}$ involves a negative energy. This fact is so unusual that it is natural to ask if it accounts for Gold being the most noble of noble metals. That is, does Gold rarely make compounds because Gold so strongly ionizes with itself that it rarely ionizes with other elements?

Platinum is not like Gold in this regard. So if Platinum is good as the anode in a CF cell, then Gold probably is *not* good in that role. So let us consider yet another candidate metal for the anode. Tungsten looks interesting because it is certainly more economical than Platinum, and it already has an important role in the history of electrical technology: it was the light bulb filament of the twentieth century. How would Tungsten now do as the anode material in a twenty-first century CF electrolytic cell?

Tungsten exhibits valence values up to 6, so the relevant information about Tungsten goes to 6 charges:

$_{74}W \rightarrow {_{74}W^+}$: 6.1633 eV

$_{74}W \rightarrow {_{74}W^{2+}}$: $6.1633 + 6.0952 = 12.2585$ eV

$_{74}W \rightarrow {_{74}W^{3+}}$: $12.2585 + 6.0281 = 18.2866$ eV

$_{74}W \rightarrow {_{74}W^{4+}}$: $18.2866 + 6.7740 = 25.0606$ eV

$_{74}W \rightarrow {_{74}W^{5+}}$: $25.0606 + 5.6662 = 30.7268$ eV

$_{74}W \to {}_{74}W^{6+}$: $30.7268 + 5.6232 = 36.3500$ eV

$_{74}W \to {}_{74}W^{-}$: -6.2327 eV

$_{74}W \to {}_{74}W^{2-}$: $-6.2327 - 5.8536 = -12.0863$ eV

$_{74}W \to {}_{74}W^{3-}$: $-12.0863 - 6.3479 = -18.4342$ eV

$_{74}W \to {}_{74}W^{4-}$: $-18.4342 - 6.4214 = -24.8556$ eV

$_{74}W \to {}_{74}W^{5-}$: $-24.8556 - 5.3578 = -30.2134$ eV

$_{74}W \to {}_{74}W^{6-}$: $-30.2134 - 5.3555 = -35.5689$ eV

and from that we have

$2W^0 \to W^+ + W^-$: $6.1633 - 6.2327 = -0.0694$ eV

$2W^0 \to W^{2+} + W^{2-}$: $12.2585 - 12.0863 = 0.1722$ eV

$2W^0 \to W^{3+} + W^{3-}$: $18.2866 - 18.4342 = -0.1776$ eV

$2W^0 \to W^{4+} + W^{4-}$: $25.0606 - 24.8556 = 0.205$ eV

$2W^0 \to W^{5+} + W^{5-}$: $30.7268 - 30.2134 = 0.5134$ eV

$2W^0 \to W^{6+} + W^{6-}$: $36.3500 - 35.5689 = 0.7811$ eV

Observe that all of these numbers are small, and many of them are positive. This clearly means that Tungsten is unenthusiastic about ionizing.

The very lethargic ionization behavior of Tungsten helps explain its success as the filament material in twentieth-century light bulbs. Recall the association between the microscopic states of ionization and the macroscopic states of matter discussed in Chapter 3. It implies that if Tungsten doesn't ionize much, then it doesn't melt fast, and it doesn't boil fast either; it just remains solid, even while glowing hot. Tungsten in fact has the highest melting point of any metal.

All of this is good for application as a filament in a light bulb. But in that environment, Tungsten is immersed in an inert gas and sealed in the glass bulb, protected from the local atmosphere. How would Tungsten do in water? Remember the last Chapter, about the oxygen gas molecules dissolved in the water that provided a reaction much more plausible than the one originally given in the referenced textbook. In the CF problem, oxygen molecules dissolved in the electrolyte might attack the electrodes.

Tungsten ore is mainly the oxide WO_3, so consider the reaction

$$2W + 3O_2 \rightarrow 2WO_3 \ .$$

The likely ionic configurations are

$$2W^0 + 3(O^{2+} + O^{2-}) \rightarrow 2(W^{6+} + 3O^{2-}) \ ,$$

which evaluates to

$$2 \times 0 + 3 \times [17.5613 + (-27.3788)] \rightarrow$$
$$2 \times [36.3500 + 3 \times (-27.3788)]$$

or

$$-29.4525 \text{ eV} \rightarrow -91.5728 \text{ eV} \ .$$

This reaction would certainly transpire.

Of course, a similar analysis could be developed about Platinum too. A commercially used form of Platinum is the oxide PtO. So consider the reaction:

$$2Pt + O_2 \rightarrow 2PtO$$

with ionic configurations

$$2Pt^0 + (O^{2+} + O^{2-}) \rightarrow 2(Pt^{2+} + O^{2-}) \ ,$$

that evaluate to

$$2 \times 0 + 17.5613 + (-27.3788) \rightarrow 2[12.3288 + (-27.3788)] \ ,$$

or

$$-9.8175 \text{ eV} \rightarrow -30.1000 \text{ eV} \ .$$

This reaction too would certainly transpire. So there is little qualitative difference between these two choices for anode material: both are prone to attack by molecular oxygen.

And as a matter of fact, the same is true about, Palladium, the cathode material. Palladium occurs as PdO, and also as PdO_2 and Pd_2O_3, with water. So consider the reactions

$$2Pd + O_2 \rightarrow 2PdO ,$$

$$Pd + O_2 \rightarrow PdO_2 ,$$

$$4Pd + 3O_2 \rightarrow 2Pd_2O_3$$

with possible ionic configurations

$$2Pd^0 + (O^{2+} + O^{2-}) \rightarrow 2(Pd^{2+} + O^{2-})$$

$$Pd^0 + (O^{2+} + O^{2-}) \rightarrow Pd^{2+} + 2O^-$$

$$4Pd^0 + 3(O^{2+} + O^{2-}) \rightarrow 2(2Pd^{3+} + 3O^{2-})$$

that evaluate to

$$2 \times 0 + [17.5613 + (-27.3788)] \rightarrow 2[12.8863 + (-27.3788)]$$

$$1 \times 0 + [17.5613 + (-27.3788)] \rightarrow 12.8863 + 2 \times (-12.4354)$$

$$4 \times 0 + 3 \times [17.5613 + (-27.3788)] \rightarrow$$
$$2 \times [2 \times 19.1149 + 3 \times (-27.3788)]$$

or

-9.8175 eV $\rightarrow -28.985$ eV

-9.8175 eV $\rightarrow -11.9845$ eV

-29.4525 eV $\rightarrow -87.8132$ eV $_{eV}$

Clearly, all of these reactions transpire.

So Tungsten is not uniquely vulnerable to oxidation; Tungsten just makes us think of the oxidation issue because of Tungsten's historical role in electrical technology. Except for Gold, all of the metals of interest in CF technology present the oxidation issue.

The oxidation issue means that the heavy-water electrolyte for a CF cell needs to be prepared properly for duty: *i.e.*, left for some long time to outgas a lot of dissolved molecular oxygen. It has been known for some time that a period of some weeks is needed for what is known in the trade as 'incubation'. But there hasn't been a plausible suggestion before as to what the incubation might be all about.

3. WHAT HAPPENS IN THE ELECTROLYTIC SOLUTION

The electrolytic solution begins with heavy water. There are several molecules possible: H_2O, DHO, D_2O, HTO, DTO, T_2O. By far, H_2O dominates the mix as a whole, and DHO dominates among the heavy species. The D_2O and the other molecules are probably negligible.

In [1], Gordon talked about 'co-deposition', meaning co-introduction of Palladium and Deuterium into the CF cell, both in solution form. The added Palladium is in the form of the salt $PdCl_2$, which presumably dissociates into the ions Pd^{2+} and $2Cl^-$. Once liberated, the Palladium must form a maximum-entropy distribution of various ionization states.

The Deuterium is, I believe, like normal water in that it does not dissociate very much, but instead stays as a molecule, with tetrahedral shape and ionic structure $(D^- + H^- + O^{2+})$. Can the Palladium ions work on the normal DHO molecule to convert it to linear form, $D \bullet O \bullet H$, with ionic structure $(D^+ + O^{2-} + H^+)$? Consider the required reaction:

$$(D^- + H^- + O^{2+}) \rightarrow (D^+ + O^{2-} + H^+).$$

The left side of the reaction embodies

$$-45.70125 - 91.4025 + 17.5613 = -119.54245 \text{ eV}.$$

The right side of the reaction embodies

$$7.125 - 27.3788 + 14.250 = -6.0038 \text{ eV}.$$

So the reaction needs energy:

$$-6.0038 + 119.54245 = 113.53865 \text{ eV}.$$

Palladium ions in the electrolytic solution can provide the needed energy. Here are several viable reaction pathways, given for illustration of the principle involved. First:

$$(D^- + H^- + O^{2+}) + 10Pd^{2+} + 20e^- \rightarrow (D^+ + O^{2-} + H^+) + 10Pd \ .$$

This provides $10 \times 12.8863 = 128.863$ eV. Next:

$$(D^- + H^- + O^{2+}) + 5Pd^{4+} + 20e^- \rightarrow (D^+ + O^{2-} + H^+) + 5Pd \ .$$

This provides $5 \times 25.1512 = 125.756$ eV. Next:

$$(D^- + H^- + O^{2+}) + 10Pd + 20e^- \rightarrow (D^+ + O^{2-} + H^+) + 10Pd^{2-} \ .$$

This provides $-10 \times (-13.2639) = 132.639$ eV. Finally:

$$(D^- + H^- + O^{2+}) + 5Pd + 20e^- \rightarrow (D^+ + O^{2-} + H^+) + 5Pd^{4-} \ .$$

This provides $-5 \times (-25.8257) = 129.1285$ eV.

Observe that all of these examples of Palladium-catalyzed reactions use some free electrons. That is why they must occur in the pathway of the electrical current in the electrolytic cell.

Consider that there must also exist many other reactions, involving other Palladium ions, that come to the same kind of result: a linear-isomer water molecule with a heavy isotope of Hydrogen in it. Indeed, there must exist the kind of riotous profusion of possibilities that is typical of catalysis schemes generally, as discussed in Chapter 6.

For the sake of completeness, we can look at D_2O as well. The required reaction is:

$$(2D^- + O^{2+}) \rightarrow (2D^+ + O^{2-}) \ .$$

The left side of this reaction embodies

$$2 \times (-45.70125) + 17.5613 = -73.8412 \text{ eV} \ .$$

The right side of the reaction embodies

$$2 \times 7.125 - 27.3788 = -13.1288 \text{ eV} \ .$$

So the reaction needs energy

$$-13.1288 + 73.8412 = 60.7124 \text{ eV} \ .$$

This requirement is more modest than the one for HDO. Here are a couple of ways to meet it:

$$(2D^- + O^{2+}) + 5Pd^{2+} + 10e^- \rightarrow (2D^+ + O^{2-}) + 5Pd \ .$$

This provides $5 \times 12.8863 = 64.4315$ eV.

$$(2D^- + O^{2+}) + 5Pd + 10e^- \rightarrow (2D^+ + O^{2-}) + 5Pd^{2-} \ .$$

This provides $-5 \times (-13.2639) = 66.3195$ eV.

There are many more ways to meet the requirement. As always with catalysis, there is the riotous profusion of possibilities.

4. WHAT HAPPENS AT THE CATHODE

Once in a while, one sees a Gold cathode in a CF related experiment, but usually the cathode in a CF cell is Palladium. The choice of Palladium rather than Gold is important. The cathode is a solid surface, so if Palladium rather than Gold, it can accomplish a trapping reaction. Trapping reactions were first discussed in Chapter 6 for the catalytic converters cleaning up after hydrocarbon combustion. There, the reaction was:

$$C \bullet O + 2Pd \rightarrow Pd \bullet C + Pd \bullet O$$

Here, the needed trapping reaction is:

$$D \bullet O \bullet H + 4Pd \rightarrow Pd \bullet D + Pd \bullet O \bullet Pd + Pd \bullet H$$

The ionic configurations here are

$$(D^+ + O^{2-} + H^+) + 4Pd^0 \rightarrow (D^+ + Pd^-) + (O^{2-} + 2Pd^+) + (H^+ + Pd^-)$$

The right side of the trapping reaction has incremental energy corresponding to

$$2(Pd^+ + Pd^-) = 2 \times (6.4899 - 6.5845)$$
$$= 2 \times (-0.0946) = -0.1892 \text{ eV}$$

This energy is negative, even if not by very much. So this trapping reaction will occur. And when it does, the stage will be set for the arrival of another linear isomer of a heavy water molecule. Then we can get a reaction like

$$(D^+ + O^{2-} + H^+) + (D^+ + Pd^-) + 2Pd \rightarrow$$
$$(He^{2+} + 2e^-) + (O^{2-} + 2Pd^+) + (H^+ + Pd^-)$$

If this reaction were a purely chemical, it would not transpire spontaneously, because it would need energy in the amount corresponding to $2Pd \rightarrow 2Pd^+$, or $2 \times 6.4899 = 12.9798$ eV. The missing energy could, of course, be supplied by adding in more Palladium ions acting as catalysts. But what is important about the basic reaction is that it is *not* purely chemical. The two naked deuterons have fused to make a Helium nucleus. This is cold fusion. It releases an energy that is tremendous by chemical standards.

Naturally, the first thing that happens is that the electrolytic solution boils. To protect the whole experiment from melting, this loss of fluid has to be countered by adding more fluid. People do worry about using up the expensive resource that heavy water is. But think about it: the way that heavy water is made in the first place is by boiling natural sea water to drive off the light H_2O. So what comes off the CF electrolytic fluid is also mostly light H_2O, and can be replaced almost entirely with light H_2O.

CONCLUSION

Let us review the basic requirements for fusion of any kind:
1) Removal of shielding electrons from target nuclei.
2) Confinement of the naked nuclei within the intended fusion volume.
3) Defeat of natural Coulomb repulsion forces between naked nuclei.
4) Handover to the nuclear attraction forces between naked nuclei.

Traditional Hot Fusion meets these requirements in a brute-force way. It heats the input heavy water species to plasma state, it confines the plasma with huge magnetic forces, and compresses the plasma until separations between the heavy isotopes of Hydrogen are small enough for nuclear forces to come into play and dominate the Coulomb repulsion forces between them.

By contrast, Cold Fusion meets the requirements with subtlety and guile. It removes electrons from heavy Hydrogen isotopes by subjecting heavy water molecules to conditions that create linear isomers of those molecules. It traps the heavy isotopes in reactions with Palladium. It sequesters them on the surface of the cathode. Despite the trapped nuclei, the cathode still looks very negative to heavy isotopes arriving subsequently. It attracts them too. Eventually, a heavy isotope arriving at the surface of the cathode meets another heavy isotope already trapped there, and cold fusion ensues.

A Project for Readers

Some of you may have on your shelves some books about the twenty-first century energy problem and/or proposed solutions thereof. Try re-reading them, this time armed with the techniques of Algebraic Chemistry. You will see a great many places in need of quantitative arguments based on numerical analysis.

Acknowledgments

I am grateful to the Natural Philosophy Alliance and to Infinite Energy Magazine for providing a forum for discussions of this subject matter. The first written description appeared under the title "A New Theory for Important New Technologies", in the 2011 Proceedings of the Natural Philosophy Alliance, vol. 8, pages 700-708, 2011. The second, and more widely disseminated, written description appeared under the title "New Theory Applied to Important New Technologies", in Infinite Energy Magazine, Vol. 17, issue 101, pages 14-21, 2012.

Reference

[1] Gordon, F. *How Hot is Cold Fusion?* power-point presentation to NPA 2010; for copy or update, e-mail fgordon@san.rr.com.

Part II. Chemistry as Quantum Mechanics

PROLOG TO PART II

What does it mean to explain some phenomenon that occurs in the domain of Chemistry? Today, the gold standard is considered to be an explanation that is based in Quantum Mechanics (QM). That is because QM occupies a position so central in relation to the rest of modern mathematical physics.

Before QM we had Newton launching mathematical physics as a discipline, with his three laws of mechanics for idealized point mass particles. We also had Maxwell developing mathematics for what came to be known as electromagnetic fields, which exist in the space between point particles, and evolve as waves. Also, we had Fluid Mechanics, Thermodynamics, and we acquired Statistical Mechanics, all to help us think about large populations of atoms or molecules in terms of gross motions, probabilities of particular states, and average values of parameters.

These pre-QM approaches have different domains of utility. Newton's laws are best for small numbers of particles. In fact, two particles define the limit for exact solution with Newton's laws. Systems more complicated than that are treated approximately, in a hierarchical way, with planets orbiting a star, moons orbiting the planets, *etc.* Maxwell theory has singularities where any charged particles are located, so it is limited to situations where that is not a problem. Fluid mechanics, Thermodynamics and Statistical Mechanics are about large numbers of particles, and aren't so helpful for just a handful of particles.

None of these pre-QM domains of utility quite accommodates the kinds of problems that Chemistry presents, which typically involve atoms in numbers of a few, to a few hundred, sometimes a few thousand. QM comes closer to the right domain of applicability because it can begin with single-electron states and put them together to emulate a system with as many electrons as may be needed.

Where did QM really come from? Maxwell theory might have been right-sized enough for Chemistry, but it seemed to indicate a problem in the simplest possible atom: the Hydrogen atom. That atom was imagined to consist of a proton nucleus with an orbiting electron. But an orbiting electron is an accelerating electron, and from Maxwell theory, it follows that an accelerating electron must radiate energy. So how could stable atoms even co-exist with Maxwell theory? There was a clue produced by the experimental discovery of the photoelectric effect. Something unexpected happened, and it suggested the existence of a new fundamental constant of Nature; namely, Planck's constant, and that in turn invited the introduction of a Postulate to found a new theory for a new domain of applicability distinct from that of Maxwell theory; namely, Quantum Mechanics.

After Quantum Mechanics, we do have a great flowering of physics at all scales, and a great deal of interest in the foundations of QM in particular. The recent anthology compiled by Pahlavani [1] gives a good picture of the diversity of the investigations. But is QM as yet adequately developed for the kinds of applications desired in Chemistry? There does exist a large body of current research developing a purpose-built Quantum Chemistry (QC) [2], and that fact implies that we still have a long way to go. What are the biggest problems in QC today? Mainly, they have to do with computational burdens. These problems are the main motivation for a recent journal Special Issue complied by Betsy Rice [3].

Algebraic Chemistry offers a computationally modest approach for QC. Part II of this Book collects mathematical analyses that are relevant for developing a mental picture of atoms that comports with the numerical facts we have at hand. Part II Chapter 9 revisits Hydrogen. Chapter 10 applies the same approach to other charge pairs. Chapter 11 generalizes from charge pairs to charge rings and ultimately to complex structures that model electron populations in atoms. Chapter 12 comments on ionization events and chemical reactions of a more dramatic kind than any of those detailed in the Chapters of Part I.

REFERENCES

[1] *Theoretical Concepts of Quantum Mechanics*, Ed. Mohammad Reza Pahlavani (InTech, Rijeka, Croatia, 2012).

[2] *Quantum Frontiers of Atoms and Molecules*, Ed. Mihai V. Putz (Nova Science Publishers, New York, 2009).

[3] *Quantum Mechanical Foundations of Multiscale Modeling*, Ed. Betsy M. Rice, Special Issue of Advances in Physical Chemistry, June 2012.

Chapter 9

HYDROGEN AS THE PROTOTYPICAL ATOM

ABSTRACT

This Chapter re-examines the currently prevailing theory of atoms, and by extension molecules and reactions, which originates in the Quantum Mechanics of the early 20th century. Assumptions from that time are recalled and re-examined. Where assumptions can be questioned, situations are re-analyzed in terms of Classical Electrodynamics instead. One important assumption was that energy loss by radiation would be the only electrodynamic phenomenon affecting the energy balance in a system like the Hydrogen atom. Here, two other important phenomena that affect energy balance are identified and analyzed. One of them makes the radiation problem greatly worse, but the other one completely balances the radiation problem with a mechanism that causes a compensating energy gain.

INTRODUCTION

In the early 20th century, the Hydrogen atom was the prototype atom for the development of Quantum Mechanics (QM) for all atoms. That history is recounted, for example, by Born [1]. The Hydrogen atom is again the prototype atom here. This Chapter revisits the development, this time without some of the seemingly benign, but nevertheless unnecessary, assumptions embedded in the original development. The assumptions are targeted because they might not be so benign after all. They might account for the enduring perception of conflict between the pre-QM classical electrodynamics (CED), based on Maxwell's electromagnetic theory, and the facts about atoms in general, and Hydrogen in particular. Removing unnecessary assumptions might allow us to better reconcile the QM of atoms with CED.

The referenced result from CED is that any accelerating charge should radiate electromagnetic energy. This result is documented, for example, in Jackson [2]. The Hydrogen atom has nominally one, and actually two, accelerating charges: an electron and a proton. Viewed that way, the Hydrogen atom ought to lose orbit energy and collapse very quickly. But in fact, it doesn't collapse at all. That fact left the world ready for a complete break from CED, and development of a disconnected new theory, QM.

The first development in QM had been the discovery of the photoelectric effect. When incident on a metal surface, light of a frequency high enough can cause the ejection of an electron. The QM of atoms is founded in part on the numerical characterization of the photoelectric effect in terms of Planck's constant. The rescuing postulate for the Hydrogen situation was that Planck's constant simply intervenes, and fixes discrete stable states for the Hydrogen atom, between which instantaneous transitions occur, and emit the observed discrete spectrum. This was the Bohr atom [Ref. 1, Chapt. V].

The present Book Chapter restructures the problem of Hydrogen so as to remove the rescuing Postulate, and replace it with some additional CED results. That way, we can view QM less as a break with the past, and more as a bridge to the future. And of course, the future includes the relatively easy calculations used in Algebraic Chemistry.

Section 1 revisits the basic problem: radiation from charges that accelerate. It does not use the Born-Oppenheimer approximation; namely, that the proton, being so massive in comparison to the electron, does not move. So both particles accelerate, and that of course makes the radiation problem slightly worse.

Section 2 revisits the Coulomb attraction acting between the electron and the proton. It does not assume *a-priori* the turn-of-the 20th century Liénard-Wiechert solution for the Maxwell potentials and fields arising from a moving source. Instead, it seeks a solution that fits the boundary conditions needed for the Hydrogen atom. That makes a slightly different solution available. It implies a second mechanism that works in opposition to radiation, a mechanism for energy gain in opposition to energy loss. This second mechanism is an internal torquing within the atom.

Section 3 shows how the newly discussed torquing mechanism also makes the problem from the radiation mechanism a lot worse. That happens because a system without Newtonian instantaneous-action-at-a-distance cannot provide Newtonian immobility of the center of mass. But then Section 4 shows that the final balance between all the mechanisms at work turns out to be about right for explaining the Hydrogen atom ground state.

Section 5 discusses the excited states of the Hydrogen atom. It offers a new interpretation for the very term 'excited state': it is understood as construct involving several atoms, not an attribute of a single atom alone. Section 6 investigates a phenomenon known as sub-states of Hydrogen.

1. RADIATION FROM ACCELERATING CHARGES

The classical Larmor formula for total radiation power P due to acceleration a of a charge q (in electrostatic units) is $P_q = 2q^2 a^2 / 3c^3$. In the classical electron-proton system, we have the electron with charge $q = -e$ and the proton with charge $q = +e$, with the electron orbiting the center of mass at relatively large radius r_e, and the proton orbiting at relatively small radius r_p, both at the same frequency Ω. Most of the radiation comes from the electron, and amounts to

$$P_e = 2e^2(r_e\Omega^2)^2 / 3c^3 .$$

The radiation power from the proton is

$$P_p = 2e^2(r_p\Omega^2)^2 / 3c^3 .$$

This is smaller than P_e by a factor of $(r_p/r_e)^2 = (m_e/m_p)^2$, where the mass ratio m_e/m_p is about $1/1800$. That makes P_p negligible, so the total radiation power is well estimated as

$$P_R = P_e + P_p \approx P_e = 2e^2(r_e\Omega^2)^2 / 3c^3 .$$

2. TORQUING IN THE HYDROGEN ATOM

Since well before the 20th century, it has been assumed that light signals propagate just like infinite plane waves propagate; *i.e.* without evolution in shape, and with fixed speed c with respect to whatever receiver will eventually see the signal, and conveys information from the source at position retarded in time by $\Delta t = \ell/c$, where ℓ is the distance from the receiver at reception time t to the source at time $t - \Delta t$.

This assumption is not confirmable by giving a finite-energy light pulse as input to Maxwell's differential equations, and then following its evolution. Something different happens. Part III, Chapter 1, of this book details the situation. For present purposes, it suffices to say that the signal pulse evolves into a wavelet, and the wavelet begins its journey at speed c with respect to the source, but concludes its journey at speed c with respect to the receiver, and the source information it brings (position, speed, acceleration) characterizes the source at half retarded time $t - \Delta t/2$.

If the commonly believed assumption were true, then the electron and the proton in the Hydrogen atom would each be experiencing Coulomb attraction from the other one as if at essentially its present position. The Coulomb forces would be almost central. The system would be almost Newtonian. Almost, ... although not quite. In principle, a system running on Maxwell's equations is simply not a Newtonian system.

And if the more justifiable assumption is taken, then the system is very clearly not at all Newtonian. The Coulomb forces are not even nearly central. As a result, there is a significant torque within the system. It amounts to

$$\mathbf{T} = \mathbf{r}_e \times \mathbf{F}_e + \mathbf{r}_p \times \mathbf{F}_p .$$

and

$$\mathbf{r}_e \times \mathbf{F}_e = \mathbf{r}_P \times \mathbf{F}_P ,$$

so

$$\mathbf{T} = 2\mathbf{r}_e \times \mathbf{F}_e .$$

The angle between $\mathbf{r}_e$ and $\mathbf{F}_e$ is

$$\theta = r_P \Omega_e / 2c = (m_e / m_P)(r_e \Omega / 2c) .$$

So the torque magnitude is

$$T = 2\theta r_e F_e = (m_e / m_P)(r_e \Omega / c)\left[e^2 / (r_e + r_P) \right]$$

The torque provides a mechanism for energy gain. The rate of energy gain is the power

$$P_T = T\Omega = (m_e / m_P)\left[r_e \Omega^2 / c \right]\left[e^2 / (r_e + r_P) \right] ,$$

and with $r_e \Omega^2 = e^2 / m_e (r_e + r_P)^2$, that reduces to

$$P_T = e^4 / m_P c (r_e + r_P)^3 .$$

3. EVEN MORE RADIATION

The model for the Hydrogen atom that is developing here has yet another non-Newtonian feature: the Coulomb forces do not cancel. So the system as a whole, as represented by its center of mass, moves: it traverses a small circle at angular frequency Ω. This is an additional source of acceleration, and hence an additional source of radiation.

A fact that is relevant to the situation at hand was first uncovered in 1927. L.H. Thomas was studying the then-anomalous magnetic moment of the electron: half its expected value. He showed that coordinate frame attached to a particle traversing a circular path naturally rotates at half the imposed revolution rate. [3]

Because the Thomas rotation effect was discovered in connection with a study of sequential non-parallel Lorentz transformations, it was originally thought to be relativistic in nature. But it is not; it just the nature of three-dimensional space working with one-dimensional time, and it arises just as clearly with sequential non-parallel Galilean transformations.

The upshot of Thomas rotation is that when the atomic two-body system is driven in a circle at a rate equal to its internal rotation rate Ω, it looks to the outside observer to be rotating at rate 2Ω. That scales up the far-field radiated power, from

$$P_R = 2e^2(r_e\Omega^2)^2/3c^3 ,$$

to

$$P_R = 2^4 \times \left[2e^2(r_e\Omega^2)^2/3c^3\right] = 2^5 e^2(r_e\Omega^2)^2/3c^3 .$$

and with $r_e\Omega^2 = e^2/m_e(r_e + r_p)^2$, that reduces to

$$P_R = 2^5 e^6/3m_e^2 c^3 (r_e + r_p)^4 .$$

4. BALANCE AT THE GROUND STATE

Let us now ask: in what circumstance is the rate of energy loss due to radiation just balanced by the rate of energy gain due to torquing? That is, under what circumstance does $P_T \equiv P_R$? The balance requires

$$e^4/m_p c(r_e + r_p)^3 \equiv 2^5 e^6/3m_e^2 c^3 (r_e + r_p)^4 .$$

That boils down to the formula

$$(r_e + r_p) \equiv 32 m_p e^2/3m_e^2 c^2 ,$$

and that evaluates to

$$(r_e + r_p) = 5.5 \times 10^{-9} \text{ cm} .$$

This result is fairly consistent with the accepted value of $r_e = 5.28 \times 10^{-9}$ for the Hydrogen atom.

In the conventional approach to QM, r_p is taken to be zero, and r_e is expressed in terms of Planck's constant h:

$$r_e = h^2 / 4\pi^2 \mu e^2$$

where μ is the so-called reduced mass, defined by $1/\mu = 1/m_e + 1/m_p$, which makes $\mu \approx m_e$. Taking $(r_e + r_p) \approx r_e$, we have

$$h \approx (\pi e^2 / c)\sqrt{128 m_p / 3 m_e}$$

This expression evaluates to 6.77×10^{-34} Joule-sec, which is close to the accepted value 6.626176×10^{-34} Joule-sec. The implication here is that Planck's constant is not actually an independent constant of Nature, but rather a function of other constants of Nature.

5. EXCITED STATES

The standard interpretation of QM is that the Hydrogen atom has multiple stable states, each with energy determined by a principle quantum number n according to the formula $E_n = E_1 / n^2$, where E_1 is the energy of the ground state. Since the ground state is a bound state, E_1 is negative, so E_n is less negative, and hence higher than E_1. An electron can reside in an upper state ($n > 1$), but only rather precariously, and when it teeters and falls back to $n = 1$, a photon is emitted with energy

$$\Delta E_{n \to 1} = |E_1|(1 - 1/n^2) .$$

When the electron falls, but not all the way to the ground state, the photon has a lesser energy, of the form

$$\Delta E_{n \to m} = |E_1|(1/m^2 - 1/n^2) .$$

where $m < n$ identifies the finishing state. Considering all possible values of n and m, a rich spectrum of possible photon energies results. Indeed there is a countable infinity of possible spectral lines.

But with only two constituent particles, the Hydrogen atom is a very simple system. It seems to lack enough degrees of freedom to support so much complexity. So consideration of an alternative interpretation seems inviting. Suppose that the 'excited state' is not really an attribute of a single atom, but rather *the* attribute of a system involving several atoms.

Consider the possibility that the nth excited state of the Hydrogen atom is actually a construct involving n Hydrogen atoms – not a molecule, but rather a super atom, with a cluster of n electrons and a cluster of n protons. When this construct of n atoms falls apart, the photon with energy ΔE_n can be emitted. Or a cascade of several photons of lesser energy, adding up to a total energy not exceeding ΔE_n, can be emitted, along with some waste heat. In this way, the whole rich spectrum of Hydrogen can be accounted for.

CONCLUSION

Taking Hydrogen as a prototype atom leads to some novel mental images about atoms in general. For example, the light spectrum emitted by any element has discrete lines, not because a single atom is describable by a differential equation with a countable set of solutions, but rather because a spectral line comes from an atomic system containing a countable number of atoms.

But the Chapter is about options, and not about mandates. It offers a picture that is so far a Model, and not a Theory. The Model is a useful way of thinking about things. Subsequent Chapters will develop it further, to begin to explain the numerical regularities that exist in chemical data. But already the Model doesn't comport with some mandates asserted in the context of 20th century QM. The Model is about discrete particles, and not about probability waves. So it lives uneasily, in parallel with standard QM Theory.

The first requirement for any such Model to be considered potentially viable as the beginning of a real Theory is an explanation for how charged particles can form into charge clusters. That is the mission of the next Chapter.

REFERENCES

[1] Born, M. *Atomic Physics*, Sixth Edition, Hafner Pub. Co., New York, Blackie and Son, London, Glasgow, Canada, India, 1957.

[2] Jackson, J.D. *Classical Electrodynamics*, third edition, John Wiley and Sons, New York, 1962.

[3] Thomas, L.H. The Kinematics of an electron with an Axis, *Phil. Mag.* S. 7, Vol. 3 No. 13, pp. 1-22 (1927).

Chapter 10

GENERAL CHARGE PAIRS

ABSTRACT

The Hydrogen atom treated in the last Chapter is just one example of a charge pair forming a two-body system. This Chapter investigates the full range of electrically bound two-body systems that form the basis for the development of the Algebraic Chemistry of atoms and molecules in general. It shows the many similarities, and the key differences, between the different types of two-body system.

INTRODUCTION

The analysis of Hydrogen has so far had some approximations in it, and removing them sets the stage for analyzing other charge pair two-body systems. One other such charge pair is the well-known construct known as 'positronium'. This system consists of one electron and one positron. Positronium is of interest because it is like Hydrogen, but without the extreme mass disparity. Another construct is the proton pair. The possible existence of this one supports the idea from the last Chapter that the terminology 'excited state' could characterize something other than a single atom; namely, a system of several atoms that is not a molecule, but rather a sort of 'super atom'. Finally, there is the well-known electron pair, the Cooper pair. This one is the first in a series of electron systems that are treated in the next Chapter, all of them being needed for developing the idea of an electron subsystem in hierarchical model for an atom.

The following Sections detail each of these mentioned two-body systems.

1. HYDROGEN

The analysis of Hydrogen in the last Chapter had two 'small-angle' approximations in it. In the expression for energy loss rate by radiation, there are vector projections, and hence angle cosines, which were approximated by unity. In the expression for energy gain rate due to torquing, there are vector cross products, and hence angle sines, which were approximated by the angle values in radians.

We suppose the electron and the proton both execute circular orbits about their system center of mass. There is a simple geometry describing what each charge experiences. Let α be the angle measured at the center of mass between the electron present position and the electron half-retarded position that is experienced by the proton. The angle α is essentially electron orbit speed V_e divided by light speed c. The relationship between V_e and r_e is Newton's law connecting mass times acceleration to Coulomb force:

$$m_e a_e = m_e V_e \Omega = m_e V_e^2 / r_e = e^2 / (r_e + r_p)^2 .$$

So, approximately,

$$V_e \approx e / \sqrt{m_e r_e} .$$

And by the same logic,

$$V_p \approx e / \sqrt{m_p r_p} .$$

Let β be the angle measured at the electron position between the proton present position and the proton half retarded position that is experienced by the electron. Then $\alpha - \beta$ will be the angle measured at the proton position between the electron present position and its half retarded position as experienced by the proton.

The angles $\alpha - \beta$ and β are the angles by which the two attractive forces are tilted off center. When $V_e \ll c$, the angle $\alpha - \beta$ is small, and β is even smaller. The analysis in the last Chapter made the approximations that sines of the angles are equal to the angles, and cosines of the angles are equal to unity. When those approximations are removed, the energy loss rate due to radiation becomes

$$P_R = \left[2^5 e^6 / 3 m_e^2 c^3 (r_e + r_p)^4 \right] \times \left[\cos(\alpha - \beta) + (m_e / m_p) \cos(\beta) \right]^2$$

and the energy gain rate due to torquing becomes

$$P_T = \left[e^4 / m_p c (r_e + r_p)^3 \right] \times \frac{1}{2} \left[\frac{\sin(\alpha - \beta)}{(\alpha - \beta)} + \frac{\sin(\beta)}{\beta} \right]$$

Observe that both of these expressions are oscillatory. But where P_R is confined to non-negative values, P_T is not. So while the radiation P_R always represents an energy loss mechanism, and the torquing mechanism P_T often acts as an energy gain mechanism, it sometimes acts as an energy loss mechanism.

The behavior of P_R and P_T as a function of r_e is displayed in Figure 1. The horizontal axis is the system radius. The vertical axis is energy loss rate for P_R or energy gain rate for P_T, on a logarithmic scale, so situations with energy loss rate due to torqiung are not displayed. The familiar ground state for Hydrogen occurs way off the Figure, to the right.

The P_R and P_T curves clearly cross each other repeatedly, so a whole sequence of additional balance points exists below the ground state. These solutions occur at very small electron orbit radii r_e, corresponding to very high electron speeds. The angle $\alpha = V_e / 2c$ is an odd multiple of π, so V_e takes values

$$V_e = 2\pi c, 6\pi c, 10\pi c, ...$$

etc. All such solution speeds are dramatically superluminal. Given Einstein's Special Relativity Theory (SRT), superluminal speeds of any magnitude may seem quite unbelievable. But consider that if such speeds do occur, they do so *deep* inside an atom – hardly the kind of 'free-space' environment for which Einstein's particle-speed limit c was legislated.

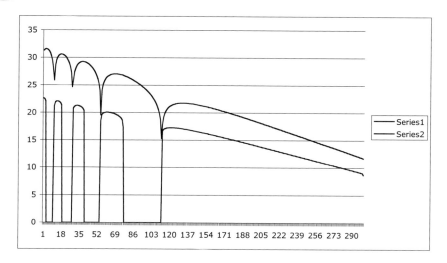

Figure 1. Behavior of P_R and P_T in the Hydrogen atom.

And this issue is just one of a list of issues that exist between QM and SRT. Another one is the existence of instantaneous correlations at macroscopic distances, predicted and demonstrated in the quantum domain. Another one is the big conceptual difference between the photon as imagined in QM and the electromagnetic fields as described by Maxwell's equations, which are today regarded as an integral part of SRT. And of course, there is that problem worried over so early in the development of QM: the immortality of the Hydrogen atom despite radiation reaction.

The superluminal speed solutions can be characterized by the formula

$$V_e \approx 2\pi c(1+2n), \quad n = 0,1,2,\ldots \text{ etc.}$$

We see that kinetic energies, if proportional to V_e^2, must advance with n^2. This behavior is like having a radial quantum number that is fractional. There exists an extensive experimental literature on the spectroscopy of Hydrogen, and some of it seems inexplicable in terms of traditional QM. In particular, the work of Dr. Randall L. Mills and associates shows some extreme UV lines that are suggestive of Hydrogen atom states with fractional radial quantum numbers. Dr. Mills tells that story quite comprehensively in [1].

2. POSITRONIUM

The refined Hydrogen analysis provides a stepping-stone to the analysis of systems in which the participant charges are not of very dissimilar mass, as in Hydrogen, but rather of equal mass, such as in the electron cloud in a trans-Hydrogenic atom. This Section explores positronium, they system consisting of one electron and one positron. The mass symmetry makes for orbit radius equality:

$$r_p = r_e,$$

and angle symmetry:

$$\alpha - \beta \equiv \beta \equiv \alpha/2.$$

The energy loss rate due to radiation becomes

$$P_R = \left[2^5 e^6 / 3m_e^2 c^3 (2r_e)^4\right] \times \left[2\cos(\alpha/2)\right]^2.$$

The energy gain rate due to torquing becomes

$$P_T = \left[e^4 / m_e c(2r_e)^3\right] \times \left[\frac{\sin(\alpha/2)}{(\alpha/2)}\right].$$

The analysis then follows:

1) The condition defining the angle α becomes

$$\alpha = (V_e/2c) \times (2r_e)/r_e = V_e/c$$

2) The formal condition for defining angle β becomes

$$\tan(\beta) = \frac{\sin(\alpha)}{1+\cos(\alpha)} = \frac{2\sin(\alpha/2)\cos(\alpha/2)}{1+\cos^2(\alpha/2)-\sin^2(\alpha/2)}$$
$$= \frac{\sin(\alpha/2)\cos(\alpha/2)}{\cos^2(\alpha/2)} = \tan(\alpha/2)$$

This relationship is indeed satisfied by $\beta = \alpha/2$.

3) The condition defining V_e becomes

$$m_e V_e^2/r_e = e^2/(2r_e)^2 .$$

The solution is then

$$V_e \approx e/\sqrt{m_e(2r_e) \times (1/2)} = e/\sqrt{m_e r_e} .$$

4) The factor for removing the small-angle approximation from the energy loss rate due to radiation, P_R, simplifies, making the expression into:

$$P_R = \left[2^5 e^6 / 3 m_e^2 c^3 (2r_e)^4\right] \times \left[2\cos(\alpha/2)\right]^2 .$$

5) Similarly, the expression for the energy gain rate due to torquing P_T simplifies to:

$$P_T = \left[e^4 / m_e c (2r_e)^3\right] \times \left[\frac{\sin(\alpha/2)}{(\alpha/2)}\right] .$$

Being formally analogous to Hydrogen, Positronium also has a plot somewhat similar to Figure 1, with 'ground-state', low-speed solution, plus a family of high-speed sub-ground solutions. But the peaks in radiation power are twice as broad, and the intercepts with the torquing power occur in the middle of torquing peaks, and so produce pairs of solution points. These solution pairs are located around V_e at odd multiples of $2\pi c$, or

$$V_e = 2\pi c, 6\pi c, 10\pi c,...$$

, *etc.*, or

$$V_e \approx 2\pi c \times (1+2M), \quad M = 0,1,2,\ldots \text{ etc.}$$

(Here M means 'multiple', not 'nuclear mass' as it did in earlier Chapters.) Figure 2 shows all this behavior.

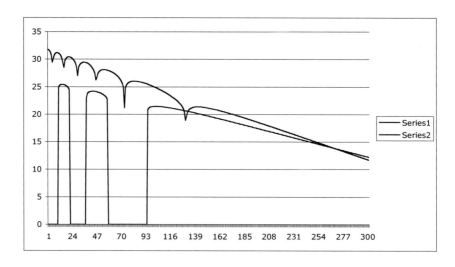

Figure 2. Behavior of P_R and P_T in the positronium system.

3. THE PROTON PAIR

The system formed by two protons is almost the same as the positronium system, except that 1) the mass involved, call it m_P, is much larger; 2) the radiation is not dipole, but rather quadrupole; 3) the torquing effect has the opposite sign everywhere. We have:

$$P_R = \left[2^4 \times \text{quadrupole radiation power}\right] \times \left[2\cos(\alpha/2)\right]^2$$

and

$$P_T = \left[-1 \times \text{torquing gain power}\right] \times \left[\frac{\sin(\alpha/2)}{(\alpha/2)}\right].$$

The condition for balance is still

$$P_T = P_R.$$

Because of the negative sign in P_T, the plot analogous to Figure 2 has peaks where there were gaps, and gaps where there were peaks. So there is no low-speed crossing indicating a 'ground-state' solution. Figure 3 shows this.

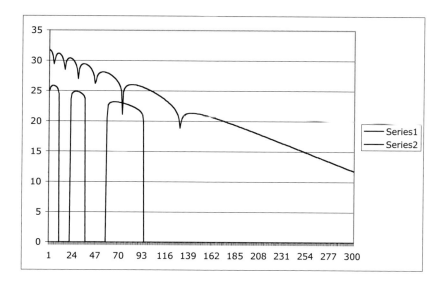

Figure 3. Behavior of P_R and P_T in same-charge binary systems.

There is also no solution at the first radiation zero, which occurs at $V_e = 2\pi c$, or any subsequent odd multiple of $2\pi c$. The only solutions occur interleaved with the high speed solutions on Figure 2; that is, at V_e equal to even multiples of $2\pi c$, or

$$V_P \approx 4\pi c,\ 8\pi c,\ 12\pi c,\ 16\pi c$$

etc., i.e.:

$$V_P \approx 2\pi c \times (2N),\ N = 1, 2, 3, 4,\ etc.$$

(Here N just means 'number', but it turns out later to be related to the N parameter introduced earlier in association with the periods on the Periodic Table.)

The behavior illustrated by Figure 3 applies also for the electron pair, discussed next.

4. THE ELECTRON PAIR

The electron pair is famous in solid-state physics under the name 'Cooper pair'. It is also obviously important in Chemistry, as witnessed by the high ionization potentials of Helium $_2$He, and, to a lesser extent, other similar elements. In the language of QM, these elements have a filled s state, *i.e.* a filled $l = 0$ state, *i.e.* a spin 'up' electron and a spin 'down' electron. They are Beryllium $_4$Be, Magnesium, $_{12}$Mg, Calcium $_{20}$Ca, Strontium $_{38}$Sr, Barium $_{56}$Ba, Radium $_{88}$Ra.

The mathematical description for the electron pair is really quite similar to that for the proton pair, except that it is written with the electron mass m_e. So the only solutions occur at speeds

$$V_e \approx 4\pi c, 8\pi c, 12\pi c, 16\pi c$$

etc., *i.e.*:

$$V_e \approx 2\pi c \times (2N), \ N = 1,2,3,4, \textit{etc.}$$

The least of these, $V_e = 4\pi c$, is the one that is most likely to be found in a two-electron ring like the Helium atom has.

Beryllium $_4$Be presents a situation fairly similar to that of Hydrogen. It has a moderately high ionization potential, and that fact suggests that its electron population forms two very stable electron pairs. One can picture these pairs stacked together, inasmuch as each pair is like a microscopic current loop, and in macroscopic physics, parallel current loops attract each other magnetically. These microscopic current loops are different in lacking a positive metallic matrix to travel in, so they do present Coulomb repulsion, but it is insignificant compared to the magnetic attraction of electrons moving in parallel at a superluminal speed.

The rest of the elements mentioned have not only electron pairs, but also rings with odd numbers of electrons: 3, 5, and 7. Some of these are detailed in the next Chapter.

Conclusion

We have learned in this Chapter that binary charge systems exhibit an important difference between opposite-sign charges and same-sign charges. Opposite-sign binary charge systems have an orbit solution analogous to the ground state of Hydrogen, plus an infinite set of sub-ground solutions with super-luminal orbit speeds. Same-sign binary charge systems do not have any analog to the Hydrogen ground state; they only have an infinite set of solutions with super-luminal orbit speeds. The super-luminal speeds involved are the same, regardless of the mass, or any other parameter, involved. That fact makes it easy to proceed to model electron systems with more than two electrons. Eventually, we can model systems of any number of electrons. That is the mission of the next Chapter.

REFERENCE

[1] Mills, R.L. The Nature of the Chemical Bond Revisited and an Alternative Maxwellian Approach, *Physics Essays* 2004, 17 (3) 342-389.

Chapter 11

ELECTRON RINGS AND STRUCTURES THEREOF

ABSTRACT

This Chapter develops a model for the electron population in all atoms. The model is based on the concept of electron rings, a generalization of the electron pairs introduced in the last Chapter. The present Chapter extends the notion of electron rings to rings of various numbers of electrons, as is required to correspond to the various sets of single-electron quantum states that make up the known electron population of an atom of any element. This exercise produces some straightforward explanations for some otherwise baffling behaviors of different elements.

INTRODUCTION

The concept of charge rings actually goes back a long way. Some of the early work pertains to the problem of modeling the electron itself. There has always been a problem about the electron; namely, why don't internal Coulomb forces make the electron self-destruct? J.P. Wesley reviewed a lot of the history about this subject in [1]. Early twentieth-century efforts by Abraham and Loentz did not get to a ring at all, and ended up with the energy wrong for the mass by a factor of 4/3. Slightly later models had the spinning ring, but had the angular momentum wrong by a factor of 2. Late in the twentieth century, Bergman and Wesley developed an electron ring model that fits the known mass, magnetic moment, and spin. It answers the original question, about avoiding self-destruction, in terms of magnetic pinch force.

But all this work is about a single electron by itself. For Chemistry, we are less interested in the possible ring structure of individual electrons, as elementary particles, and more interested in rings with multiple electrons, as parts of atoms. The fundamental issue for the electron ring in an atom is the same as it was with the single electron: why doesn't the imagined ring construct self-destruct? The answer turns out to be more easily found for the ring of several discrete electrons than it can be for a single electron shaped as a continuous ring.

This Chapter builds on the now-preliminary treatment of some of the issues in [2]. What was preliminary there, and is updated here, concerns electron orbit speeds. Recall from the last Chapter that, with two electrons, we have balance between energy loss by radiation and energy gain by torquing at orbit speeds

$$V_e \approx 4\pi c, 8\pi c, 12\pi c, 16\pi c$$

etc., i.e.:

$$V_e = 2\pi c \times (2N), \ N = 1,2,3,4 \text{, } etc.$$

With more than two electrons, we have a different electron count, a different radiation angular pattern and amplitude, and a different amount of energy gain rate caused torquing. *But we still have the same set of orbit-speed solutions.* That is because radiation and torquing amplitudes do not matter; only the fact that they cross each other, and the fact that they balance at the crossing points, matters. And those crossing points are characterized entirely by orbit speeds. So variations on those orbit speeds that I considered in [2] are now omitted.

Different members of the given orbit-speed solution set occur in different situations. Some elements are in very comfortable situations, and some are in very distressed situations. Section 1 discusses elements that are very comfortable. Electron rings explain why. Section 2 discusses elements that are very distressed. Electron rings explain why. Section 3 discusses elements that are just plain peculiar. Electron rings explain those peculiarities.

1. COMFORTABLE ELEMENTS

The champion of all comfortable elements is of course Helium $_2$He, with one two-electron ring. Also very comfortable are Beryllium $_4$Be, Magnesium, $_{12}$Mg, Calcium $_{20}$Ca, Strontium $_{38}$Sr, Barium $_{56}$Ba, Radium $_{88}$Ra. These all apparently have two rings of two electrons each. Every two-electron ring represents a filled s state, *i.e.* a filled $l = 0$ state, *i.e.* a spin 'up' electron and a spin 'down' electron. The comfortable elements beyond Berylium also have additional filled rings along with their two-electron rings.

1.1. A Ring of Three Electrons

The number 3 triggers one to think first of Lithium $_3$Li. Its three electrons might indeed form a ring, but if they do, then that ring is not anywhere near as stable as the two-electron ring that a Helium $_2$He atom would have. Why is that?

The evidence is that a three-electron ring can be comfortable only if there are enough two-electron rings along with it. The first element that appears to be comfortable and to allow a three-electron ring is Nitrogen $_7\text{N}$. It appears to have a Beryllium-like two rings of two electrons each, plus a three-electron ring. A plausible mental image is that the two rings of two-electrons each form a magnetic containment volume, within which three-electron ring can reside. The magnetic confinement concept is the same thing that one sees on a macroscopic scale in a hot fusion reactor. The three-electron ring is like the hot plasma in the fusion reactor. It can be confined if its radius is comfortably smaller than the radius of the confinement volume. Its radius will indeed be comfortably smaller if the three electrons are moving, not at the first solution speed $V_e = 4\pi c$, but rather at the second solution speed, $V_e = 8\pi c$. Why? Because, in an attractive system, high velocity means low potential energy, and hence very negative total energy, and hence very small orbit radius.

1.2. Two Rings of Three Electrons

Consider Neon $_{10}\text{Ne}$. It apparently has the Beryllium-like two rings of two electrons each, forming a magnetic containment volume, containing in turn two Nitrogen-like rings of three electrons each. The two two-electron rings circulate at the first solution speed $V_e = 4\pi c$, while the two three-electron rings circulate at the second solution speed, $V_e = 8\pi c$.

Consider also $_{12}\text{Mg}$. It has three two-electron rings and two three-electron rings. So there are two magnetic confinement volumes available for the two three-electron rings. So which way are the two three-electron rings deployed? Are they deployed as singletons, into separate magnetic confinement volumes? Or are they deployed together, into one magnetic confinement volume, thus leaving the other magnetic confinement volume totally empty? This sort of question recurs from here on throughout the Periodic Table. A clue about the answer comes only later, in Sect. 3.

1.3. A Ring of Five Electrons

Consider Manganese $_{25}\text{Mn}$. It appears to have four two-electron rings, presumably all circulating at the first solution speed, $V_e = 4\pi c$. They make three magnetic confinement volumes. These three confinement volumes contain altogether four three-electron rings, all of them smaller because they are circulating at the second solution speed, $V_e = 8\pi c$. Again we have the sort of ambiguity noted above for $_{12}\text{Mg}$: two of the magnetic confinement volumes may contain one three-electron ring, and one of them, possibly the middle one, may contain two three-electron rings. Or two of the magnetic confinement volumes may each contain two three-electron rings, with the third magnetic confinement volume being left totally empty.

In any event, any pair of three-electron rings in turn makes a stronger magnetic confinement volume, which can contain a five-electron ring. This ring can be smaller still because it can circulate at the third solution speed, $V_e = 12\pi c$.

At this point, it becomes helpful to introduce some more distinguishing language. Think of the magnetic confinement volume made by two two-electron rings as a 'magnetic bottle', and think of the stronger magnetic confinement volume made by two three-electron rings as a 'magnetic thermos jug': a magnetic container too, but a more substantial one than the magnetic bottle.

1.4. Two Rings of Five Electrons

Consider $_{30}Zn$. Like $_{25}Mn$, it must contain four two-electron rings, presumably all of them circulating at the first solution speed, $V_e = 4\pi c$, which make three magnetic confinement volumes. Each of these confinement volumes can contain one or two three-electron rings, circulating at the second solution speed, $V_e = 8\pi c$. Two of the confinement volumes may contain one such ring, and one of them, possibly the middle one, may contain two such rings. Or just two of them may contain two such rings each. In any event, any pair of three-electron rings makes a stronger magnetic confinement volume, which can contain one or two five-electron rings circulating at the third solution speed, $V_e = 12\pi c$.

With two such five-electron rings, we have an even stronger magnetic confinement volume. Think of it as a 'magnetic Dewar flask'.

1.5. A Ring of Seven Electrons

Consider Europium, $_{63}Eu$. It appears to have six two-electron rings, all circulating at the first solution speed, $V_e = 4\pi c$. These six circulating rings make five weak magnetic confinement volumes. There are eight three-electron rings distributed among these five weak confinement volumes, one or two to an individual volume, and circulating at the second solution speed $V_e = 8\pi c$. So at least three of these weak confinement volumes have two of these three-electron rings. These in turn create three stronger magnetic confinement volumes that can hold five-electron rings circulating at the third solution speed $V_e = 12\pi c$. There are four of these five-electron rings, and hence at least one pair of them among the three stronger confinement volumes. The pair of five-electron rings creates an even stronger magnetic confinement volume. This can hold a seven-electron ring circulating at the fourth solution speed, $V_e = 16\pi c$.

The electrons of the seven-electron ring are thus very deep buried, and hardly ever become involved in reactions to make compounds. So Chemistry cannot do much to distinguish and separate the 28 different elements in which seven-electron rings are being filled.

This explanation for the difficulty of distinguishing and separating these elements can be contrasted with the explanation given by the usual shell model of the atom.

In terms of the notation commonly used in QM, these electrons are said to have orbital quantum number $n = 6$ or $n = 7$, and angular momentum quantum number $l = 3$. Despite the fact that n is large, which suggests great distance from the nucleus, the large l is said to intervene, and work against that, and keep these electrons deep buried near the nucleus.

1.6. Two Rings of Seven Electrons

Consider Ytterbium, $_{70}Yb$. It has the same six two-electron rings, all circulating at the first solution speed, $V_e = 4\pi c$, making the same five weak magnetic confinement volumes.

There are the eight three-electron rings distributed among these, circulating at the second solution speed $V_e = 8\pi c$, making at least three stronger magnetic confinement volumes.

There are the four five-electron rings circulating at the third solution speed, $V_e = 12\pi c$, making at least one pair that in turn makes an even stronger confinement volume. Both seven-electron rings reside there, circulating at the fourth solution speed $V_e = 16\pi c$. With two such seven-electron rings, we again have an even stronger magnetic confinement volume.

But fortunately we don't yet have to give this kind of confinement volume a memorable name, because Chemistry has not yet revealed anything to be confined within it!

2. UNCOMFORTABLE ELEMENTS

The set of uncomfortable elements begins with the elements that begin periods. The first one is of course Hydrogen $_1H$, which would rather be in a molecule, or else be in the ions of a hydrogen plasma, than remain a neutral atom. Then we have the alkali metals: Lithium $_3Li$, Sodium $_{11}Na$, Potassium $_{19}K$, Rubidium $_{37}Rb$, Cesium $_{55}Cs$, and Francium $_{87}Fr$. They all have ionization potentials that are even lower than that of Hydrogen $_1H$. That means they are all trying even harder to dump an electron.

A similar situation arises with Boron $_5B$, Aluminum $_{13}Al$, Gallium $_{31}Ga$ Indium, $_{49}In$, Thallium $_{81}Tl$, and with Oxygen $_8O$, Sulfur $_{16}S$, Selenium $_{34}Se$, Tellurium $_{52}Te$, and Polonium $_{84}Po$: a three-electron ring beginning, but only one vanguard electron there to establish the ring.

And another similar situation occurs with the brittle metals Scandium $_{21}$Sc, Yttrium $_{39}$Y, and Lutetium $_{71}$Lu and Lawrencium $_{103}$Lr, and with the ductile metals Iron $_{26}$Fe, and Rubidium $_{44}$Ru, Osmium $_{76}$Os and Hassium $_{108}$Hs: a five-electron ring starting, but only one vanguard electron present.

And yet another similar situation transpires with rare earths Lanthanum $_{57}$La and Actinium $_{89}$Ac, and with Gadolinium $_{64}$Gd, Curium $_{96}$Cm: a seven-electron ring starting, but only one electron as yet present.

In every case, the ionization potential for the first electron of a ring is lower than that for the last electron in that ring. In between, there is the gradual rise in ionization potential. With the idea of the electron rings, one can visualize the cause of the discomfort that is driving this behavior. When there is only one electron, that isn't even a ring at all. It has only the magnetic moment of the electron itself to make it at all responsive to magnetic confinement. The one electron may not properly integrate into the atomic charge cluster at all. Instead, it may just make the atom highly reactive with other atoms. Hydrogen and the alkali metals are definitely like that. But the others aren't so clear: the brittle metals Scandium and Yttrium aren't so famously reactive, and Lutetium and Lawrencium aren't so fully studied yet. Lanthanum and Actinium and Gadolinium and Curium appease the one electron by pairing it with another electron robbed from an existing five-electron ring. That is a peculiar response. And it is not the only case.

3. Peculiar Elements

There are presently 19 elements that are known to be somewhat peculiar in terms of what single-electron quantum states are filled to complete the atom. Lanthanum and Actinium and Gadolinium and Curium were mentioned above; the full list of known peculiar elements with their diagnoses goes as follows:

Chromium $_{24}$Cr robs one electron from a two-electron ring to complete a five-electron ring.

Copper $_{29}$Cu robs one electron from a two-electron ring to complete a five-electron ring.

Niobium $_{41}$Nb robs one electron from a two-electron ring to complete a five-electron ring.

Molybdenum $_{42}$Mo robs one electron from a two-electron ring to complete a five-electron ring.

Rubidium $_{44}$Ru robs one electron from a two-electron ring to complete a five-electron ring.

Rhodium $_{45}$Rh robs one electron from a two-electron ring to complete a five-electron ring.

Palladium $_{46}$Pd completely consumes a two-electron ring to complete a five-electron ring.

Silver $_{47}$Ag robs one electron from a two-electron ring to complete a five-electron ring.

Lanthanum $_{57}$La puts an electron in a five-electron place instead of a seven-electron place.

Cerium $_{58}$Ce puts an electron in a five-electron place instead of a seven-electron place.

Gadolinium $_{64}$Gd puts an electron in a five-electron place instead of a seven-electron place.

Platinum $_{78}$Pt robs one electron from a two-electron ring to complete a five-electron ring.

Gold $_{79}$Au robs one electron from a two-electron ring to complete a five-electron ring.

Actinium $_{89}$Ac puts an electron in a five-electron place instead of a seven-electron place.

Thallium $_{90}$Th puts an electron in a five-electron place instead of a seven-electron place.

Protactinium $_{91}$Pa puts an electron in a five-electron place instead of a seven-electron place.

Uranium $_{92}$U puts an electron in a five-electron place instead of a seven-electron place.

Neptunium $_{93}$Np puts an electron in a five-electron place instead of a seven-electron place.

Cerium $_{96}$Cm puts an electron in a five-electron place instead of a seven-electron place.

But recall from Part I, Chapter 1, that I said there were *no* special cases in Chemistry. So how can all this peculiarity be part of the same story? The answer lies in the numbers. The ionization potentials that all these elements present do not stray from the overall numerical pattern followed by all of the elements. It is only the details of the actual realizations that show the peculiarities. Those realization details may be accompanied by energy details, but the energy details are just too small for us humans to notice.

The big questions that arise from this list of peculiarities are:

1) Why does the frequent robbery from a two-electron ring to complete a five-electron ring occur?
2) Why does the complete consumption of a two-electron ring to complete a five-electron ring ever occur?

3) Why does the placement of one or two electrons to start a five-electron ring instead of a seven-electron ring occur?

The answers based on the present model using electron rings are as follows:

1) In a two-electron ring, each electron has orbit energy $-m_e V_e^2 / 2 = -m_e (4\pi c)^2 / 2$, for a total of $-m_e (4\pi c)^2$. If the ring gets broken, and one electron is left adrift while the other gets relocated to a growing five-electron ring, then their total orbit energy becomes $0 - m_e (12\pi c)^2 / 2$. This is more negative by a factor of $9/2$. This is significant for the two electrons, although not so significant for the atom overall, since even the least of the peculiar atoms has 22 other electrons. These other electrons can just stay in place if the robbed two-electron ring is not a participant in forming a magnetic confinement volume that is holding any of them. So here we have a clue about the ambiguity questions raised earlier about $_{12}$Mg, $_{25}$Mn, *etc.*: do electron rings get deployed as pairs, leaving some magnetic confinement volumes totally empty? Apparently, the answer is: yes, they do.

2) When a two-electron ring is totally consumed to complete a five-electron ring, both of the two electrons change orbit speed from $4\pi c$ to $12\pi c$, and so change negative energy by a factor of 9. So the real question is not why this happens once, for Palladium, but rather why it does *not* happen several times, for several elements. Evidently, negative energy is not the only consideration. The other consideration frequently seen in Chemistry is geometry. For $_{46}$Pd, total consumption of a two-electron ring allows an extremely symmetric electron-ring stack: 2, 3, 5,5, 3, 2, 3,3, 2, 3, 5,5, 3, 2. Its neighbor $_{47}$Ag can get this same nice stack, plus one stray electron, which makes it a better conductor than $_{46}$Pd is. As for $_{45}$Rh, it can get a stack almost as nice: 2, 3, 5,4, 3, 2, 3, 4,5, 3, 2, plus one stray electron. It, too, is a better conductor than $_{46}$Pd is.

3) In all cases where electrons are diverted to start a five-electron ring instead of a seven-electron ring, it is just one or two electrons that are being diverted, by themselves, and not with others. Since a new five-electron ring will soon start legitimately, the confinement volume for it is certainly available, there between two three-electron rings. And that location is just easier to reach than the confinement volume appropriate for a seven-electron ring, which is buried between two already-existing five-electron rings.

CONCLUSION

In this Chapter the concept of electron rings has been developed to model any atom. Electron rings are like current loops, so they can align and stack like so many magnets.

Electron rings circulate at discrete super-luminal speeds that occur in multiples of $4\pi c$. Higher orbit speeds correspond to smaller orbit radii. So electron rings with higher orbit speeds can nestle into magnetic confinement volumes created by other electron rings running at lower orbit speeds.

The resulting iconic image of the atom is *not* a *sphere* of electron *shells* nestled *around* the nucleus, with larger electron counts *outside* of smaller ones. It is instead a *cylinder* of stacked electron *rings*, located at a nominal orbit radius *away from* the nucleus, with larger electron counts nestled *inside* of magnetic containment volumes created by smaller ones.

Such stacking and nestling is how electron rings can, for any atom, form an electron cloud that is somewhat confined, and acts as a sub-system within the atom. It can orbit around the nucleus as a unit. This is a mental image quite different from that of all the electrons being spread all around the nucleus all at once.

Nature's nominal algorithm for filling electron rings appears to be this: Always build and store an electron ring for the largest number L of electrons possible, where 'possible' means having a suitable magnetic confinement volume available to fit into, and where 'suitable' means created by two electron rings with smaller electron count, and not yet filled with two L-electron rings. Sometimes only a new $L = 2$ electron ring is possible, and that is what starts a new period in the Periodic Table. The parameter N introduced in Part I, Chapter 1 can be re-characterized here as the number of different L-values occurring in a period.

APPENDIX

It may be helpful to have some abbreviated language for distinguishing the types of completed electron rings. In [2] I suggested 'binar' for the two-electron ring, but electronic word processors always change 'binar' to 'binary'. So now I suggest using 'duo' for two electrons. And then 'tert' is for three electrons, 'quint' is for five electrons, and 'sept' is for seven electrons.

This language fits with the language introduced earlier for the different levels of magnetic confinement that pairs of electron rings provide. In short, in the 'magnetic bottle' between two 'duos' can hold one or two 'terts', the 'magnetic thermos jug' between two 'terts' can hold one or two 'quints', and the 'magnetic Dewar flask' between two 'quints' can hold one or two 'septs'.

The concept of electron rings, and all the language that goes with it, can be helpful in explaining why the filling of single-electron states occurs in the way that it does across the whole Periodic Table, including even the occasional cases where exceptions occur in violation of expectations based on traditional quantum numbers alone. Richard Scerri devotes a Chapter in [3] to the problem: traditional quantum numbers reveal the pattern in the way single-electron states fill, but do not offer a whole lot in the way of explanation for it. The electron ring concept and its language may help fill this gap.

Stacked and orbiting electron rings: is that *really* how atoms are? We do not know. And we cannot know - just as we cannot know that the usual Model consisting of electron shells enclosing the nucleus is really the way atoms are. But it doesn't matter, because Science is more about predicting the observables than about actually knowing the *un*observables.

A Project for Readers

Each of you can make for yourself a study of some, or all, of the 19 presently known peculiar elements. Develop your own opinion as to why each one that you study does its own particular peculiar thing. We can then all collaborate to create a comprehensive study. I will be happy to collect your results to publish with proper credit in a future edition of this book.

Acknowledgments

I thank Richard Scerri for directing me onto the subject of single-electron-state filling order early in my struggles.

References

[1] Wesley, J.P. *Selected Topics in Advanced Fundamental Physics*, Chapt. 11, Benjamin Wesley, Blumberg, Germany, 1991.
[2] Whitney, C.K. Visualizing Electron Populations in Atoms, *Int. J. Chem. Model.* Vol. 1, pp. 245-297, 2008.
[3] Scerri, R. *The Periodic Table, Its Story and Its Significance*, Chapter 9, Oxford University Press, 2007.

Chapter 12

EXPLOSIONS AND EXPLANATIONS

ABSTRACT

This Chapter revisits the numerical regularities developed in Part I, Chapter 1, in light of the particular view of Chemistry as Quantum Mechanics developed in this Part II, with the goal of developing some explanations for those numerical regularities. In addition, some of those regularities haven't even been exploited yet, as we have been looking only at reactions involving gentle, one-at-a time exchanges of electrons. But there are also reactions of a more violent type, in which electrons move as large groups: explosions.

INTRODUCTION

Quantum Mechanics (QM) is generally viewed as the legitimate source for deep explanation of all phenomena in Chemistry. Among those phenomena, the numerical regularities that exist in chemical data certainly stand in need of such deep explanation. But the numerical regularities are on a pretty synoptic scale, whereas QM has traditionally been rather particular-case oriented, mainly because of computational-load issues.

It is to be hoped that the ideas about the QM of atoms that are developed in this Part II of this Book imply computations simple enough that they can help produce some useful degree of explanation for the numerical regularities from Part I.

The blatant questions include the following:

1) Why do all the plots of ionization potentials on a log scale apparently consist of straight-line segments?
2) What story are the plots of higher-order ionization potentials trying to tell us?
3) What do electron rings explain about the pattern established by first-order ionization potentials?
4) What is the information about higher-order ionization potentials really good for?

1. LOG-LINEARITY OF IONIZATION POTENTIALS

Let us recall Figure 1 from Part I, Chapter 1:

The most pervasive fact revealed by Figure 1 is that all sub-periods, beyond the first one, which has only two electrons, clearly have two halves. In standard QM language, the two sub-period halves correspond to two spin states, $s = 1/2$ and $s = -1/2$. In the language of electron rings, the two sub-period halves correspond to the filling of two electron rings, each one containing 3 electrons, or 5 electrons, or 7 electrons.

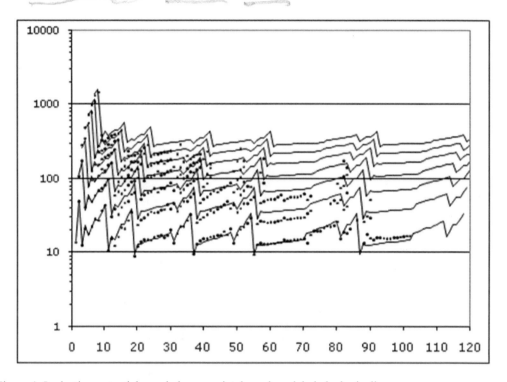

Figure 1. Ionization potentials, scaled appropriately and modeled algebraically.

Within each half sub-period, *i.e.* within the filling of each of two spin states, *i.e.* within the development of each of two electron rings, the ionization potential increases linearly on the log scale. Such linear rises imply power laws. Why do we have power laws here?

Consider the situation from the viewpoint of electron-to-electron bonds. The quantum number l specifies the maximum number of electrons in a ring as $2l+1$. Suppose the actual number of electrons in an incomplete ring is a. The number of electron-to-electron bonds in the incomplete ring is $a(a-1)/2$.

It grows from a minimum of zero, up to a maximum $(2l+1)2l/2 = (2l+1)l$, practically as a^2. So no wonder it produces an energy effect that looks like a power law.

2. ON THE MEANING OF HIGHER-ORDER IONIZATION POTENTIALS

Let us recall Figure 2 from Part I, Chapter 1, relating the IP's with higher ionization order, $IO > 1$, to the first-order IP's, with $IO = 1$:

The general pattern is revealed in $IP_{IO,IO+3}$. The first term is universal. The second term is period specific. The third term is element specific.

Now knowing empirically what all these terms apparently are, we can pause for a moment and study them. The many facts displayed seem to be trying very hard to communicate a coherent story. Presumably, the story is not just *what* the facts are, but *why* they are what they are. The following paragraphs try to capture that story.

$$IP_{IO,IO} = 2 \times IP_{1,1} \times IO^2$$

$$IP_{IO,IO+1} = IP_{1,1} \times 2 \times IO^2 + \frac{1}{2} \times IP_{1,1} \times IO + \frac{1}{2} \Delta IP_{1,2} \times IO$$

$$IP_{IO,IO+2} = \frac{1}{2} IP_{1,1}(IO^2 + IO) + \frac{1}{2} \Delta IP_{1,3} \times (IO^2 + IO)$$

$$IP_{IO,IO+3} = \frac{1}{2} IP_{1,1}(IO^2 + IO) + \frac{1}{2} \Delta IP_{1,3}(IO^2 - IO) + \Delta IP_{1,4} IO, \text{ etc..}$$

$$\Delta IP_{1,3} \to \Delta IP_{1,IO+N} \text{ at } IP_{IO,IO+N} \text{ for } IO + N = 11, 19, 37, 55, 87$$

Figure 2. Behavior of higher-order IP's.

2.1. Observations

Observe that the leading terms representing any IP are proportional to IO^2. This dependence suggests that the physical process generating the data involves removing IO electrons all at once, and not removing just the single electron left after $IO - 1$ other electrons have already been removed, or possibly just skipped over and left in place. This distinction about removing IO electrons all at once, *vs.* any scenario that removes just one electron, is obvious from the mathematical factor IO^2, but it is not obvious from a description by a typical text phrase, such as 'third-order ionization potential', for example. So when reading the existing literature on ionization potentials, always watch out for the possibility of confusion arising from inadequate language.

The secondary terms representing any arbitrary IP are linear in IO. These terms implement the 'shift right' behavior seen in Figure 1. They also determine the numerical pattern that the period rises follow. For $IO \equiv 1$, the period rises are *all 7/2*. For all $IO > 1$, all but the first period are the same, but less than $7/2$, gradually approaching unity as IO

increases. The formula for the period rises can be determined by looking at the ratio $IP_{IO,IO+10}/IP_{IO,IO+2}$:

$$\frac{\frac{1}{2}IP_{1,1}(IO^2+IO)+\frac{1}{2}\Delta IP_{1,3}(IO^2-IO)+\Delta IP_{1,10}IO}{\frac{1}{2}IP_{1,1}(IO^2+IO)+\frac{1}{2}\Delta IP_{1,3}\times(IO^2+IO)}$$

Numerical values for $IP_{1,1}$ and all ΔIP's are detailed the Appendix to Part I, Chapter 1. For now, we note that $\Delta IP_{1,3} = -\frac{1}{8}IP_{11}$ and $\Delta IP_{1,10} = \left(\frac{7}{8}\times\frac{7}{2}-1\right)IP_{1,1} = \frac{33}{16}IP_{1,1}$. From these values, one finds that, except for the first one, period rises follow the simple formula $(IO+6)/(IO+1)$, which reduces to $7/2$ for $IO=1$.

2.2. Explanations

The exception for the rises in first periods when $IO > 1$ demands an explanation. To begin developing the explanation, note the factor of 2 in the formula for $IP_{IO,IO}$ for $IO > 1$:

$$P_{IO,IO} = 2 \times IP_{1,1} \times IO^2 .$$

Why does this formula not come out, as it must for $IO \equiv 1$, simply $IP_{IO,IO} \equiv IP_{1,1} \times IO^2$? A possible explanation is that, unlike first-order ionizations, these *total* ionizations are *not* best characterized as 'removal of electrons from the atomic system'; they are *better* characterized as 'removal of the *nucleus* from the atomic system'. Note that electrons have a lot of kinetic energy; on average, an amount equal to half the magnitude of their (negative) potential energy in the atomic system. By contrast, the nucleus has almost *no* kinetic energy. So removing the nucleus from the atomic system takes essentially *twice* the energy that removing all of the electrons *together* from the atomic system would take. But the fragility of the electron subsystem probably prevents the latter scenario.

Next, consider the formula for the scenario that starts with $IO+1$ electrons, and ends up with 1 electron:

$$IP_{IO,IO+1} = IP_{1,1} \times 2 \times IO^2 + \frac{1}{2}\times IP_{1,1}\times IO + \frac{1}{2}\Delta IP_{1,2}\times IO .$$

The first term, $IP_{1,1} \times 2 \times IO^2$, is identical to the formula for starting with IO electrons and ending up with no electrons. So this part of the scenario looks like the blow-the-nucleus-out scenario, but not with just the nucleus alone; instead, this nucleus takes one electron with it, so that what is blown out is a nucleus-plus-electron system that is like a nucleus with charge IO instead of $IO+1$. That leaves an electron system with IO electrons still to dispose of.

The second term, $\frac{1}{2} \times IP_{1,1} \times IO$, defines what it means to 'dispose of' IO electrons. Each electron is blown away with enough energy, $\frac{1}{2} \times IP_{1,1}$, so that any pair of them has energy $IP_{1,1}$, sufficient to keep them away from each other.

The third term, $\frac{1}{2}\Delta IP_{1,2} \times IO$, suggests a possible structure for the system of IO electrons. If the system is a ring, or several rings, it takes IO binary cleavages to destroy the system. So there is the factor of IO multiplying $\Delta IP_{1,2}$. But why is there also the factor of $\frac{1}{2}$ multiplying $\Delta IP_{1,2}$? Well, consider that when every electron gets energy $\frac{1}{2}\Delta IP_{1,2}$, then any pair of them has enough energy, $\Delta IP_{1,2}$, to resist reuniting as a pair.

Next, consider the scenario that starts with $IO+2$ electrons, and ends up with 2 electrons. We have:

$$IP_{IO,IO+2} = \frac{1}{2}IP_{1,1}(IO^2 + IO) + \frac{1}{2}\Delta IP_{1,3} \times (IO^2 + IO).$$

The first term, $\frac{1}{2}IP_{1,1}(IO^2 + IO)$, contains $\frac{1}{2}IP_{1,1}IO^2$, which is quite different from the $IP_{1,1} \times 2 \times IO^2$ seen in the cases of IO and $IO+1$: it has the overall factor of $\frac{1}{2}$ instead of 2. This means that there is no blowing-out of a nucleus-like system consisting of the nucleus plus two electrons, which would have charge IO instead of $IO+2$. Instead, IO electrons are blown away from the atom, leaving the nucleus-like system behind. The factor of $\frac{1}{2}$ on $\frac{1}{2}IP_{1,1}IO^2$ mean the nucleus-like system and the blown-away electron system have between them enough energy, $IP_{1,1}IO^2$, to keep them away from each other.

The first term here also contains $\frac{1}{2}IP_{1,1} \times IO$, which was present in the $IO+1$ case as the second term there. The $\frac{1}{2}IP_{1,1} \times IO$ means the same thing here: each electron is blown away with enough energy, $\frac{1}{2} \times IP_{1,1}$, so that any pair of them has energy $IP_{1,1}$, sufficient to keep them away from each other.

The second term here, $\frac{1}{2} \Delta IP_{1,3} \times (IO^2 + IO)$, is actually negative, because $\Delta IP_{1,3} = -\frac{1}{8} IP_{11}$. The second term scales down the energy increments recommended by the first term as being necessary to keep away the IO electrons, either as a group (the IO^2 term), or as individuals (the IO term). The scale-down reflects the fact that an electron system consisting of just two electrons by themselves is the most stable electron system seen anywhere in Nature: like a hundred dollar bill – too big to break!

Next, consider the scenario that starts with $IO + 3$ electrons and ends up with 3 electrons. We have:

$$\ldots IP_{IO,IO+3} = \frac{1}{2} IP_{1,1}(IO^2 + IO) + \frac{1}{2} \Delta IP_{1,3}(IO^2 - IO) + \Delta IP_{1,4} IO \ .$$

The first term is the same as it was in $IP_{IO,IO+2}$, and has the same meaning here. The second term, $\frac{1}{2} \Delta IP_{1,3}(IO^2 - IO)$, differs by the internal minus sign from the second term $\frac{1}{2} \Delta IP_{1,3}(IO^2 + IO)$ in $IP_{IO,IO+2}$. That minus sign on the already negative $\Delta IP_{1,3}$ effectively adds to the energy requirement to keep individual electrons from coming back.

The third term, $\Delta IP_{1,4} IO$, further raises the energy needed to keep individual electrons from coming back, inasmuch as $\Delta IP_{1,4}$ is positive. The message is: three electrons is a really an unstable situation; having an extra electron is like having extra small change in your pocket: it is just waiting to be left somewhere.

3. ON THE MEANING OF FIRST-ORDER IONIZATION POTENTIALS

The actual data that determines the higher-order IP's is the set of all first-order IP's. This Section sets out to explain the pattern of that data in terms of electron rings.

3.1. Full Periods

Let us recall Figure 3 from Part I, Chapter 1.

	$_1$H,	$IP_{1,1}$	→ 7/2 →	$IP_{1,2}$,	$_2$He	
7/8	↓	↓	← 1/4 ←	↵	↓	7/8
	$_3$Li,	$IP_{1,3}$	→ 7/2 →	$IP_{1,10}$,	$_{10}$Ne	
7/8	↓	↓	← 1/4 ←	↵	↓	7/8
	$_{11}$Na,	$IP_{1,11}$	→ 7/2 →	$IP_{1,18}$,	$_{18}$Ar	
7/8	↓	↓	← 1/4 ←	↵	↓	7/8
	$_{19}$K,	$IP_{1,19}$	→ 7/2 →	$IP_{1,36}$,	$_{36}$Kr	
1	↓	↓	← 2/7 ←	↵	↓	1
	$_{37}$Rb,	$IP_{1,37}$	→ 7/2 →	$IP_{1,54}$,	$_{54}$Xe	
1	↓	↓	← 2/7 ←	↵	↓	1
	$_{55}$Cs,	$IP_{1,55}$	→ 7/2 →	$IP_{1,86}$,	$_{86}$Rn	
1	↓	↓	← 2/7 ←	↵	↓	1
	$_{87}$Fr,	$IP_{1,87}$	→ 7/2 →	$IP_{1,118}$,	$_{118}$??	

Figure 3. First-order IP's: map of main highways through the periods.

The most startling facts are that the rise on *every* period is the same, and that the value is a simple ratio of integers: total rise $= 7/2$. Is there some insight about these facts available from the electron rings model?

Firstly, why are all the period rises the same? That behavior arises from the nature of the algorithm that fills single-electron states across the Periodic Table. Recall from the last Chapter that the algorithm goes: Always build and store an electron ring for the largest number L of electroms possible, where 'possible' means having a suitable magnetic confinement volume available to fill, and where 'suitable' means created by two $(L-2)$-electron rings, and not yet filled with two L-electron rings. Sometimes only a new $L = 2$ electron ring is possible, and that is what starts a new period in the Periodic Table.

Like most construction algorithms in Nature, this algorithm is basically 'fractal', a term defined by Benoit Mandlebrot [1], meaning: it does the same thing over and over, at smaller and smaller geometric scales. The rings composed of two electrons accommodate rings of three electrons, and those in turn accommodate rings of five electrons, and those in turn accommodate rings of seven electrons: four levels so far. It is tempting to wonder about what comes at a fifth level: 9 because it is an odd number, or 11 because it is a prime number. We will find out only when, or if, enough new elements can be synthesized, and then studied within their very-short lifetimes.

In any case, the fractal nature of the filling algorithm means that the situation at the beginning of any period is very much like the Hydrogen situation: one orphan electron. And the situation at the end of any period is very much like the Helium situation: some two-electron ring or rings with whatever else confined between them. So of course all the period rises are numerically the same.

And why is the numerical value of the factor for the rise on every period $7/2$? From the viewpoint of numerology, 7 is the highest electron count, and 2 is the lowest electron count, among the electron rings encountered in reality! But that is only useless numerology. To approach a real explanation, it may help to split the $7/2$, into 1, for the non-change of $IP_{1,1} = 14.250$, and $5/2$ for the change from $\Delta IP_{1,1} = 0$ to $\Delta IP_{1,2} = 35.625$. The mystery lies entirely with the $5/2$. It is not yet solved, and so stands as a research topic.

3.2. Sub-Period Levels

Looking again to Figure 1, observe that within every period, the sub-periods occur with L decreasing. That means the filling of electron rings with higher L correlates with lower ionization potential. So seven-electron rings are correlated with low ionization potentials. So filling of seven-electron rings, $l = 3$ single-electron states, or so-called f states, is correlated with low ionization potentials. The subsequently filled l states generally correlate with higher ionization potentials.

In the standard literature, electron states are imagined in terms of the standard shell-around-nucleus model for atoms. The f states are imagined to lie closer to the nucleus than the subsequently filled l states. This asserted proximity to the nucleus is imagined to give the f states lower energy than the subsequently filled l states, thus accounting for their being filled first. But what can one say about the ionization potentials? Liberating an electron in from a lower-energy f state ought to require a *higher* ionization potential than liberating an electron from a higher-energy subsequently filled l state.

This situation constitutes a troubling contradiction. But there is generally no comment about it in the standard literature.

The observed behavior makes more sense in terms of the stacked-electron-ring model. There, the energy of a state being filled has precious little to do with the nucleus, its particular identity or its proximity, and a lot more to do with the other electrons. The seven-electron rings are very delicate because of the many electron-electron interactions. Any slight disturbance of geometry can change the situation to where an electron has positive rather than negative potential energy in the electron-electron interactions. This delicacy makes for the low ionization potentials.

3.3. Sub-Period Slopes

Let us recall also Figure 4 from Part I, Chapter 1. We had

incremental rise = total rise × fraction

with the fraction being given by the Figure 4 reproduced below:

N	l	fraction	l	fraction	l	fraction	l	fraction
1	0	1						
2	0	1/2	1	3/4				
2	0	1/3	1	3/4				
3	0	1/4	2	5/18	1	2/3		
3	0	1/4	2	5/18	1	2/3		
4	0	1/4	3	7/48	2	5/16	1	9/16
4	0	1/4	3	7/48	2	5/16	1	9/16

Figure 4. First-order IP's: map of local roads through the periods.

For first sub-periods, the fractions are empirical numbers; for subsequent sub periods, they fit the empirical formula:

$$\text{fraction} = \left[(2l+1)/N^2\right]\left[(N-l)/l\right].$$

We can re-express the Table and the formula in terms of the number L of electrons in the ring being formed:

N	L	fraction	L	fraction	L	fraction	L	fraction
1	2	1						
2	2	1/2	3	3/4				
2	2	1/3	3	3/4				
3	2	1/4	5	5/18	3	2/3		
3	2	1/4	5	5/18	3	2/3		
4	2	1/4	7	7/48	5	5/16	3	9/16
4	2	1/4	7	7/48	5	5/16	3	9/16

and

$$\text{fraction} = \left[L/N^2\right]\left[(2N-L)/(L-1)\right].$$

This is only a small step, and one can reasonably hope that with more investigation more insight can be gained.

4. ON THE UTILITY OF HIGHER-ORDER IONIZATION POTENTIALS IN CHEMISTRY

The present Book has so far been about chemical reactions that are gentle: that is, where exchanges of electrons occur one at a time. So the information required has been just the set of first-order ionization potentials. But there are also important reactions that are not at all gentle.

Let us consider super fast events, occurring not as reactions among several molecules, but rather within individual molecules, where participant atoms are very close at hand, and exchanges of electrons can happen all at once. That scenario amounts to detonation of an explosion. The information about higher-order ionization potentials can be relevant here.

Consider dynamite. The active ingredient can be nitroglycerin, which is also known by the alternative name, trinitroxypropane. That alternative name gives a good clue about the real structure of the molecule. It suggests starting with propane, and replacing Hydrogen atoms with radicals providing the same overall ionization states. From Part I, Chapter 5, propane is

$$\begin{array}{ccc} H & H & H \\ \cdot & \cdot & \cdot \\ H \cdot C \cdot C \cdot C \cdot H \\ \cdot & \cdot & \cdot \\ H & H & H \end{array}$$

with possible ionic configurations

$$\begin{array}{ccc} H^+ & H^- & H^+ \\ \cdot & \cdot & \cdot \\ H^+ \cdot C^{4-} \cdot C^{4+} \cdot C^{4-} \cdot H^+ \\ \cdot & \cdot & \cdot \\ H^+ & H^- & H^+ \end{array}$$

and

$$\begin{array}{ccc} H^- & H^+ & H^- \\ \cdot & \cdot & \cdot \\ H^- \cdot C^{4+} \cdot C^{4-} \cdot C^{4+} \cdot H^- \\ \cdot & \cdot & \cdot \\ H^- & H^+ & H^- \end{array}$$

So for nitroglycerin we could have:

$$H^-$$
$$H^+ \cdot C^{4-} \cdot C^{4+} \cdot C^{4-} \cdot H^+$$
$$O^{3+} \quad O^{2-} \quad O^{3+}$$
$$O^+ \cdot N^{3-} \quad N^{3+} \quad N^{3-} \cdot O^+$$
$$O^+ \quad O^- O^- \quad O^+$$

The ions displayed require the energies:

$2C^{4-}$: $2 \times (-54.6304) = -109.2608$ eV and $1C^{4+}$: 22.187 eV ;

$2H^+$: $2 \times 14.1369 = 28.2738$ eV and $1H^-$: -90.6769 eV ;

$2N^{3-}$: $2 \times (-47.6706) = -95.3412$ eV and $1N^{3+}$: 23.6535 eV ;

$2O^{3+}$: 26.6007 eV and $1O^{2-}$: -27.3788 eV

$4O^+$: $4 \times 7.9399 = 31.7596$ eV

$2O^-$: $2 \times (-12.4354) = -24.8708$ eV

This all adds up to

$$-109.2608 + 22.187 + 28.2738 - 90.6769 - 95.3412 + 23.6535$$
$$+ 26.6007 - 27.3788 + 31.7596 - 24.8708 = -215.0539 \text{ eV}$$

This negative result is indicative of a stable molecule. Or we could have the charge mirror image molecule:

$$H^+$$
$$H^- \cdot C^{4+} \cdot C^{4-} \cdot C^{4+} \cdot H^-$$
$$O^{3-} \quad O^{2+} \quad O^{3-}$$
$$O^- \cdot N^{3+} \quad N^{3-} \quad N^{3+} \cdot O^-$$
$$O^- \quad O^+ O^+ \quad O^-$$

The ions displayed require the energies:

$2C^{4+}$: $2 \times 22.187 = 44.374$ eV and $1C^{4-}$: -54.6304 eV

$2H^-$: $2 \times (-90.6769) = -181.3538$ eV and $1H^+$: 14.1369 eV

$2N^{3+}$: $2 \times 23.6535 = 47.307$ eV and $1N^{3-}$: -47.6706 eV

$2O^{3-}$: $2 \times (-15.0672) = -30.1344$ eV and $1O^{2+}$: 17.5613 eV

$4O^-$: $4 \times (-12.4354) = -49.7416$ eV and

$2O^+$: $2 \times 7.9399 = 15.8798$ eV

This all adds up to

$44.374 - 54.6304 - 181.3538 + 14.1369 + 47.307 - 47.6706$
$-30.1344 + 17.5613 - 49.7416 + 15.8798 = -224.2718$ eV

This result is more negative than the previous one, and so is the more likely one. But the difference is slight, so nitroglycerin is likely to be present in both of these ionic configurations.

So what makes nitroglycerin explosive? To understand that phenomenon, we need to consider the possible products. Among them are the simple diatomic molecules: H_2, N_2, O_2, CO, NO, and a few tri-atomic ones: H_2O, NO_2, CO_2, and even a four atom one: H_2CO.

Here is one possible set of products that completely uses exactly one molecule of nitroglycerin:

1CO, 1CO_2, 1HCO, 3NO, 1H_2, 1O_2.

The most likely ionic configurations of these molecules were discussed in Part I, Chapter 5. They are:

$CO : C^{2+} + O^{2-}$, $CO_2 : C^{4+} + 2O^{2-}$, $HCO : H^- + C^{2+} + O^-$,

$NO : N^{3-} + O^{3+}$, $H_2 : H^+ + H^-$, $O_2 : O^{2+} + O^{2-}$

with energy requirements

$$1 \times (C^{2+} + O^{2-}): 1 \times (14.2505 - 27.3788) = -13.1283 \text{ eV}$$

$$1 \times (C^{4+} + 2O^{2-}):$$
$$1 \times [22.187 + 2 \times (-27.3788)] = 1 \times (22.187 - 54.7576) = -32.5706 \text{ eV}$$

$$1 \times (H^- + C^{2+} + O^-):$$
$$1 \times (-90.6769 + 14.2505 - 27.3788) = -103.8052 \text{ eV}$$

$$3 \times (N^{3-} + O^{3+}):$$
$$3 \times (-47.6706 + 26.6007) = 3 \times (-21.0699) = -63.2097 \text{ eV}$$

$$1 \times (H^+ + H^-): 1 \times [14.1369 + (-90.6769)] = -76.54 \text{ eV}$$

$$1 \times (O^{2+} + O^{2-}): 1 \times (17.5613 - 27.3788) = -9.8175 \text{ eV}$$

Making the whole set of them takes energy

$$-13.1283 - 32.5706 - 103.8052 - 63.2097 - 76.54 - 9.8175 =$$
$$-299.0713 \text{ eV}$$

This is more negative than either of the nitroglycerin numbers.

And there is more. Here is a different, and simpler, set of products that one nitroglycerin molecule can produce:

3HCO, 3NO$_2$

Again, the most likely ionic configurations of these molecules were discussed in Part I, Chapter 5. They are:

$$\text{HCO}: H^- + C^{2+} + O^-, \quad \text{NO}_2: N^{3-} + O^+ + O^{2+}$$

The energy requirements are then

$$3 \times (H^- + C^{2+} + O^-):$$

$$3 \times (-90.6769 + 14.2505 - 27.3788) = -311.4156 \text{ eV}$$

and

$$3 \times (N^{3-} + O^{+} + O^{2-}):$$
$$3 \times (-47.6706 + 7.9399 + 17.5613) = -66.5082 \text{ eV}$$

So making all the molecules takes

$$-311.4156 - 66.5082 = -377.9238 \text{ eV}.$$

This number is even more negative than the first product set was in relation to either of the nitroglycerin numbers. So this is the more likely product set.

Let us focus on the more likely ionic configurations for the nitroglycerin molecules and the product molecules, and look at the overall reaction. There are 18 atoms involved. Some of them lose electrons, and some gain electrons. One H^+ gains two electrons. Two C^{4+} gain two electrons each to become two C^{2+}, and one C^{4-} loses six electrons to become a C^{2+}. Two N^{3+} gain six electrons each to become two N^{3-}. These are all transactions with two electrons, or multiples of two electrons. The six Oxygen ions can do a variety of things, but as a set, they change from charge -6 to charge $+6$; $i.e.$ they lose twelve electrons, again a multiple of two.

Given that everything is happening within the small spatial extent of a single molecule, changes in electron loading on atoms are not necessarily going to occur in the orderly one-at-a-time process used throughout most of this Book. Instead, they can occur in the all-at-once kind of process here associated with the higher-order ionization potentials. In particular, pairs of electrons are likely to change allegiance together. The energy numbers, up and down, are larger. The situation is an explosion.

This story about nitroglycerin means that the molecule does not need a partner reagent molecule to react with. At most, it needs a percussive rap, and then it can simply self-detonate. And when it self-detonates, it releases a lot of energy all at once.

CONCLUSION

This Chapter identified five blatant questions to answer:

1) Why do all the plots of ionization potentials on a log scale apparently consist of straight-line segments?
2) What story are the plots of higher-order ionization potentials trying to tell us?
3) What do electron rings explain about the pattern established by first-order ionization potentials?
4) What is the information about higher-order ionization potentials really good for?

The answers developed in this Chapter are the following:

1) The straight lines mean power laws, and the variable raised to a power appears to be the number of electrons in an electron ring.
2) The higher-order ionization potentials appear to be telling a very complicated story, one essential feature of which is events that affect multiple electrons all at once.
3) The electron rings form a structure that is essentially fractal, so that all periods are basically alike, and just have various self-similar sub-periods inserted.
4) Detonation of nitroglycerine illustrates a type of chemical scenario where the rather elaborate information about higher-order ionization potentials appears very relevant for a proper numerical analysis. Detonation involves just one molecule self-destructing, rather than several molecules reacting with each other.

REFERENCES

[1] Mandelbrot, B. *The Fractal Geometry of Nature*, Macmillan, New York, 1983.

Part III. Quantum Mechanics as Electrodynamics

PROLOG TO PART III

In the twentieth century, it was generally accepted that Quantum Mechanics (QM) represented a real break with earlier branches of Classical Physics (CP). Newtomian mechanics was about idealized point particles, whereas QM was about waves. Electrodynamics was about waves, but only light waves, whereas QM was about matter waves. Statistical mechanics was about uncertainties and probabilities describing an ensemble of particles, whereas QM was about uncertainty and probability inherent even to a single particle. And so on. Connections were not sought; in fact, they were actively disparaged.

Post has argued in [1] that there is good reason to revisit the possibilities for connections between QM and the other areas of CP with a different attitude: ensembles of multiple systems, instead of single systems. This Part III of my own Book adopts that view, and sets out to establish some of the possible connections. Chapter 13 is about the connections between photons and Maxwell fields. This is needed in order to establish the directionality of forces that occurred in the models for the Hydrogen atom, and positronium, and the various charge pairs and rings, which were developed in Part II. Chapter 14 is about the invariance of Maxwell's equations, not just under Lorentz transformations, as is currently believed by many people, but also under Galilean transformation, or any other invertible transformation. This freedom is needed to justify the superluminal speeds used in Part II.

The result of Chapter 13 is a photon model that is harmonious with Maxwell fields, but not with the assumption that was used without question going into the twentieth century. That assumption concerned both the photons for QM and the signals for Einstein's Special Relativity Theory (SRT). The assumption was that such entities would travel like bullets, with shape constant, and with speed c, a constant relative to any observer in general, and their target in particular. That key assumption was common, but unstated, in the 19th century. It was first truly formalized in the 20th century by Einstein, in his Second Posulate for SRT.

The result of Chapter 14 is an expansion of tensor notation to allow a clear representation for any kind of transformation, not just a space-time symmetric one, like the Lorentz transformation is. The expansion amounts to a doubling of the tensor notation. It is not unlike other doublings that have marked the development of Mathematics: inclusion of negative numbers, inclusion of imaginary numbers, and more.

REFERENCE

[1] Evert Jan Post, Quantum Reprogramming, Boston Studies in Philosophy of Science (Kluwer Academic Publishers, Dordrecht, Boston, London, 1995).

Chapter 13

PHOTONS AND MAXWELL'S EQUATIONS

ABSTRACT

This Chapter aims to reconcile the two ideas of light: 1) as a stream of photons, and 2) as a set of propagating electromagnetic fields. The approach is to construct a model for the photon expressed in terms of fields that are solutions to Maxwell's equations, chosen from among all the possible solutions to combine in a way that satisfies some boundary conditions that are designed to represent the effect of the source of the photon, and the receiver of the photon. The result is a photon model that, when applied to the so-called virtual photons that communicate electromagnetic forces, *i.e.* applied to electromagnetic signals, explains why the attractive forces in an atom are not necessarily central. This is the key to modeling atoms for the purposes of Algebraic Chemistry.

INTRODUCTION

The photoelectric effect, and the concept of the photon that came from it, triggered the development of Quantum Mechanics (QM), which is today the basis for the development of Quantum Chemistry (QC). But the photon concept is still very enigmatic. The idea of light as a stream of photons seems completely different from the idea of light as traveling Maxwell fields. The photon is apparently discrete and particle-like, rather than continuous and wave-like, and it has a total energy that depends on frequency, rather than having a local energy density that depends on field intensities.

So the introduction of the photon, especially when followed with the perceived problem about the immortality of the Hydrogen atom, was thought to require a complete departure from Maxwell's classical electrodynamics (CED). And so we got 20th century development of Quantum Mechanics (QM). But the original assessment of CED may have been too extreme. After all, the ideas of total integrated energy and local energy density are not inherently incompatible. We can have a photon that is not an ideal point particle, but rather a somewhat extended field construct, and it can indeed have an integrated total energy related to frequency, and a local energy density that depends on Maxwell field amplitudes.

This Chapter develops such a model for the photon. Section 1 describes the vital role of boundary conditions. Section 2 illustrates one of the many similar parts of the problem. Section 3 shows how a field pulse evolves into a field wavelet. Section 4 studies the energy density profile of wavelets, and describes how the total energy density profile of the photon model evolves over time. Section 5 analyzes what happens when there is relative motion between the source and the receiver. A contradiction in our current understanding of this problem is revealed and resolved.

1. APPROACH

Patrick Cornille [1] spells out the historical problem about a wave packet: it may spread. When we allow a dispersive medium, we apparently have to invoke a non-linearity to prevent spreading. Cornille was speaking about a wave packet representing an electron, but something similar can be said about a wave packet representing a photon. Even without a dispersive medium, a photon wave packet will spread due to the action of Maxwell's equations. By the term 'Maxwell's equations', I mean the first-order field equations, *not* the second-order wave equations that result by inserting one field equation into the equation for the other to eliminate coupling. When both fields are considered, each field profile has its edges, and at these edges it generates the other field.

So does modeling a photon with a wave packet require some sort of non-linearity like modeling an electron with a wave packet apparently requires? No, it does not; we can exploit a much more familiar idea from Applied Mathematics. Here is the approach:

1) Recall that differential equations always offer a family of possible solutions, and that real-world problems always involve particular boundary conditions that have to be fit. We always construct the particular solution we want by making a combination of several possible solutions that, when taken together, make a sum solution that fits the boundary conditions.

2) Note that a photon is not a self-sufficient thing; to come into being, it requires a source, and to affect anything, it must encounter a receiver. So a mathematical model for a photon must involve not only Maxwell's differential equations, to describe its evolution, but also two boundary conditions, one to represent its source, and one to represent its receiver. Some aspect of the mathematical function representing the photon must be defined as zero at the two boundaries. Let that aspect be the electric field $\mathbf{E}$, leaving the magnetic field $\mathbf{B}$ to respond however it must. This choice makes the source and the receiver analogous to mirrors in a laser cavity: they keep the electromagnetic energy confined within the space between them.

3) So a realistic model solution requires at least three parts. The first part is the main part, traveling from the source to the receiver like the photon does. The other two parts can be shaped like the main part, but traveling in the opposite direction, to make appropriate cancellations at the source and the receiver. The solution we want is then the sum of the three parts, evaluated within the spatial domain between the source and the receiver.

4) And be prepared: each of the three solution parts has four sub parts. That is because it has both an $\mathbf{E}$ field and a $\mathbf{B}$ field (the second field being needed to make the photon travel), and it has a second pair of $\mathbf{E}$ and $\mathbf{B}$ fields located at $90°$ in space and a quarter cycle in time away from the first pair (the second pair of fields being

needed to make the photon spin). So far, there are a staggering twelve parts to the complete solution. And really there are infinitely more parts, because each new part added to fix one boundary condition slightly affects the other boundary condition, thereby demanding yet another tiny correction by addition of yet another tiny part, leading to an infinite sum of increasingly inconsequential parts. But fortunately, we won't have to study any of these individual parts in any detail.

2. E'S AND B'S FOR ONE OF THE TWO ORIENTATIONS

The main steps for implementing the three-solution model for either one of the two $90°$ orientations are the following:

1) Observe that Maxwell's equations imply that the wave travel direction for each solution part is determined by the cross product of the electric field $\mathbf{E}$ and magnetic field $\mathbf{B}$ assigned to it. So concentrate in the $\mathbf{E}$ fields, and let the $\mathbf{B}$ fields follow accordingly to give the propagation direction required. Call the propagation direction x, and call the $\mathbf{E}$ direction y, and call the $\mathbf{B}$ direction z.

2) Note that all three parts of the solution have to be initialized with field pulses. Let all pulses have the same Gaussian shape: rounded on top, sloping down similarly on both sides, and tailing off toward zero. Let the $\mathbf{E}$ pulse for the main part of the solution be positive, and let it be located at the source, on the path to the receiver.

3) To provide $\mathbf{E}$-field cancellation at the source, let $\mathbf{E}$ pulse for the second part of the solution be negative, and locate it at the source. To provide $\mathbf{E}$-field cancellation at the receiver, let the $\mathbf{E}$ pulse for the third part of the solution be negative, and locate it at double distance after the receiver.

4) To provide the desired propagation directions, assign all the $\mathbf{B}$ fields to be positive. Observe that, while $\mathbf{E}$ fields are cancelled at the source and receiver, $\mathbf{B}$ fields are doubled at the source and receiver. The zero $\mathbf{E}$ fields at the source and receiver mean zero Poynting vectors $\mathbf{P} = \mathbf{E} \times \mathbf{B}$ there, trapping the photon energy between the source and receiver. The double $\mathbf{B}$ fields at the source and receiver mean the photon energy begins piled up in front of the source, and ends piled up behind the receiver.

3. WAVEFORM EVOLUTION

The evolution of the initial field pulses under the action of Maxwell's first-order differential equations for fields exhibits the following main features:

1) As the differential operators in Maxwell's field equations work on the Gaussian $\mathbf{E}$ and $\mathbf{B}$ pulses, they generate Hermite polynomials of higher and higher order. These multiply their generating Gaussians, and the resulting product functions look like wavelets. So the scenario always has a mix of $\mathbf{E}$ and $\mathbf{B}$ field wavelets that are dying, and $\mathbf{E}$ and $\mathbf{B}$ field wavelets that are developing.

2) The field wavelets travel because of Poynting vector cross products between dying **E** field wavelets and dying **B** field wavelets, and between developing **E** field wavelets and developing **B** field wavelets. Travel is consistent because wavelets both of which are dying, or developing, are the same function, except for field vector direction.
3) The field wavelets spread because of Poynting vector cross products between dying **E** field wavelets and developing **B** field wavelets, and between developing **E** field wavelets and dying **B** field wavelets. Spreading happens because one of these wavelets is an even function while the other wavelet is an odd function, so the vector cross product always directs some energy forward, and an equal energy backward.
4) No matter how many zero-crossings develop, the separations between zero always crossings remains very similar. The wavelets always appear to have half wavelength equal to the width of the original Gaussian.

Figure 1 illustrates this behavior at the stage where an input Gaussian has spread and has developed five peaks (four zero crossings). Series 1 is the original input Gaussian function, Series 2 is the Gaussian after the overall spreading has developed to this point, and Series 3 is the wavelet that has emerged in the process; *i.e.* the spread-out Gaussian times the fourth-order Hermit polynomial generated.

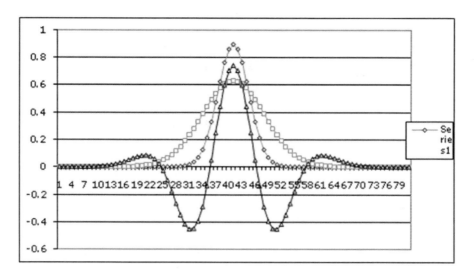

Figure 1. A wavelet develops when an EM pulse is acted upon by Maxwell's equations. [2].

4. WAVEFORM ENERGY DENSITIES

Figure 2 presents the information from Fig. 1 squared. Waveform amplitudes squared represent energy density, and integrated over space they represent total energy. Observe that when squared and integrated, generated wavelets have the same total energy as do their generating Gaussian squared and integrated.

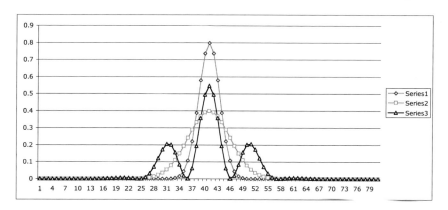

Figure 2. The squared amplitudes of the wavelet, its generating Gaussian, and the original Gaussian. [2].

1) The problem at hand has three solution parts with four field wavelets per part, making twelve wavelets altogether, so it is quite complicated. But viewed from the perspective of total energy, the situation is much simpler.

2) At the beginning of the scenario, we have two squared Gaussians at the source, one moving toward the receiver, and the other moving further away from the receiver, and we have one squared Gaussian at double distance beyond the receiver, and moving toward it. Within the space between the source and the receiver, the sum of all these Gaussians is essentially just a double height half Gaussian in front of the source.

3) At the end of the scenario, all three squared Gaussians have spread out. In addition, the one moving away from the receiver is now double distance away. The two moving toward the receiver have arrived at the receiver. Within the space between the source and the receiver, the sum of all these Gaussians is essentially a spread out, double height, half Gaussian before the receiver.

4) The beginning of the scenario and the end of the scenario are qualitatively similar, but quantitatively different on account of the spreading that has occurred. This spreading shows that Maxwell's coupled field equations naturally produce an important real-world physical phenomenon: the so-called 'arrow of time'.

5. RELATIVE MOTION

Now suppose that there is relative motion between the source and the receiver. We are immediately faced with the central question of Special Relativity Theory: with respect to what reference should the travel speed of any of these Gaussians be asserted to be light speed c? For the present model, the governing proposal is the following:

1) The two Gaussians that are placed outside of the domain of interest to enforce the two zero boundary conditions are not physical things, but only ideas. They can move however they must to accomplish their job.

2) The one Gaussian that travels from the source to the receiver must start at speed c relative to the source, and finish at speed c relative to the receiver. The transition

from one condition to the other must be symmetric with respect to the mid-point of the scenario.
3) The mid-point of the scenario is defined by the condition that both the source and the receiver are at that instant equidistant from that position, and they are moving at equal speeds in opposite directions with respect to that position.
4) Further details don't actually matter. It only matters that the mid-point of the scenario exists, and whatever information about the source position, velocity, and acceleration the virtual photon from the source may convey, that information is time-tagged to the mid-point of the scenario.

I believe the above model responds appropriately to an issue discussed at length by McCausland [3], Phipps [4], and many others they cite: what controls the photon: the source, the receiver, or the medium? My response is that the medium determines the numerical parameter c, the source provides the initial reference for c, and the receiver provides the final reference for c.

6. IMPLICATIONS FOR FIELDS DELIVERED

The received wisdom about this subject traces to the turn of the 20th century, when A. Liénard [5] (1898) and E. Wiechert [6] (1901) developed models for the potentials and fields created by rapidly moving charges. Although Liénard and Wiechert worked independently, they made the same assumption, and they got the same results, and so confirmed each other.

Expressed in Gaussian units, the Liénard-Wiechert (LW) scalar and vector potentials at position **r** and time t are

$$\Phi(\mathbf{r},t) = e\left[1/\kappa R\right]_{\text{retarded}} \text{ and } \mathbf{A}(\mathbf{r},t) = e\left[\vec{\beta}/\kappa R\right]_{\text{retarded}}$$

where $\kappa = 1 - \mathbf{n}\cdot\vec{\beta}$, $\vec{\beta}$ is source velocity normalized by c, and $\mathbf{n} = \mathbf{R}/R$ (a unit vector), and $\mathbf{R} = \mathbf{r}_{\text{source}}(t - R/c) - \mathbf{r}$ (an implicit definition for the terminology 'retarded'). The LW fields obtained from those potentials are then

$$\mathbf{E}(\mathbf{r},t) = e\left[\frac{(\mathbf{n}-\vec{\beta})(1-\beta^2)}{\kappa^3 R^2} + \frac{\mathbf{n}}{c\kappa^3 R}\times\left((\mathbf{n}-\vec{\beta})\times\frac{d\vec{\beta}}{dt}\right)\right]_{\text{retarded}}$$

and $\mathbf{B}(\mathbf{r},t) = \mathbf{n}_{\text{retarded}} \times \mathbf{E}(\mathbf{r},t)$.

The LW fields have some interesting properties. The $1/R$ fields are radiation fields, and they make a Poynting vector (energy flow per unit area per unit time) that lies along $\mathbf{n}_{\text{retarded}}$:

$$\mathbf{P} = \frac{c}{4\pi} \mathbf{E}_{radiative} \times \mathbf{B}_{radiative} =$$

$$\frac{c}{4\pi} \mathbf{E}_{radiative} \times \left(\mathbf{n}_{retarded} \times \mathbf{E}_{radiative} \right) = \frac{c}{4\pi} (E_{radiative})^2 \mathbf{n}_{retarded}$$

But the $1/R^2$ fields are Coulomb-Ampère fields, and the Coulomb field does *not* lie along $\mathbf{n}_{retarded}$ as one might naively expect; instead, it lies along $(\mathbf{n} - \vec{\beta})_{retarded}$. Assume that $\vec{\beta}$ does not change much over the total field propagation time, in which case $(\mathbf{n} - \vec{\beta})_{retarded}$ is virtually indistinguishable from $\mathbf{n}_{present}$. So then the Coulomb field and the radiation are arriving to the observer from different directions.

This result seems very unbelievable. Can it mean there may exist imperfections in our common heritage of results in the subject of Electrodynamics? Yes, that is always possible. Indeed, the journal Galilean Electrodynamics has featured such problems for over two decades.

So suppose we try the same exercise with the present photon model instead of the LW assumption. Then the potentials become

$$\Phi(\mathbf{r},t) = e\left[1/\kappa R \right]_{half\ retarded} \quad \text{and} \quad \mathbf{A}(\mathbf{r},t) = e\left[\vec{\beta}/\kappa R \right]_{half\ retarded}$$

The fields become:

$$\mathbf{E}(\mathbf{r},t) = e\left[\frac{(\mathbf{n} - \vec{\beta})(1 - \beta^2)}{\kappa^3 R^2} + \frac{\mathbf{n}}{c\kappa^3 R} \times \left((\mathbf{n} - \vec{\beta}) \times \frac{d\vec{\beta}}{dt} \right) \right]_{half\ retarded}$$

and $\mathbf{B}(\mathbf{r},t) = \mathbf{n}_{half\ retarded} \times \mathbf{E}(\mathbf{r},t)$.

The Poynting vector $\mathbf{P}(\mathbf{r},t)$ becomes:

$$\frac{c}{4\pi} \mathbf{E}_{radiative} \times \mathbf{B}_{radiative} =$$

$$\frac{c}{4\pi} \mathbf{E}_{radiative} \times \left(\mathbf{n}_{half\ retarded} \times \mathbf{E}_{radiative} \right) = \frac{c}{4\pi} (E_{radiative})^2 \mathbf{n}_{half\ retarded}$$

Observe that now the direction of the Coulomb field is

$$(\mathbf{n} - \vec{\beta})_{half\ retarded} \approx (\mathbf{n}_{present})_{half\ retarded} \stackrel{\Delta}{=} \mathbf{n}_{half\ retarded}$$

and the direction of the Poynting vector is $\mathbf{n}_{\text{half retarded}}$ too. So now, the Coulomb field and the Poynting vector are reconciled to the same direction.

7. Implications for Relativity Theory

The assumption that underlies the Liénard-Wiechert formulation of potentials and fields from a moving source is the same one that Einstein later formalized as his Second Postulate for Special Relativity Theory (SRT) [7]. The assumption comes from the reduction of Maxwell's four coupled field equations to two uncoupled wave equations. This is accomplished by inserting two Maxwell equations into the two other Maxwell equations. The four coupled field equations are

$$\nabla \cdot \mathbf{B} = 0, \ \nabla \cdot \mathbf{D} = 4\pi\rho,$$

$$\nabla \times \mathbf{E} + \frac{1}{c} \partial \mathbf{B}/\partial t = 0, \quad \nabla \times \mathbf{H} - \frac{1}{c} \partial \mathbf{D}/\partial t = \frac{4\pi}{c}\mathbf{J}$$

The two uncoupled wave equations are

$$\nabla^2 \mathbf{E} - \frac{1}{c^2}\partial^2 \mathbf{E}/\partial t^2 = 0 \text{ and } \nabla^2 \mathbf{B} - \frac{1}{c^2}\partial^2 \mathbf{B}/\partial t^2 = 0$$

Observe that the time derivatives in the coupled field equations are first order, whereas the time derivatives in the wave equations are second order. That means the wave equations obscure the evidence of the arrow of time. As a result, the uncoupled wave equations allow a larger set of solutions than the coupled field equations allow. Some solutions to the uncoupled wave equations do *not* also solve Maxwell's coupled field equations. As you have seen, the electromagnetic pulse that travels without spreading longitudinally is one of those solutions to the un-coupled wave equations only. And it was used as the representative of Maxwell's coupled field equation in the development of SRT. So caution about SRT is warranted. At the very least, SRT does not do what it was meant to do; namely, capture the essence of Maxwell's electrodynamics for SRT. That goal was indeed a worthy one, and the fact that it was not fully accomplished a hundred years ago ought not preclude the renewed attempt today.

Conclusion

I believe the above exercise usefully removes a contradiction from our common narrative of Physics. But it won't be the last one. Many other authors, for example Russian mathematician J.G. Klyushin [8], have written about similar problem areas. A.A. Nassikas has argued in [9] that, on grounds of Logic and Linguistics, there will *always* be such contradictions! And indeed, fixing this one contradiction just leads to another one, which will be discussed in the next Chapter. Nevertheless, removing this one contradiction is useful. As a result of the reconciliation between directions for the Coulomb field and the Poynting vector, we have tangential forces and unbalanced forces in the Hydrogen atom. The corrected Coulomb field in the Hydrogen atom produces the torquing and the enhanced radiation that lead in turn to the balance between energy loss by radiation and energy gain by torquing that is discussed in Part II, Chapter 9.

ACKNOWLEDGMENTS

This all seems fairly straightforward now, but it took a long time to develop, with more than one person thinking about it. This Chapter has an important Appendix, and I did not write it. My teacher and lifelong friend, the late Laszlo Tisza, wrote it. He was my Ph.D. thesis supervisor at M.I.T. in the late 1960's. In the 1970's, he talked to me, and wrote in his notebooks, about the photon. Remembering those conversations, and reading those passages from his notebooks, I have realized again how much each generation really inspires the next generation. I thank Prof. Tisza's wife Magda for the chance to read the notebooks and share the information quoted here.

APPENDIX

I have retyped the following passage from the late Prof. Laszlo Tisza's research notebooks, which are now resident in the archives of the Science Library at the Massachusetts Institute of Technology. I accept responsibility for responding to any unclear points that readers may detect in my presentation of Prof. Tisza's text. Square brackets in the text indicate words restored, replaced, or added, or comments added, by me. *CKW*.

[A] Theory of the Photon

It is an undisputed postulate of modern physics that light, like other fundamental entities of physics, has [both] corpuscular and wave properties These two concepts are inconsistent within classical theory, and the development of a consistent and empirically satisfying [modern] theory had to follow one of two options: 1) Start from classical mechanics of particles (CMP), and modify it guided by experimental fact; or 2) Do the same thing [starting from classical electrodynamics (CED)].

In the course of the actual historical development, much more emphasis was put on Option 1. The purpose of this discussion is to show that valuable insights can be gained by exploring Option 2. Although the one-sided emphasis of the past is undoubtedly connected with the Newtonian bias of classical physicists, Option 1 could have been favored on pragmatic [grounds] because of its simplicity. Option 2 is undoubtedly more complicated; however, this complication is commensurate with the nature of the problem. At any rate, the present discussion has only a heuristic character, though it should be sufficient to direct attention to this other Option.

The Concept of Convergence

The fundamental facts connected with the emission and absorption of light can be summed up as follows:

1) CED is symmetric with respect to future and past; we have both diverging and converging waves.[1] However, in the macroscopic theory, only the initial conditions for the diverging wave can be realized for a source in free space. This situation leads to unlimited [spatial] dilution of the radiative energy.
2) This picture has to be modified on the microscopic quantum scale on at least two counts:

   a) Whereas the radiation in the macro theory is symmetric with respect to inversion, on the micro scale the radiation is directed. The emitter suffers a recoil, and the transmitted momentum imposes a direction on the exiting wave. Actually, the emitter transmits energy, momentum, and angular momentum. We [can] refer to this situation briefly as 'momentum transfer'.
   b) The 'needle radiation' of a) would still seem to become indefinitely diluted, whereas quantum observations at low intensity reveal that the radiative energy remains at all times available in a small region of space-time.

Taking care of points a) and b) calls for two 'quantum postulates'. The first is the conventional one, although [I] prefer to put it in a slightly novel form: [I] postulate that the photon carries an angular momentum h, and this implies by an elementary phenomenological argument that it carries a four-momentum hk.

The second postulate is new. We require that the diverging beam should 'naturally' refocus itself. Before going into any quantitative elaboration, let me at first explain the meaning of this statement.

The conception underlying the second postulate can be compared with Bohr's quantization of the [Hydrogen] atom. [Bohr] assumed that the atom can exist in discrete states provided by CMP, while [transitions] between these states [are] non-classical. In the present situation, we assume that the photon goes through phases of diverging and converging radiation as provided by CED, but the refocusing [that takes] the diverging [beam] into the converging beam is non-classical. At the same time, the present situation compares most favorably with the one prevailing in the atom. The traditional quantum transition defies intuitive understanding, and even from a formal point of view, the transition is described in terms [of] perturbation theory; *i.e.*, an approximation that is unbecoming for describing a fundamental phenomenon.

---

[1] The diverging spherical wave may have an angular dependence of the radiated energy depending on the symmetry of the source. However, the radiated energy is always symmetric with respect to inversion at the source.

Although the refocusing here postulated is non-classical, in the sense that it is not predicted by CED, it is consistent with our intuition; refocusing is a most common phenomenon in optical instruments, and it is also subject to precise mathematical description. We have, of course, the major problem of explaining why and how such a refocusing takes place under 'natural' conditions.

Note that the quantization described under a) involves the harmonious cooperation of relativity and quantum theory: it is a fundamental contribution of relativity that we have an expression for the four-momentum equally valid for particles with finite and with zero rest mass.

As we shall see [eventually], the situation is much more involved with respect to b). Quite apart from the 'why' and 'how', we wish to point out that the picture given is in qualitative agreement with what we know about light.

Note: Speaking of focusing: we cannot have, and [do not] need, perfect imaging; the radiant energy will always occupy a finite region in space and time. The important point is that this volume does not increase in the course of time. This statement is merely an expression of the stability of the photon. It is a much weaker assumption than the traditional statement that the photon is a particle.

Eventually, we shall have a description of the photon in [position-momentum] phase space, although the detailed features will be different from the conventional one. The constancy of the volume is essentially an application of Liouville's theorem.

[Compare] with the traditional theory: by asserting that the propagation consists of a succession of divergent and convergent phases of the photon, we completely avoid the difficulties of the wave –particle dualism. In its spread-out state, the photon can interact with a spatially periodic structure (grating), whereas in the concentrated phase, it can interact with a small system exhibiting resonance, *i.e*, temporarily periodic structure [an atom]. *There [exists] no phenomenon to support the present view, according to which the photon propagates as a rigid wave packet.* [emphasis added; following paragraph break added].

The [traditional] theory has, however, an elaborate formalism with excellent quantitative predictions. It is plausible that these predictions should be, by and large, consistent with the present viewpoint. The question is: what kind of formalism can be assigned to the new aspects of the ideas here advanced? [This is a question addressed by the present paper].

In the present approach, we are dealing with divergent and convergent beams; *i.e.* with a measure of convergence, or curvature of the wave front. This concept is certainly physically meaningful, but it is completely lost in the [traditional] method, which deals with plane waves. (There is a curious hubris involved in ignoring geometrical optics in fundamental theory, presumably because it is elementary.

Yet the lens formula essentially expresses a conservation law for curvature.). In other words, the traditional theory established the wave particle analog in terms of plane waves only, thereby emasculating the wave theory of some of its fundamental features. [Prof. Tisza's text continues into the subject of elementary particles. The Archives of the Science Library at the Massachusetts Institute of Technology contain this text, and many other interesting research materials].

REFERENCES

[1] Cornille, P. *Advanced Electromagnetism and Vacuum Physics*, Chapter 7, The Wave Packet Concept, World Scientific, New Jersey, London, Singapore, Hong Komg, 2003.

[2] Whitney, C.K. Maxwell's Maxima, *Proceedings of the NPA*, Vol. 6, No. 2, pp. 374-382, 2009.

[3] McCausland, I. *A Scientific Adventure: Reflections on the Riddle of Relativity*, Appendix III, On Faraday's Ray-Vibrations and the Postulates on the Velocity of Light, Apeiron, Montreal, Canada, 2011.

[4] Phipps, T.E. *Heretical Verities: Mathematical Themes in Physical Description*, Chapter 3, The Remarkable Behavior of Light, Classic Non-Fiction Library, Urbana, IL, 1996.

[5] Liènard, A. Champ Electrique et Magnétique produit par une Charge Electrique Concentrée en un Point et Animée d'un Movement Quelconque, *L'Eclairage Electrique*, Vol. XVI No. 27, pp. 5-14; No. 28, pp. 53-59, No. 29, pp. 106-112, 1898.

[6] Wiechert, E. Elektrodynamische Elementargesetze, *Archives Néerlandesises des Sciences Exactes et Naturelles*, série II, Tome IV, 1901.

[7] Einstein, A. On the Electrodynamics of Moving Bodies (1905), as translated in *The Collected Papers of Albert Einstein*, Vol. 2, pp. 140-171, Princeton University Press, Princeton, New Jersey, 1989.

[8] Klyushin, J.G. *Fundamental Problems in Electrodynamics and Gravidynamics,* Updated English-Language Edition, Galilean Electrodynamics, Arlington, Massachusetts, 2009.

[9] Nassikas, A.A. *Minimum Contradictions Everything*, Hadronic Press, Palm Harbor, Florida, 2008.

Chapter 14

ON THE INVARIANCE OF MAXWELL'S EQUATIONS

ABSTRACT

This Chapter confronts one more contradiction that exists in our common narrative of Physics. This one concerns the status of Maxwell's equations as tensor equations and the status of Lorentz Transformations as the unique choice of coordinate transformation allowable for use in Physics.

INTRODUCTION

It is generally believed that Maxwell's equations are invariant under Lorentz transformations, and not invariant under Galilean transformations, and so it is believed that they provide decisive evidence in favor of Einstein's Special Relativity (SRT) Theory over Newtonian Mechanics (NM). But it is also generally believed that Maxwell's equations are tensor equations. Now the whole purpose of tensor theory is to render mathematical statements that are 'coordinate free', *i.e.* independent of the choice of coordinate frame. So there is no room there to prohibit any kind of coordinate transformations, except for ones that contain singularities, and so are not invertible.

What is going on here? It sounds like a contradiction. Either Maxwell's equations are *not* themselves tensor equations, and therefore are *not* qualified to endorse SRT as a candidate for the status of tensor theory, and hence its Lorentz Transformations for exclusive use in physics, or else Maxwell's equations really *are* tensor equations, and therefore do *not* endorse any kind of limitation on choice of coordinate frame, and hence on the kind of transformations allowable, and hence in favor of SRT for exclusive use in physics. This Chapter is about resolving this contradiction.

A very few people, R.M. Kiehn [1] being the only one I personally know, have written contrary to the general beliefs, and supported the position that Maxwell's equations really are invariant under arbitrary coordinate transformations, so long as the transformations are invertible. I agree. But I think we need a more precise notation to demonstrate the point. Section 1 here develops a suitably expanded tensor notation, and records explicit demonstrations of its meaning in particular cases that can be displayed using only 2×2

matrices. Section 2 then lays out the transformation of Maxwell's equations in the general case, which requires 4×4 matrices.

1. EXTENDED TENSOR NOTATION, WITH MATRIX DEMONSTRATIONS

Tensor notation uses the position of an index to convey information about what transformation to apply. For an arbitrary tensor A, there are four possible index positions:

$$A_\alpha, \; A^\alpha, \text{ and } {}_\alpha A, \; {}^\alpha A .$$

Only the first two are in common use today. They are called 'covariant' and 'contravariant'. The other two can be called 'transcovariant' and 'transcontravariant', with the prefix 'trans' reminding us about the index position being left instead of right.

The introduction of ${}_\alpha A$ and ${}^\alpha A$ amounts to a doubling of the tensor algebra. There is a long and venerable history of such doublings in the development of Mathematics. One example is the doubling of Arithmetic to include not only positive numbers, but also negative ones, and even zero. Another is the doubling of arithmetic to include not only real numbers, but also imaginary ones. And there are many more such doublings. Rowlands [2] tells the whole fascinating story up to the present time. There is every reason to believe that Mathematics will continue to exploit doublings in the future too.

Indeed, there has been some prior work thinking about tensors more generally than has been common so far. Moon and Spencer [3] use the term 'holor' to cover many useful generalizations of the idea of 'tensor'. So if the left-side indices introduced here seem at all upsetting, then it may be appropriate to consider the objects defined as holors, rather than as tensors, and to consider the algebra developed as the corresponding holor algebra, rather than as the extended tensor algebra.

For purposes of unambiguous discussion, we can represent arbitrary tensors, or holors, by examples that can be displayed as column vectors. The problems of interest are four dimensional, but two dimensions are enough to start with. Call them n and m, with particular numbers in the range 0 to 3 to be specified later. We have

$$A_\alpha \to \begin{bmatrix} A_n \\ A_m \end{bmatrix}, \; A^\alpha \to \begin{bmatrix} A^n \\ A^m \end{bmatrix}, \text{ and } {}_\alpha A \to \begin{bmatrix} {}_n A \\ {}_m A \end{bmatrix}, \; {}^\alpha A \to \begin{bmatrix} {}^n A \\ {}^m A \end{bmatrix} .$$

All of these tensors are subject to transformations. Because the different transformations have subtle distinctions, it is useful to represent them as 2×2 matrices. That allows four parameters, but we only need two, because we are only interested in transformations that are 'unimodular'; *i.e.* have unit determinant, *i.e.* don't change the intrinsic size of anything. We can have

$$\begin{bmatrix} A'_n \\ A'_m \end{bmatrix} = \frac{1}{\sqrt{1-BC}} \begin{bmatrix} 1 & C \\ B & 1 \end{bmatrix} \begin{bmatrix} A_n \\ A_m \end{bmatrix}, \quad \begin{bmatrix} A'^n \\ A'^m \end{bmatrix} = \frac{1}{\sqrt{1-BC}} \begin{bmatrix} 1 & -C \\ -B & 1 \end{bmatrix} \begin{bmatrix} A^n \\ A^m \end{bmatrix},$$

and

$$\begin{bmatrix} {}_nA' \\ {}_mA' \end{bmatrix} = \frac{1}{\sqrt{1-BC}} \begin{bmatrix} 1 & B \\ C & 1 \end{bmatrix} \begin{bmatrix} {}_nA \\ {}_mA \end{bmatrix}, \quad \begin{bmatrix} {}^nA' \\ {}^mA' \end{bmatrix} = \frac{1}{\sqrt{1-BC}} \begin{bmatrix} 1 & -B \\ -C & 1 \end{bmatrix} \begin{bmatrix} {}^nA \\ {}^mA \end{bmatrix}.$$

For the record, we should also write down the matrix transpose versions of the same relationships:

$$A_\alpha \to [A_n \ A_m], \quad A^\alpha \to [A^n \ A^m],$$

and

$$_\alpha A \to [{}_nA \ {}_mA], \quad {}^\alpha A \to [{}^nA \ {}^mA],$$

with

$$[A'_n \ A'_m] = [A_n \ A_m] \frac{1}{\sqrt{1-BC}} \begin{bmatrix} 1 & B \\ C & 1 \end{bmatrix},$$

$$[A'^n \ A'^m] = [A^n \ A^m] \frac{1}{\sqrt{1-BC}} \begin{bmatrix} 1 & -B \\ -C & 1 \end{bmatrix}.$$

and

$$[{}_nA' \ {}_mA'] = [{}_nA \ {}_mA] \frac{1}{\sqrt{1-BC}} \begin{bmatrix} 1 & C \\ B & 1 \end{bmatrix},$$

$$[{}^nA' \ {}^mA'] = [{}^nA \ {}^mA] \frac{1}{\sqrt{1-BC}} \begin{bmatrix} 1 & -C \\ -B & 1 \end{bmatrix}.$$

There are two special cases of interest,

$C = B$ and $C = -B$

The first special case is good for Lorentz transformations, in which case we can make $n = 0$, $m = 1$. The second special case is good for rotations, in which case we can make $n = 2$, $m = 3$. For the special cases, we don't have all four different transformation matrices; we only have

$$\frac{1}{\sqrt{1-B^2}}\begin{bmatrix} 1 & B \\ B & 1 \end{bmatrix} \text{ and } \frac{1}{\sqrt{1+B^2}}\begin{bmatrix} 1 & B \\ -B & 1 \end{bmatrix},$$

$$\left(\text{or the transpose } \frac{1}{\sqrt{1+B^2}}\begin{bmatrix} 1 & -B \\ B & 1 \end{bmatrix}\right).$$

So we have no reason *not* to make

$$_\alpha A = A_\alpha \text{ and } {}^\alpha A = A^\alpha.$$

That is, we do not need four tensor index positions to express pure Lorentz transformations or three-space rotations, or indeed any combinations thereof, which make up the whole Lorentz group of transformations.

But we do have reason to worry that we may not live in a Universe limited to these special cases. That is why we need to keep open the four-index-position notational option that is available to us. Suppose, for example, that we want to look into the behavior of Galilean transformations. That would require $B = -V/c$, $C = 0$. We would then need all four cases:

$$\begin{bmatrix} A'_0 \\ A'_1 \end{bmatrix} = \begin{bmatrix} 1 & 0 \\ -V/c & 1 \end{bmatrix}\begin{bmatrix} A_0 \\ A_1 \end{bmatrix}, \begin{bmatrix} A'^0 \\ A'^1 \end{bmatrix} = \begin{bmatrix} 1 & 0 \\ +V/c & 1 \end{bmatrix}\begin{bmatrix} A^0 \\ A^1 \end{bmatrix},$$

$$\begin{bmatrix} {}_0 A' \\ {}_1 A' \end{bmatrix} = \begin{bmatrix} 1 & -V/c \\ 0 & 1 \end{bmatrix}\begin{bmatrix} {}_0 A \\ {}_1 A \end{bmatrix}, \begin{bmatrix} {}^0 A' \\ {}^1 A' \end{bmatrix} = \begin{bmatrix} 1 & +V/c \\ 0 & 1 \end{bmatrix}\begin{bmatrix} {}^0 A \\ {}^1 A \end{bmatrix},$$

or, expressed in the transpose,

$$\begin{bmatrix} A'_0 & A'_1 \end{bmatrix} = \begin{bmatrix} A_0 & A_1 \end{bmatrix}\begin{bmatrix} 1 & -V/c \\ 0 & 1 \end{bmatrix},$$

$$\begin{bmatrix} A'^0 & A'^1 \end{bmatrix} = \begin{bmatrix} A^0 & A^1 \end{bmatrix} \begin{bmatrix} 1 & +V/c \\ 0 & 1 \end{bmatrix},$$

$$\begin{bmatrix} {}_0A' & {}_1A' \end{bmatrix} = \begin{bmatrix} {}_0A & {}_1A \end{bmatrix} \begin{bmatrix} 1 & 0 \\ -V/c & 1 \end{bmatrix},$$

$$\begin{bmatrix} {}^0A' & {}^1A' \end{bmatrix} = \begin{bmatrix} {}^0A & {}^1A \end{bmatrix} \begin{bmatrix} 1 & 0 \\ +V/c & 1 \end{bmatrix}$$

3. GALILEAN TRANSFORMATION OF MAXWELL'S EQUATIONS

Tensor transformation is typically represented with 'tensor contraction', which means repeated indexes, up and down. How can that type of representation be applied here? There are four cases to distinguish, and we can use all four possible index positions to do that job clearly. We can, for example, capture the distinctions by w riting

$$A'_\beta = ({}^\alpha T_\beta) A_\alpha, \quad A'^\beta = ({}_\alpha T^\beta) A^\alpha,$$

and $\quad {}_\beta A' = ({}_\beta T^\alpha)_\alpha A, \quad {}^\beta A' = ({}^\beta T_\alpha)^\alpha A$

Maxwell's equations in current tensor notation read:

$$\partial_\alpha F^{\alpha\beta} = \frac{4\pi}{c} J^\beta \text{ and } \partial_\alpha D^{\alpha\beta} = 0^\beta.$$

The two-index tensors $F^{\alpha\beta}$ and $D^{\alpha\beta}$ refer to the electromagnetic field and the 'dual' thereof. The electromagnetic field tensor $F^{\alpha\beta}$ contains the three-dimensional electric and magnetic field vectors, **E** and **B**. The $D^{\alpha\beta}$ is the dual to $F^{\alpha\beta}$, which contains components of **B** and $-\mathbf{E}$. The one-index tensors J^β and ∂_α refer to the source charge-current density vector and the differential operator vector. The indexes α and β each take four values: $0,1,2,3$.

The seeming limitation of Maxwell's equations to invariance only under Lorentz transformation arises entirely from the differential operator being written as a covariant vector. In the extended tensor algebra, this operator is identified as <u>trans</u>covariant, and then Maxwell's equations look like:

$$(_\alpha\partial)F^{\alpha\beta} = \frac{4\pi}{c}J^\beta \text{ and } (_\alpha\partial)D^{\alpha\beta} = 0^\beta.$$

Expanded into matrix notation, $_\alpha\partial$ becomes a row vector, and takes its transformation matrix on its right, and for motion in the direction 1, that matrix is the 4×4 extension of $\begin{bmatrix} 1 & 0 \\ -V/c & 1 \end{bmatrix}$. Both $F^{\alpha\beta}$ and $D^{\alpha\beta}$ are square matrices, and take transformations on both sides, the one on the left being the 4×4 extension of $\begin{bmatrix} 1 & 0 \\ +V/c & 1 \end{bmatrix}$. So what occurs between the differential operator $_\alpha\partial$ and its operand $F^{\alpha\beta}$ or $D^{\alpha\beta}$ is the 4×4 extension of the matrix product

$$\begin{bmatrix} 1 & 0 \\ -V/c & 1 \end{bmatrix}\begin{bmatrix} 1 & 0 \\ +V/c & 1 \end{bmatrix} = \begin{bmatrix} 1 & 0 \\ 0 & 1 \end{bmatrix}.$$

So written with $_\alpha\partial$, Maxwell's equations are manifestly form invariant, not only under Lorentz transformation, but also under Galilean transformation, and in fact any invertible transformation.

CONCLUSION

The present Chapter challenges the long-standing belief that only Lorentz transformations of space-time coordinates are valid for use in physics. This belief comes from the fact that SRT requires Lorentz transformations, and SRT is presently accepted as valid. One reason for its early acceptance was that its mandated Lorentz transformations were demonstrably in harmony with Maxwell's equations. But we see in the present Chapter that, when written with clear enough tensor notation, *all* invertible transformations are in harmony with Maxwell's equations. So, actually, Maxwell's equations don't provide any evidence whatsoever as to what sort of coordinate transformations Physics ought to use. So they neither confirm, nor deny, the validity of SRT. They just stand on their own, apart from SRT.

This challenge comes on top of the challenge to SRT raised by the last Chapter. Einstein had founded SRT on his Second Postulate, which says that that light speed is the same consant c with respect to any observer. With this Postulate, he formalized the assumption made by Liénard and Wiechert, and many other authors of that time. One is always encouraged to challenge any Assumption, but one does not really feel invited to challenge a Postulate. So the status of Postulate long protected the Assumption from successful scrutiny. But we saw in the last Chapter that the Second Postulate could rationally be challenged, because an alternative more compatible with Quantum Mechanics could be constructed from particular solutions to Maxwell's coupled field equations. And rest assured: although not discussed here, many more challenges to SRT do exist.

What do such challenges mean for the subject of Electrodynamics? Are we recommending upheaval there? Not exactly. A Newtonian conception of Electrodynamics is alive and well, and always has been, within the world of electrical technology. Why? Because the technology is not about point particles or inertial motions; it is about circuits, and coils, and oscillations. My friend Peter Graneau has been writing about Newtonian Electrodynamics for years, and his book with his son Neal [4] has exactly that title.

The same is true of Chemistry: it is not about point particles or inertial motions. Its smallest meaningful system is an atom, and the essence of an atom is closed-path motion of its parts. I believe we are now at a point in time where it is completely acceptable for Chemistry to consider some ideas that seem forbidden from the viewpoint of SRT, especially ideas concerning events that occur deep inside of atoms and hence far from the idealized free space that SRT presumes, and especially ideas that help us explain and quantify the many chemical processes that we can actually observe.

REFERENCES

[1] Kiehn, R.M. *Non-Equilibrium Systems and Irreversible Processes*, Vol. 4, Adventures in Applied Topology, especially Sect. 2.2.3; http://www22.pair.com/csdc/kok/ebookvol4.pdf, or Lulu Enterprises, Inc., 3131 RDU Center, Suite 210, Morrisville, NC 27560, 2009.

[2] Rowlands, P. *Zero to Infinity – The Foundations of Physics*, Chapter 16, The Factor 2 and Duality, World Scientific, New Jersey, London, *etc.*, 2007.

[3] Moon P., and Spencer, D.E. *Theory of Holors: A generalization of tensors*, Cambridge University Press, London, New York, *etc.*, 1986.

[4] Graneau, P., and Graneau, N. *Newtonian Electrodynamics*, World Scientific, Singapore, New Jersey, London, Hong Kong, 1996.

CONCLUSION

There is a moral to this Book. In Science, you must always read both widely and critically. Even the most exalted of standard textbooks may contain some unexamined dogma, and even the most excoriated of fringe doggerel may report some interesting phenomenon. You must use your own mind to sort things out. Algebraic Chemistry is one tool to use for that task.

Throughout Science in general, the sorting task creates a tension between conservatism and radicalism. Science is inherently conservative most of the time: before we accept any new Theory, we try to disprove it. But the process is not always as thorough as it should be. We should *also* always ask if we have inadvertently overlooked anything important about the Theory we had *already*. This part is sometimes skipped over. That problem is illustrated in Part III of the present Book. Why does it ever happen? The following paragraphs explore the etiology.

1. SOME HISTORY TO RECALL

Thomas Kuhn described in [1] how revolutions occur in Science. Once in a while, because of great ennui or perceived crisis, its practitioners come to want a bit of revolutionary spirit, and that initiates the process. That was the scenario in the late 19th century, when electrical phenomena led to investigations by many researchers, and to the ultimate unified formulation by Maxwell. That development brought to Physics something new that hadn't been part of Newtonian mechanics: the unified electromagnetic field. The new knowledge led to important technological advances, and that development was undeniably good.

The welcome for revolutionary spirit existed again at the turn of the 20th century. There was a great morass of data about atoms, especially spectroscopic data, and there was a great frustration about light, especially in relation to the believed ideas about 'ether' and the observed non-confirming behavior of light in the Michelson-Morley interferometer. And then there was the photoelectric effect. That did it. There was a warm welcome waiting for the 20th century developments: Quantum Mechanics (QM), in its several incarnations, and Relativity Theory (RT), in its two parts, Special and General (SRT and GRT).

The new QM and RT looked sophisticated, modern, and deep. So there was great hope that we were getting closer to the physical truth of things. QM and RT were embraced as the two embodiments of an important Scientific Revolution.

RT as developed was a departure from Newtonian mechanics, and QM as developed was a departure from Maxwell's electromagnetic theory. But the present Book stands to testify that all this emphasis on 'departure' was not so good.

Today we are certainly not satisfied with what we have for Theory. What we have is a big unsolved problem: QM and RT are departures, not only from earlier Theories, but apparently also from each other. Recall that QM features wave functions and instantaneous distant correlations, whereas RT features point particles and finite signal speed. This contradictory situation leaves many people wondering what it all means. Is one of the two pillars of 20th century science flawed? Could both be flawed? What a quandary!

So maybe conservatism should have remained a bit stronger in those early days. The new QM and RT did look a bit mystical. That aspect might have triggered some caution. And the several 20th century formulations of QM were all considered departures from Maxwell's Classical Electrodynamics (CED). Considering the phenomenal practical success of CED, that too could have triggered some caution. And the new RT was supposed to have been founded on CED, so there *had* to be some sort of discord between the QM and the RT. That too could have raised more questions at the time.

So, can we nevertheless re-assert some conservatism today? The requirements of Chemistry may be just the prod needed to help make that happen. Chemistry has reactions. Reactions embody the 'arrow of time'. Maxwell's four coupled field equations provide the arrow of time. The two uncoupled wave equations that describe pulses that propagate without any evolution into wavelets do not provide the arrow of time. The no-evolution traveling pulse is *the* basis for Special Relativity Theory (SRT), and hence *a* basis for General Relativity Theory (GRT) too. So all of RT may not have any useful role in problems that involve Chemistry.

However, RT can only be ignored, and not invalidated. Puzzlement about SRT in particular is sporadically rekindled by various observations, like the recent ones from CERN concerning neutrinos. [2] But such puzzlement is always quickly put to rest. That is because it is really impossible to test SRT by experimental observations, since the assumption that needs testing, *i.e.* the postulated light speed c, is always embedded somewhere in the algorithm for processing the data taken. Philosophically, we are faced with an unavoidable tautology. That makes SRT, and by extension all of RT, not exactly a scientific Theory, according to the meaning that Philosophy of Science currently attributes to the term, which requires that a scientific Theory be falsifiable [3]. So RT is more a Convention for human thinking than a Theory about physical things.

QM is more robust than RT, although it is still a work in progress. A lot of current deep thought is included in a recent book edited by Jonathan P. Groffe [4]. The topics treated there are numerous and diverse, but there is some clustering around certain issues that also appear in the present Book. These issues include:

1) What is the minimum system content that one can meaningfully talk about? In [4], this issue underlies references to 'entanglement', 'non-separability', 'distant correlation', 'irreducible holistic property', *etc*. The same issue came up many times in the present Book; for example, in the basic Hydrogen atom Model, which has two moving parts and three important phenomena.

2) What is the minimum time history that one can meaningfully recount? In [4], this issue underlies references to 'measurement', 'wave function collapse', *etc*. The same

issue came up in the present Book in connection with the photon model, which involved emission from a source in the form of **E** and **B** field pulses, development into a spreading wavelet, and regression back to a less spread waveform for absorption by a receiver.

3) Can Physics become more unified than it presently seems to be? In [4], this issue underlies references to 'interpretation', 'philosophical implication', 'domain of application', and juxtapositions like 'QM and GRT', 'QM and SRT', 'phases and topological effects', *etc*. The same issue came up in the present Book in connection with the photon model, which went back to Maxwell for its foundation.

4) What really matters for technology development? In [4], this issue is answered in references to 'numerical methods', 'quantum computation', 'biological membranes', *etc*. The same issue lies at the heart of Algebraic Chemistry. AC renders practical assistance in the business of technology.

2. THE TASK TO ADDRESS

The fact that the most fundamental Theory to underpin AC is Maxwell's Electromagnetic Theory means the best pre-existing Theory for upgrading the status of the present AC Model actually predates both of the two major pillars of 20th century Physics, QM and RT. I believe the Maxwell connection can put Chemistry into the position of a third pillar, equal in importance to the other two, and possibly a bit *more* important. Will you pursue this kind of work?

How can you do it? It is widely believed that progress in Science depends on crucial Experiments. The vision is that, at any point in history, there exists a dominant scientific paradigm, and its shortcomings are revealed when it fails in an experiment. But this is not really how things go. There would have to be, not just one experiment, but rather overwhelmingly *many* experiments, *all* demonstrating failures, before conceptual shortcomings could even be acknowledged to exist. Each such experiment being both difficult and expensive to perform, the requirement for an overwhelming number of them is not possible to fulfill.

There is a chicken-and-egg problem here. There has to be a major loss of faith already, before challenging experiments can even be funded and carried out. And there also has to be some alternative paradigm on offer, to replace the old one.

And it has to make some interesting predictions for comparison to new experimental results. That is, progress requires a three-legged stool. Without all three legs in place, all at once, nobody will stand on the stool. So experiment is *not* a promising way to accomplish scientific advance.

In [5], Imre Lakatos spoke of the corresponding situation in Mathematics. He made an important distinction, between 'discovery' and 'justification'. That distinction is important for the situation in physical science too. Discovery can occur in a variety of ways, but justification is rigorously mathematical. That is why this Book is so mathematical in character. Math is a core value that you can take away from this Book. Regardless of what your specific undertaking may turn out to be: 'Master the Math'.

3. Specific Tools to Use

Some specific techniques for pursuing the mathematical approach to re-examining conventional wisdom, and possibly challenging conventional wisdom, have been demonstrated in the present Book. In case you have a similar mission in mind for your own future, here is a list of these techniques:

1) Observe the structure of the present Book: Math theory in Part III, Physics Theory in Part II, and Chemistry Applications in Part I. You can guess that the present Book is written in an order often reversed from the order in which the research was actually done. The practical applications are presented first, and any upsetting information about conventional wisdom comes later. That approach seems to work, and in my experience, no other approach has worked anywhere near as well. So for presentation purposes, it can help to 'Reverse the Research'.
2) In Part I, the upsetting Chapters 6 and 7 about Catalysis and Cold Fusion embody the following script: "It is generally believed that (fill in the blank) occurs (or doesn't occur). But numerical analysis shows that the general belief is not very likely to be true, whereas an alternative scenario (fill in the blank) is really more likely." This script works well. So always remember: 'Nail the Numbers'.
3) In Part II, Chapter 9 about the Hydrogen atom reflects a typical engineer's question: what is the minimum system we can talk about? The Hydrogen atom involves acceleration, so it has to have at *minimum* two charges. Early 20th century analyses focused on *one* charge. That approach left many troubling mysteries, such as exploding electrons and run-away trajectories. So it is a good idea to 'Restrain the Reductionism'.
4) In Part III, Chapter 13 about photons, you see a long-ignored quirkiness noted (*i.e.* the Liénard-Wiechert formulae), and then tracked down, and then resolved. You should always 'Question any Quirkiness' you may detect, and then chase it to some sort of resolution.
5) In Part III, Chapter 14 about tensor analysis, you see a new mathematical technique that was eagerly adopted for a new branch of Physics (*i,e*, SRT) being turned back to illuminate the shaky foundations of that new branch of Physics. So given any new math tool always, 'Review Earlier Research'.
6) The application of all such tools together, and any other ones you may think of, constitutes 'Due Diligence'. Whatever you do in life, you should always perform Due Diligence to the best of your ability.

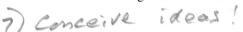

Acknowledgments

This Book developed out of journal articles, and the journal articles developed in turn out of conference presentations. That is how research seems to work. I want to thank all the organizers of all the conferences. There are too many. But I can mention the conferences of the Natural Philosophy Alliance, and the efforts of its long-time President, Prof. Domina Eberle Spencer. And I can mention the conferences titled "Physical Interpretations of

Relativity Theory", launched by Prof. Michael C. Duffy, and now carried forward by Dr. Peter Rowlands. And I can mention the conferences of the International Association for Relativistic Dynamics, to which Dr. Ruggero M. Santilli introduced me following a PIRT presentation. He also opened the pages of Hadronic Journal to this line of research.

This Author had many teachers to whom this long-developing book should be dedicated. Most especially I thank my MIT thesis supervisors Laszlo Tisza, Martin Schetzen, and Walter Thorson, and my high-school physical science teachers Richard Mason and Edwin Cooper.

REFERENCES

[1] Kuhn, T.S. *The Structure of Scientific Revolutions*, Second Edition, Enlarged, University of Chicago Press, 1970.
[2] CERN, OPERA Collaboration Team, http://arxiv.org/abs/1109.4897v2.
[3] Popper, K. *The Logic of Scientific Discovery*, Hutchinson, London, 1934.
[4] *Quantum Mechanics*, Ed. Jonathan P. Groffe, Nova Science Publishers, New York, 2009.
[5] Lakatos, I. *Proofs and Refutations*, J. Warrol and E. Zahar Eds., Cambridge University Press, 1976.

INDEX

A

Actinium, 8, 157, 177, 322, 323
Alchemists, 13
Alchemy, 1, 14
Aluminum, 5, 6, 59, 254, 321
Americium, 8, 179
Antimony, 6, 120
Argon, 5, 71, 170, 172, 180
Aristotle, 1, 15, 159
Arsenic, 6, 95, 176
Astatine, 8, 178

B

Balance, 303
Barium, 7, 314, 318
Berkelium, 8
Berylium, 318
Bismuth, 8, 147
Bohrium, 8
Boiling Points, 179
Boltzmann Factors, 186
Boron, 5, 41, 321
Bromine, 6, 97, 174, 195, 202
Butane, 221

C

Cadmium, 6, 113
Calcium, 5, 6, 314, 318
Californium, 8
Carbon, 5, 44, 176, 177, 196, 200, 207, 208, 210, 211, 220, 226, 235, 240, 245, 246, 247, 248, 255, 256, 257, 260
Catalysis, v, 263, 276, 277, 370
catalytic converter, 3, 15, 237, 251, 254, 256, 260, 291
Cerium, 7, 254, 323
CERN, 368, 371
Cesium, 7, 125, 166, 169, 179, 195, 199, 321
Charge Pairs, v, 307
chemical explosions, 4
Chlorine, 5, 70
Chromium, 5, 77, 263, 266, 276, 322
Classical Electrodynamics, 299, 305, 368
Cobalt, 5, 81, 196
Cold Fusion, 15, 279, 280, 292, 293, 370
Combustion, 207, 238-241, 244, 245, 247-250
Comfortable Elements, 318
Convention, 368
Copernicus, 13
Copper, 6, 84, 322
Curium, 8, 322

D

Darmstadtium, 9
Deuterium, 279, 281, 289
Dubnium, 8
Dysprosium, 7

E

Einstein, A., 358
Einsteinium, 8
Electron Pair, 313
Electron Rings, v, 317
electron-positron system, 4
Erbium, 7
Ethane, 216, 217
Europium, 7, 320
Excited States, 304

F

Fermium, 8
First-Order Ionization Potentials, 25
Fluorine, 5, 51, 177, 195, 198, 199, 202
Foundations of Physics, 24, 365
Francium, 8, 155, 166, 169, 177, 179, 321

G

Gadolinium, 7, 322, 323
Gallium, 6, 89, 195, 202, 204, 205, 321
General Relativity Theory, 368
Germanium, 6, 92
Gold, 7, 134, 204, 284, 285, 289, 291, 323

H

Hadronic Journal, 371
Hafnium, 7
Halogens, 199, 202
Hassium, 8, 322
Helium, 5, 8, 36, 161, 162, 170, 171, 177, 180, 195, 196, 197, 198, 201, 202, 279, 280, 292, 313, 314, 318, 333
Hexane, 244
Holmium, 7
Holors, 365
Hot Fusion, 279, 280, 292
Hydrocarbons, 213, 234, 244

I

Indium, 6, 114, 321
Infinite Energy Magazine, 293
International Journal of Chemical Modeling, 277
International Journal of Molecular Sciences, 25, 193
Iodine, 6
Ionization Potentials, v, 19, 20, 22, 29, 160, 328, 329, 332, 336
ionization states, 3, 29, 34, 159, 160, 164, 165, 166, 167, 170, 180, 183, 184, 185, 186, 191, 193, 205, 206, 207, 233, 235, 270, 271, 283, 285, 289, 336
Iridium, 7
Iron, 5, 79, 196, 254, 264, 276, 277, 322

J

Jackson, J.D., 305

K

Kepler, 13, 14
Klyushin, J.G., 358
Krypton, 6, 100, 170, 172, 180

L

Lanthanum, 7, 128, 322, 323
Lawrencium, 8, 177, 322
Lead, 8, 144
Lithium, 5, 37, 162, 163, 166, 167, 177, 179, 318, 321
Lutetium, 7, 177, 322

M

Magnesium, 5, 57, 314, 318
Manganese, 5, 254, 319
Maxwell fields, 4, 345, 347
Meitnerium, 8
Melting points, 166
Mendelevium, 8
Mercury, 7, 138, 174, 195, 202, 203, 204
Methane, 214, 215, 240, 241, 244
Mills, R.L., 315
Models, 2
Molybdenum, 6, 322

N

Nassikas, A.A., 358
Natural Philosophy Alliance, 293, 370
Neodymium, 7, 178
Neon, 5, 53, 170, 171, 180, 195, 201, 202, 319
Neptunium, 8, 177, 323
Nickel, 6, 196, 254
Niobium, 6, 322
Nitrogen, 5, 46, 163, 164, 195, 200, 207, 208, 210, 215, 319
Nobelium, 8, 196
Noble gasses, 200

O

Octane, 227, 247
Osmium, 7, 322
Oxygen, 5, 49, 163, 164, 195, 199, 207, 209, 210, 211, 212, 215, 220, 235, 236, 245, 246, 248, 256, 257, 260, 264, 266, 274, 321, 340

P

Palladium, 6, 109, 204, 254, 256, 257, 258, 259, 260, 282, 288, 289, 290, 291, 292, 323, 324
Peculiar Elements, 322
Periodic Arch, 176, 177, 178, 193, 195, 196
Periodic Table, 2, 3, 14, 15, 16, 19, 29, 166, 170, 174, 175, 176, 177, 179, 195, 280, 313, 319, 325, 326, 333
Periods, 25, 27, 177, 332
Phase Diagrams, 181
Philosophy of Science, 2, 346, 368
Phosphorus, 5, 64, 176, 184, 192
Platinum, 6, 7, 132, 204, 254, 258, 259, 260, 283, 284, 285, 287, 323
Plutonium, 8
Polonium, 8, 149, 321
Positronium, 307, 310, 311
Potassium, 5, 73, 166, 168, 179, 321
Praseodymium, 7
Promethium, 7
Protactinium, 8, 323
Proton Pair, 312

Q

Quantum Mechanics (QM), v, 2, 3, 4, 20, 241, 295, 297, 298, 299, 327, 343, 345, 347, 364, 367, 371

R

Radium, 8, 314, 318
Radon, 8, 153, 171, 173, 180
Reaction Steps, 269, 272
Relativity Theory, 354, 367, 371
Rhenium, 7
Rhodium, 6, 106, 196, 254, 255, 256, 257, 258, 259, 323
Rowlands, P., 365
Rubidium, 6, 103, 166, 168, 179, 321, 322
Ruthenium, 6
Rutherfordium, 8

S

Samarium, 7
Scandium, 5, 75, 177, 322
Seborgium, 8
Selenium, 6, 176, 321
Septane, 227, 245
Silicon, 5, 62, 196

Silver, 6, 111, 263, 264, 265, 269, 272, 275, 276, 277, 323
Sodium, 5, 55, 166, 168, 179, 321
Solid State, 4
Special Relativity Theory, 4, 309, 345, 351, 354, 368
States of Matter, v, 159, 164, 234
Strontium, 6, 314, 318
Sulfur, 5, 67, 176, 263, 267, 277, 321

T

Tantalum, 6, 7
Technetium, 6
Tellurium, 6, 321
Terbium, 7
Thallium, 7, 141, 321, 323
Theories, 2, 368
Thomas, L.H., 305
Thorium, 8
Thulium, 7, 178
Tin, 6, 117, 176
Titanium, 5
Torquing, 301
Tritium, 279, 281
Tungsten, 7, 129, 285, 286, 287, 289

U

Uncomfortable Elements, 321
Unnilquadium, 9
Ununbium, 9
Ununhexium, 9
Ununoctium, 9
Ununpentium, 9
Ununseptium, 9
Ununtrium, 9
Unununium, 9
Uranium, 8, 323

W

Water, 25, 159, 194, 212
Wesley, J.P., 326

X

Xenon, 6, 123, 170, 172, 180

Y

Ytterbium, 7, 196, 321
Yttrium, 6, 105, 177, 322

Z

Zinc, 6, 85
Zirconium, 6, 7